Generalized Harmonic Analysis and Wavelet Packets

Khalifa Trimèche

GORDON AND BREACH SCIENCE PUBLISHERS
Australia • Canada • France • Germany • India • Japan • Luxembourg
Malaysia • The Netherlands • Russia • Singapore • Switzerland

Amsteldijk 166
1st Floor
1079 LH Amsterdam
The Netherlands

British Library Cataloguing in Publication Data

ISBN: 90-5699-329-1

Contents

Introduction ... xi

1 The Normalized Bessel Function Of First Kind ... 1

Introduction ... 1

1.I. The Bessel function of first kind ... 2

1.I.1. Definition ... 2

1.I.2. Derivatives and differential equation of the Bessel function J_ν ... 3

1.I.3. Asymptotic formulas for the Bessel function J_ν ... 4

1.I.4. The Poisson integral representations of the Bessel function J_ν ... 5

1.I.5. The Sonine's first integral for the Bessel functions J_α and J_β ... 6

1.I.6. Addition formulas for the Bessel function J_ν ... 7

1.I.7. Product formulas for the Bessel function J_ν ... 7

1.II. The normalized Bessel function of first kind ... 8

1.II.1. Definition ... 8

1.II.2. Properties of the function $j_\alpha(\lambda x)$ ... 9

1.II.3. Integral representations of the function $j_\alpha(\lambda x)$ ... 10

1.II.3.a. The Poisson integral representations 10

1.II.3.b. The Sonine's first integral 12

1.II.4. Product formula for the function j_α 12

1.II.5. Useful formulas involving the function j_α 14

1.III. Problems 15

2 Riemann-Liouville And Weyl Integral Transforms 25

Introduction 25

2.I. The Riemann-Liouville integral transform 26

2.I.1. Definition and properties 26

2.I.2. Inversion of the operator $\mathcal{R}_\alpha$ 31

2.I.3. The Riemann-Liouville integral transform on the spaces

$L^p([0, +\infty[, dx)$, $1 \leq p \leq +\infty$ 38

2.II. The Weyl integral transform 41

2.II.1. Definition and properties 41

2.II.2. Inversion of the operator W_α 44

2.III. The Weyl integral transform on the space $\mathcal{S}_*(\mathbb{R})$ 51

2.IV. The Weyl transform on the space $\mathcal{E}'_*(\mathbb{R})$ 55

2.V. The Sonine integral transform and its dual 56

2.V.1. The Sonine integral transform 56

2.V.2. The Sonine integral transform on the spaces

$L^p([0, +\infty[, dx)$, $1 \leq p \leq +\infty$ 59

2.V.3. The dual Sonine integral transform 60

2.V.4. The Sonine transform on the space $\mathcal{E}'_*(\mathbb{R})$ 61

2.VI. Problems 62

3 Convolution Product And Fourier-Cosine Transform Of Functions, Measures And Distributions 69

Introduction 69

3.I. Convolution product of functions and distributions 69

3.I.1. The translation operator 70

3.I.2. Convolution product of functions 71

3.I.3. Convolution product of measures 73

3.I.4. Convolution product of distributions 75

3.II. The Fourier-cosine transform 75

3.II.1. The Fourier-cosine transform on $L^1([0, +\infty[, dx)$ 75

3.II.2. The Fourier-cosine transform on $\mathcal{S}_*(\mathbb{R})$ and $\mathcal{D}_*(\mathbb{R})$ 77

3.II.3. The Fourier-cosine transform on $L^2([0, +\infty[, dx)$ 78

3.II.4. The Fourier-cosine transform on $M^b([0, +\infty[)$ 79

3.II.5. The Fourier-cosine transform on $\mathcal{E}'_*(\mathbb{R})$ and $\mathcal{S}'_*(\mathbb{R})$ 80

3.III. Problems 81

4 Generalized Convolution Product Associated With The Bessel Operator 87

Introduction 87

4.I. Convolution product of radial functions 88

4.I.1. Definition and properties 88

4.I.2. Convolution product 89

4.II. Generalized translation operators associated with the Bessel operator 92

4.III. Generalized convolution product associated with the Bessel operator 97

4.III.1. Generalized convolution product of functions 97

4.III.2. Generalized convolution product of measures of $M^b([0,+\infty[)$... 106

4.III.3 Generalized convolution product of distributions.................... 107

4.IV. Problems.. 108

5 Fourier-Bessel Transform.. 115

Introduction.. 115

5.I. Fourier transform of radial functions... 116

5.II. The Fourier-Bessel transform on $L^1([0, +\infty[, d\mu_\alpha)$........................ 119

5.III. The Fourier-Bessel transform on $\mathcal{S}_*(\mathbb{R})$ and $\mathcal{D}_*(\mathbb{R})$..................... 127

5.IV. The Fourier-Bessel transform on $L^2([0, +\infty[, d\mu_\alpha)$....................... 132

5.V. The Fourier-Bessel transform on $L^p([0, +\infty[, d\mu_\alpha), 1 \leq p \leq 2$.............. 139

5.VI. The Fourier-Bessel transform on $M^b([0, +\infty[)$........................... 141

5.VII. The Fourier-Bessel transform on $\mathcal{E}'_*(\mathbb{R})$ and $\mathcal{S}'_*(\mathbb{R})$.................... 144

5.VIII. Problems.. 150

6 Infinitely Divisible Probabilities And Central Limit Theorem Associated With The Bessel Operator...................................... 173

Introduction.. 173

6.I. Dispersion of a probability measure on $[0, +\infty[$.......................... 174

6.I.1. Generalized quadratic form.. 174

6.I.2. Dispersion of a probability measure on $[0, +\infty[$.................... 176

6.I.3. Levy's theorem.. 179

6.II. Levy-Khintchine's formula.. 182

6.III. Convolution semigroups and infinitely divisible probabilities associated with the Bessel operator... 186

6.III.1. Convolution semigroups... 186

6.III.2. Infinitely divisible probabilities associated with the Bessel operator.... 187
6.IV. Central limit theorem associated with the Bessel operator.... 191

7 Continuous Wavelet Transform Associated With The Bessel Operator.... 197

Introduction.... 197
7.I. Classical continuous wavelet transform on [0, + ∞[.... 198
7.I.1. Classical wavelets on [0, + ∞[.... 198
7.I.2. Classical continuous wavelet transform on [0, + ∞[.... 201
7.II. Continuous wavelet transform associated with the Bessel operator.... 203
7.II.1. Wavelets associated with the Bessel operator.... 203
7.II.2. Continuous wavelet transform associated, with the Bessel operator.. 207
7.III. Inversion formulas for the operators $\mathfrak{R}_\alpha$ and W_α.... 218
7.IV. Inversion formulas for the operators $\mathfrak{R}_\alpha$ and W_α using wavelets associated with the Bessel operator.... 223
7.V. Problems.... 231

8 Wavelet Packets Associated With The Bessel Operator.... 239

Introduction.... 239
8.I. The P-wavelet packet transform associated with the Bessel operator..... 239
8.I.1. Plancherel and reconstruction formulas 240
8.I.2 Calderon's reproducing formula.... 245
8.II. Scale discrete scaling function associated with the Bessel operator.... 247
8.III. Modified packet associated with the Bessel operator.... 254
8.IV. S-wavelet packet associated with the Bessel operator.... 260
8.V. Multiresolution analysis by means of wavelet packets associated with the Bessel operator.... 265

8.VI. Problems........ 268

9 Continuous Linear Wavelet Transform Associated With The Bessel Operator And Its Discretization........ 273

Introduction........ 273

9.I. Linear wavelets associated with the Bessel operator........ 273

9.II. Linear wavelet packets associated with the Bessel operator........ 280

9.III. Scale discrete L-scaling function associated with the Bessel operator... 285

Bibliography........ 293

Index........ 303

Introduction

In this book we consider the Riemann-Liouville and the Weyl integral transforms associated with the Bessel operator, and we study their properties. Next using these transforms we construct a Harmonic Analysis associated with the Bessel operator (generalized convolution product, Fourier-Bessel transform, Plancherel and inversion formulas, Plancherel theorem, Paley-Wiener theorems, ...).

The objective of this book is to generalize the important subject of wavelets and wavelet packets to the case of this Harmonic Analysis.

Wavelets were introduced by J. Morlet a French petroleum engineer at ELF-Aquitaine, in connection with his study of seismic traces. The Mathematical fundations were given in a paper by A. Grossmann and J. Morlet [41]. The harmonic analyst Y. Meyer and many other mathematicians became aware of this theory and they recognised many classical results inside it (see [42] [52] [56] [63]).

Wavelets have wide applications, ranging from signal analysis in geophysics and acoustics to quantum theory and pure mathematics (see [40] [63] [88]).

We study the continuous wavelet analysis and we present a construction of wavelet packets (resp. linear wavelet packets) and the corresponding wavelet packet transforms (resp. linear wavelet packet transform) and we prove for these transforms Plancherel formulas, Calderon's reproducing formulas, and pointwise reconstruction formulas.

As applications of this theory we give the inverse of the Riemann-Liouville and Weyl integral transforms, and the Radon transform on $\mathbb{R}^n$, $n \geq 2$, using wavelets.

This book is suitable for use either as a classroom text in various mathematical courses or as a reference text for scientists.

Each chapter includes problems to make the results obtained in these chapters accessible to a broad audience.

I am grateful to Najet Sellami for her expertise, skill and endurance in producing the typescript.

Khalifa TRIMECHE

1 The Normalized Bessel Function of First Kind

Introduction.

The German astronomer F.W. Bessel (1784-1846) first achieved fame by computing the orbit of Halley's comet. In addition to many other accomplishments in connection with his studies of planetary motion, he is credited with deriving the differential equation bearing his name and carrying out the first systematic study of the general properties of its solutions (now called Bessel functions) in his famous 1824 memoir. Nevertheless, Bessel functions were first discovered in 1732 by D. Bernoulli (1700-1782), who provided a series solution (representing a Bessel function) for the oscillatory displacements of a heavy hanging chain(See [3] sec. 6.7.12. p. 282-284). Euler later developed a series similar to that of Bernoulli, which was also a Bessel function, and Bessel's equation appeared in a 1764 article by Euler dealing with the vibrations of a circular drumhead. J. Fourier (1768-1836) also used Bessel functions in his classical treatise on heat in 1822, but it was Bessel who first recognized their special properties.

Bessel functions are closely associated with problems possessing circular or cylindrical symmetry. For example, they arise in the study of free vibrations of a circular membrane. They also occur in electromagnetic theory and numerous other areas of physics and engineering. (See [3] p. 237).

We give in the first section of this chapter the definition and the important properties of the Bessel function of first kind. (See [3] [4] [5] [28] [51] [55] [62] [70] [80] [91]).

In the second section we define and study the normalized Bessel function of first kind, and we give its properties.

The main relations satisfied by this function are the Poisson integral representation and the product formula which are fundamental in the construction of the Riemann-Liouville and Weyl integral transforms and the Harmonic Analysis associated with the Bessel operator, which will be done in the following chapters.

The third section is devoted to problems in which we present proofs of some properties of the Bessel function and the normalized Bessel function of first kind. (See [3] [29] [76]).

1.I. The Bessel function of first kind.

1.I.1. Definition.

Let $\nu \in \mathbb{C}$, such that $\operatorname{Re} \nu \geq 0$. The function J_ν defined on $]0, +\infty[$ by

$$J_\nu(x) = \left(\frac{x}{2}\right)^\nu \sum_{n=0}^{\infty} \frac{(-1)^n \left(\frac{x}{2}\right)^{2n}}{n!\Gamma(n+\nu+1)}, \tag{1.I.1}$$

is called the Bessel function of first kind and order ν.

For $\nu \neq p$, with $p \in \mathbb{N}^*$, we put

$$J_{-\nu}(x) = \left(\frac{x}{2}\right)^{-\nu} \sum_{n=0}^{\infty} \frac{(-1)^n \left(\frac{x}{2}\right)^{2n}}{n!\Gamma(n-\nu+1)}, \tag{1.I.2}$$

(obtained by changing ν to $(-\nu)$ in (1.I.1)).

As the poles of the function Γ are the nonpositive integers, then for $\nu = p$, the formula (1.I.2) becomes

$$J_{-p}(x) = \left(\frac{x}{2}\right)^{-p} \sum_{n=p}^{\infty} \frac{(-1)^n \left(\frac{x}{2}\right)^{2n}}{n!\Gamma(n-p+1)}. \tag{1.I.3}$$

But if we put $n - p = k$, we obtain

$$J_{-p}(x) = (-1)^p \left(\frac{x}{2}\right)^p \sum_{k=0}^{\infty} \frac{(-1)^k \left(\frac{x}{2}\right)^{2k}}{k!\Gamma(p+k+1)},$$

thus

$$J_{-p}(x) = (-1)^p J_p(x). \tag{1.I.4}$$

Remarks.

i) From the relations (1.I.1) and (1.I.2) we obtain

(1.I.5) $$J_{\frac{1}{2}}(x) = \sqrt{\frac{2}{\pi x}} \sin x .$$

(1.I.6) $$J_{-\frac{1}{2}}(x) = \sqrt{\frac{2}{\pi x}} \cos x .$$

ii) The function

$$u(x,t) = \exp[\frac{1}{2}x(t - \frac{1}{t})], \quad t \in \mathbb{R}^*, \quad x > 0,$$

is called the generating function of the Bessel functions of integral order $J_p(x)$, since by expanding $u(x,t)$ in series involving powers of t, we show that

(1.I.7) $$u(x, t) = \sum_{p=-\infty}^{\infty} J_p(x)t^p .$$

iii) For $\nu \in \mathbb{R}$, we define the function J_ν on $]-\infty, 0[$ by

(1.I.8) $$J_\nu(-x) = e^{i\nu\pi} J_\nu(x), \quad x > 0$$

1.I.2. Derivatives and differential equation of the Bessel function J_ν .

From the series definition of the functions $J_\nu(x)$ and $J_{-\nu}(x)$ we obtain by differentiation.

(1.I.9) $$\frac{d}{dx}[x^\nu J_\nu(x)] = x^\nu J_{\nu-1}(x).$$

(1.I.10) $$\frac{d}{dx}[x^{-\nu} J_\nu(x)] = - x^{-\nu} J_{\nu+1}(x).$$

If we carry out the differentiation on the left hand sides in (1.I.9) and (1.I.10) and divide the results by the factors x^ν and $x^{-\nu}$ respectively, we deduce that

(1.I.11) $$J'_\nu(x) + \frac{\nu}{x} J_\nu(x) = J_{\nu-1}(x).$$

(1.I.12) $$J'_\nu(x) - \frac{\nu}{x} J_\nu(x) = - J_{\nu+1}(x).$$

Setting $\nu = 0$ in (1.I.12), we obtain the special case

(1.I.13) $$J'_0(x) = - J_1(x).$$

Finally, the sum and difference of (1.I.11) and (1.I.12) yields, respectively

(1.I.14) $$2J'_\nu(x) = J_{\nu-1}(x) - J_{\nu+1}(x).$$

(1.I.15) $$\frac{2\nu}{x} J_\nu(x) = J_{\nu-1}(x) + J_{\nu+1}(x).$$

This last relation is the three-term recurrence formula for the Bessel function J_ν.

From the relations (1.I.11) and (1.I.12) we deduce that $J_\nu(x)$ is a solution of the second order linear differential equation

(1.I.16) $$y'' + \frac{1}{x} y' + (1 - \frac{\nu^2}{x^2})y = 0,$$

called Bessel differential equation.

The general solution of this equation is studied in Problem 1.III.1.A.

1.I.3. Asymptotic formulas for the Bessel function J_ν .

For $\nu \in \mathbb{C}$, such that $\mathrm{Re}\, \nu \geq 0$, we deduce from the relation (1.I.1) that the function J_ν possesses the following asymptotic formula when $x \to 0^+$:

(1.I.17) $$J_\nu(x) \sim \frac{x^\nu}{2^\nu\, \Gamma(\nu+1)} .$$

On the other hand the relation (1.I.2) gives for $\nu \neq p$, $p \in \mathbb{N}^*$, the asymptotic formula for $J_{-\nu}$ when $x \to 0^+$:

(1.I.18) $$J_{-\nu}(x) \sim \frac{x^{-\nu}}{2^{-\nu}\Gamma(1-\nu)}.$$

The asymptotic formula for J_{-p}, $p \in \mathbb{N}^*$, when $x \to 0^+$ can be deduced from the relations (1.I.4) and (1.I.17).

For $\nu \in \mathbb{R}$, we have the asymptotic formula for J_ν when $x \to +\infty$:

(1.I.19) $$J_\nu(x) = \sqrt{\frac{2}{\pi x}} \cos(x - (2\nu + 1)\frac{\pi}{4}) + O(\frac{1}{x^{3/2}}).$$

1.I.4. The Poisson integral representations of the Bessel function J_ν.

Let $\nu \in \mathbb{C}$, such that $\mathrm{Re}(\nu + \frac{1}{2}) > 0$. The integral representation of the function J_ν, due to S.D.Poisson is given by

(1.I.20) $$J_\nu(x) = \frac{(\frac{x}{2})^\nu}{\sqrt{\pi}\,\Gamma(\nu + \frac{1}{2})} \int_{-1}^{1} (1-t^2)^{\nu-\frac{1}{2}} e^{ixt}\, dt.$$

(See [91] § 3, 3 (1)).

To verify this relation we start with the integral.

$$\int_{-1}^{1} (1-t^2)^{\nu-\frac{1}{2}} e^{ixt}\, dt = 2\int_{0}^{1} (1-t^2)^{\nu-\frac{1}{2}} \cos(xt)\; dt\,,$$

(1.I.21) $$\int_{-1}^{1} (1-t^2)^{\nu-\frac{1}{2}} e^{ixt}\, dt = 2\sum_{k=0}^{\infty} \frac{(-1)^k x^{2k}}{(2k)!} \int_{0}^{1} (1-t^2)^{\nu-\frac{1}{2}} t^{2k}\, dt\,,$$

where we have expressed cosxt in power series.

The integral in (1.I.21) can be evaluated in terms of the Beta function. More precisely we have

$$\int_{0}^{1} (1-t^2)^{\nu-\frac{1}{2}} t^{2k}\, dt = \frac{1}{2} B(k + \frac{1}{2},\ \nu + \frac{1}{2})$$
$$= \frac{\sqrt{\pi}\,(2k)!\,\Gamma(\nu + \frac{1}{2})}{2^{2k+1}\, k!\,\Gamma(k + \nu + 1)}\,.$$

Finally, by substituting this result in (1.I.21), we obtain from the relations (1.I.1) and (1.I.2) :

$$\int_{-1}^{1} (1-t^2)^{\nu-\frac{1}{2}} e^{ixt}\, dt = \sqrt{\pi}\,\Gamma(\nu + \frac{1}{2}) \sum_{k=0}^{\infty} \frac{(-1)^k (\frac{x}{2})^{2k}}{k!\,\Gamma(k + \nu + 1)}$$
$$= \sqrt{\pi}\,\Gamma(\nu + \frac{1}{2})\,(\frac{x}{2})^{-\nu} J_\nu(x).$$

We deduce (1.I.20) from this relation.

By making the change of variables $t = \cos\theta$ in (1.I.20) we obtain still another integral representation.

(1.I.22) $$J_\nu(x) = \frac{(\frac{x}{2})^\nu}{\sqrt{\pi}\,\Gamma(\nu+\frac{1}{2})}\int_0^\pi \cos(x\cos\theta)(\sin\theta)^{2\nu}\,d\theta\,,$$

which was originally due to F.W.Bessel.

Remarks.

i) A proof of the relation (1.I.20) is given in Problem 1.III.1. B.

ii) To obtain an integral representation of the Bessel function of integral order J_p, $p \in \mathbb{Z}$, we start with the generating function relation (1.I.7) :

$$\exp[\frac{1}{2}x(t-\frac{1}{t})] = \sum_{p=-\infty}^{\infty} J_p(x)\,t^p,\ t \in \mathbb{R}^*,\ x > 0.$$

If we set $t = e^{-i\phi}$, $\phi \in \mathbb{R}$, we obtain

$$e^{-ix\sin\phi} = \sum_{p=-\infty}^{\infty} J_p(x)\,e^{-ip\phi}\,.$$

We see that $J_p(x)$ is the Fourier coefficient of the Fourier series development of the function $e^{-ix\sin\phi}$. Then

$$J_p(x) = \frac{1}{2\pi}\int_0^{2\pi} e^{-ix\sin\phi}\,e^{ip\phi}\,d\phi$$
$$= \frac{1}{2\pi}\int_0^{2\pi}\cos(p\phi - x\sin\phi)\,d\phi.$$

This relation can also be written in the form

(1.I.23) $$J_p(x) = \frac{1}{\pi}\int_0^\pi \cos(p\phi - x\sin\phi)d\phi.$$

1.I.5. The Sonine's first integral for the Bessel functions J_α and J_β.

Let $\alpha, \beta \in \mathbb{R}$, such that $\alpha > \beta > -1$. The Sonine's first integral for the Bessel functions J_α and J_β is given by

(1.I.24) $$J_\alpha(x) = \frac{x^\alpha}{2^{\alpha-\beta-1}\,\Gamma(\alpha-\beta)}\int_0^1 (xr)^{-\beta}\,J_\beta(xr)(1-r^2)^{\alpha-\beta-1}\,r^{2\beta+1}dr.$$

(See [91] § 12, 11 (1)).

To verify this relation it suffices to remplace $J_\beta(xr)$ by its series representation and to use termwise integration.

1.I.6. Addition formulas for the Bessel function J_ν.

From [91] § 11.4 (1) and (2) (see also [5] p.35) we have the following addition formulas for the Bessel function J_ν :

- For $\nu \in \mathbb{R}, \nu \neq 0, -1, -2, \ldots$, and $x > 0, y > 0$:

(1.I.25)
$$\frac{J_\nu(\sqrt{x^2+y^2-2xy\cos\theta})}{(x^2+y^2-2xy\cos\theta)^{\nu/2}} = 2^\nu\,\Gamma(\nu)\sum_{m=0}^{\infty}(m+\nu)\,\frac{J_{m+\nu}(x)}{x^\nu}\times\frac{J_{m+\nu}(y)}{y^\nu}\,C_m^\nu(\cos\theta).$$

where C_m^ν is the Gegenbauer polynomial of degree m and order ν (see [3] p. 226-228).

- For $\nu = 0$, and $x > 0, y > 0$:

(1.I.26)
$$J_0(\sqrt{x^2+y^2-2xy\cos\theta}) = J_0(x)\,J_0(y) + 2\sum_{m=1}^{\infty} J_m(x)\,J_m(y)\cos m\theta.$$

Remark.

We consider the function

$$u(x,t) = \exp\left[\frac{1}{2}x\left(t-\frac{1}{t}\right)\right], \quad t \in \mathbb{R}^*, \quad x > 0.$$

This function satisfies the relation

(1.I.27)
$$\forall\, x > 0, y > 0, \quad u(x+y,t) = u(x,t).\,u(y,t).$$

Using the relation (1.I.7) we deduce the following addition formula for the Bessel functions of first kind and integral order J_p, $p \in \mathbb{N}$:

(1.I.28)
$$\forall\, x > 0, y > 0, \quad J_p(x+y) = \sum_{k=-\infty}^{\infty} J_k(x)\,J_{p-k}(y).$$

1.I.7. Product formulas for the Bessel function J_ν .

Let $\nu \in \mathbb{R}$, such that $\nu > -\frac{1}{2}$. By integration of the relations (1.I.25) and (1.I.26), and using the orthogonality of the polynomials $C_m^\nu(\cos\theta)$ and of the functions $\cos m\theta$, we obtain the product formula for the Bessel function J_ν :

(1.I.29) $\forall\, x > 0,\ y > 0,\ \dfrac{J_\nu(x)}{x^\nu}.\dfrac{J_\nu(y)}{y^\nu} =$

$$\frac{1}{2^\nu\,\Gamma(\nu+\frac{1}{2})\sqrt{\pi}}\int_0^\pi \frac{J_\nu(\sqrt{x^2+y^2-2xy\cos\theta})}{(x^2+y^2-2xy\cos\theta)^{\nu/2}}(\sin\theta)^{2\nu}\,d\theta.$$

(See [91] § 11, 4 (2) and [5] p. 35).

Remark.

In [51] p. 130, we find a product formula for the Bessel functions of first kind, more general than the formula (1.I.29). More precisely for $\alpha, \beta \in \mathbb{R}$, such that $\alpha > \beta > -\dfrac{1}{2}$, we have

(1.I.30) $$\frac{J_\alpha(x)}{x^\alpha}.\frac{J_\beta(y)}{y^\beta} = \frac{1}{2^\alpha\,\Gamma(\alpha+1)}\int_0^1\int_0^\pi \frac{J_\beta(\sqrt{x^2+y^2-2xy\cos\phi})}{(x^2+y^2-2xy\cos\phi)^{\beta/2}}\,dm_{\alpha,\beta}(r,\phi)\,,$$

where

(1.I.31) $$dm_{\alpha,\beta}(r,\phi) = \frac{2\Gamma(\alpha+1)}{\sqrt{\pi}\,\Gamma(\alpha-\beta)\,\Gamma(\beta+\frac{1}{2})}(1-r^2)^{\alpha-\beta-1}\,r^{2\beta+1}\,(\sin\phi)^{2\beta}\,dr d\phi\,,$$

this measure is normalized by

$$\int_0^1\int_0^\pi dm_{\alpha,\beta}(r,\phi) = 1.$$

If $\beta \to -\dfrac{1}{2}$, the measure $dm_{\alpha,\beta}(r,\phi)$ degenerates to the measure

$$\frac{\Gamma(\alpha+1)}{\sqrt{\pi}\,\Gamma(\alpha+\frac{1}{2})}(1-r^2)^{\alpha-\frac{1}{2}}(\delta(\phi)+\delta(\pi-\phi))dr\,d\phi\,.$$

The formula (1.I.30) degenerates for $\beta \to -\dfrac{1}{2}$, to

(1.I.32) $$\frac{J_\alpha(x)}{x^\alpha} = \frac{1}{2^\alpha\sqrt{\pi}\,\Gamma(\alpha+\frac{1}{2})}\int_{-1}^1 \cos(y+xr)(1-r^2)^{\alpha-\frac{1}{2}}\,dr.$$

For $y = 0$, this is the Poisson integral representation (1.I.20).

1.II. The normalized Bessel function of first kind.

1.II.1. Definition.

Let $\alpha \in \mathbb{R}$, such that $\alpha > -\dfrac{1}{2}$. The function j_α defined on $\mathbb{C}$ by

(1.II.1) $$j_\alpha(z) = \Gamma(\alpha+1) \sum_{n=0}^{\infty} \frac{(-1)^n (\frac{z}{2})^{2n}}{n!\,\Gamma(n+\alpha+1)},$$

is called the normalized Bessel function of first kind and order α.

Example.

From the relation (1.II.1) we obtain

(1.II.2) $$j_{\frac{1}{2}}(z) = \begin{cases} \dfrac{\sin z}{z} & , \quad \text{if } z \in \mathbb{C}^*, \\ 1 & , \quad \text{if } z = 0. \end{cases}$$

For $z \in]0, +\infty[$, we deduce from the relations (1.I.1) and (1.I.2) that

(1.II.3) $$j_\alpha(z) = 2^\alpha\,\Gamma(\alpha+1)\,\frac{J_\alpha(z)}{z^\alpha}.$$

From this formula and the relation (1.I.10) we obtain the following derivation formula

(1.II.4) $$\forall\, k \in \mathbb{N}, \forall\, z > 0, \left(\frac{1}{z}\frac{d}{dz}\right)^k (j_\alpha(z)) = \frac{(-1)^k\,\Gamma(\alpha+1)}{2^k\,\Gamma(k+\alpha+1)} j_{\alpha+k}(z).$$

By using the relations (1.II.3) and (1.I.15) we deduce the three-terme recurrence formula for the function j_α:

(1.II.5) $$\forall\, z \geq 0,\ z^2 j_{\alpha+1}(z) = 4\alpha(\alpha+1)(j_\alpha(z) - j_{\alpha-1}(z)).$$

In the following chapters we shall use the normalized Bessel function $j_\alpha(\lambda x)$, with $x \in \mathbb{R}$ and $\lambda \in \mathbb{C}$, which has the following series representation

(1.II.6) $$j_\alpha(\lambda x) = \Gamma(\alpha+1) \sum_{n=0}^{\infty} \frac{(-1)^n \lambda^{2n} x^{2n}}{2^{2n}\, n!\,\Gamma(n+\alpha+1)}.$$

1.II.2. Properties of the function $j_\alpha(\lambda x)$.

From the relation (1.II.6) we deduce the following properties of the function $j_\alpha(\lambda x)$:

i) For all $\lambda \in \mathbb{C}$, the function $x \to j_\alpha(\lambda x)$ is an even C^∞-function on $\mathbb{R}$.

ii) For all $x \in \mathbb{R}$, the function $\lambda \to j_\alpha(\lambda x)$ is an even entire function on $\mathbb{C}$.

iii) We have

(1.II.7) $$j_\alpha(0) = 1.$$

iv) For all $\lambda \in \mathbb{C}$, the function $x \to j_\alpha(\lambda x)$ is the unique solution of the differential equation

(1.II.8) $$\begin{cases} \ell_\alpha u(x) = -\lambda^2 u(x), \\ u(0) = 1, \ u'(0) = 0, \end{cases}$$

where ℓ_α is the Bessel operator on $]0, +\infty[$ defined by

(1.II.9) $$\ell_\alpha = \frac{d^2}{dx^2} + \frac{2\alpha+1}{x}\frac{d}{dx}.$$

For $\alpha = \frac{n-2}{2}$, $n \in \mathbb{N}$, $n \geq 2$, this operator is the radial part of the Laplacian $\Delta = \sum_{j=1}^{n} \frac{\partial^2}{\partial x_j^2}$ on $\mathbb{R}^n$.

1.II.3. Integral representations of the function $j_\alpha(\lambda x)$.

1.II.3.a. The Poisson integral representations.

Let $\alpha \in \mathbb{R}$, such that $\alpha > -\frac{1}{2}$. From the relations (1.I.20) and (1.II.3) we obtain

(1.II.10) $$\forall\, x > 0, \forall\, \lambda > 0, j_\alpha(\lambda x) = \frac{2\Gamma(\alpha+1)}{\sqrt{\pi}\,\Gamma(\alpha+\frac{1}{2})}\int_0^1 (1-t^2)^{\alpha-\frac{1}{2}} \cos\lambda xt\, dt.$$

This relation is also true for all $x < 0$, and using the relation (1.II.7), the fact that

$$\frac{2\Gamma(\alpha+1)}{\sqrt{\pi}\,\Gamma(\alpha+\frac{1}{2})}\int_0^1 (1-t^2)^{\alpha-\frac{1}{2}} dt = 1,$$

and the dominated convergence theorem, we deduce that the relation (1.II.10) is available for all $x \in \mathbb{R}$ and $\lambda > 0$. On the other hand the functions $\lambda \to j_\alpha(\lambda x)$ and $\lambda \to \int_0^1 (1-t^2)^{\alpha-\frac{1}{2}} \cos(\lambda xt)dt$ are even entire on $\mathbb{C}$. Then from the analytic extension theorem we have

(1.II.11) $$\forall\, x \in \mathbb{R}, \forall\, \lambda \in \mathbb{C},\ j_\alpha(\lambda x) = \frac{2\Gamma(\alpha+1)}{\sqrt{\pi}\,\Gamma(\alpha+\frac{1}{2})}\int_0^1 (1-t^2)^{\alpha-\frac{1}{2}} \cos(\lambda xt)dt.$$

This relation is called the Poisson integral representation of the function $j_\alpha(\lambda x)$.

We can write also this relation in the form

(1.II.12) $$\forall\ x \in \mathbb{R}, \forall\ \lambda \in \mathbb{C},\ j_\alpha(\lambda x) = \frac{\Gamma(\alpha+1)}{\sqrt{\pi}\,\Gamma(\alpha+\frac{1}{2})}\int_{-1}^{1}(1-t^2)^{\alpha-\frac{1}{2}}e^{i\lambda xt}\,dt.$$

By making the change of variables $t = \cos\theta$, we obtain

(1.II.13) $$\forall\ x \in \mathbb{R}, \forall\ \lambda \in \mathbb{C},\ j_\alpha(\lambda x) = \frac{\Gamma(\alpha+1)}{\sqrt{\pi}\,\Gamma(\alpha+\frac{1}{2})}\int_{0}^{\pi}\cos(\lambda x\cos\theta)\,(\sin\theta)^{2\alpha}d\theta.$$

Also a change of variables in the relation (1.II.11) gives the relation

(1.II.14) $$\forall\ x > 0, \forall\ \lambda \in \mathbb{C},\ j_\alpha(\lambda x) = \frac{2\Gamma(\alpha+1)x^{-2\alpha}}{\sqrt{\pi}\,\Gamma(\alpha+\frac{1}{2})}\int_{0}^{x}(x^2-t^2)^{\alpha-\frac{1}{2}}\cos\lambda t\,dt.$$

Remark.

Let $n \in \mathbb{N}$, $n \geq 2$, and $\|y\|$ the norm of $y \in \mathbb{R}^n$. The function $j_{\frac{n-2}{2}}(\|y\|)$ possesses the following integral representation

(1.II.15) $$j_{\frac{n-2}{2}}(\|y\|) = \int_{S^{n-1}} e^{i(u,y)}\,d\sigma(u),$$

where S^{n-1} is the unit sphere of $\mathbb{R}^n$, σ the normalized measure on S^{n-1} and $(.,.)$ the scalar product of $\mathbb{R}^n$.

From this formula and the analytic extension theorem we deduce the Poisson integral representations (1.II.12) and (1.II.13) for the function $j_{\frac{n-2}{2}}(\lambda x)$, $x \in \mathbb{R}$, $\lambda \in \mathbb{C}$. The method used to obtain these relations is presented in Problem 1.III.7.

Using the relation (1.II.11) we deduce the other following properties of the function $j_\alpha(\lambda x)$:

i) We have

(1.II.16) - For all $x \in \mathbb{R}, \lambda \in \mathbb{C}$, $|j_\alpha(\lambda x)| \leq e^{|\mathrm{Im}\lambda|\,|x|}$.

(1.II.17) - For all $p \in \mathbb{N}$, $x \in \mathbb{R}, \lambda \in \mathbb{C}$, $\left|\frac{d^p}{d\lambda^p} j_\alpha(\lambda x)\right| \leq |x|^p e^{|\mathrm{Im}\lambda|\,|x|}$.

(1.II.18) ii) For all $x \in \mathbb{R}$, we have $|j_\alpha(\lambda x)| \leq 1$, if and only if $\lambda \in \mathbb{R}$.

1.II.3.b. The Sonine's first integral.

Let $\alpha, \beta \in \mathbb{R}$, such that $\alpha > \beta > -\frac{1}{2}$. Using the relations (1.I.24) and (1.II.3) we obtain

(1.II.19) $\forall\, x > 0,\ \forall\, \lambda > 0,\ j_\alpha(\lambda x) = \dfrac{2\Gamma(\alpha+1)}{\Gamma(\beta+1)\Gamma(\alpha-\beta)} \displaystyle\int_0^1 (1-r^2)^{\alpha-\beta-1} j_\beta(\lambda x r) r^{2\beta+1}\, dr.$

This relation is also true for all $x < 0$, and using the relation (1.II.7), the fact that

$$\frac{2\Gamma(\alpha+1)}{\Gamma(\beta+1)\Gamma(\alpha-\beta)} \int_0^1 (1-r^2)^{\alpha-\beta-1} r^{2\beta+1}\, dr = 1,$$

and the dominated convergence theorem, we deduce that the relation (1.II.19) is available for all $x \in \mathbb{R}$, and $\lambda > 0$. On the other hand from (1.II.17) the functions $\lambda \to j_\alpha(\lambda x)$ and $\lambda \to \int_0^1 (1-r^2)^{\alpha-\beta-1} j_\beta(\lambda x r) r^{2\beta+1}\, dr$ are even entire on $\mathbb{C}$. Then from the analytic extension theorem we have

(1.II.20) $\forall\, x \in \mathbb{R},\ \forall\, \lambda \in \mathbb{C},\ j_\alpha(\lambda x) =$

$$\frac{2\Gamma(\alpha+1)}{\Gamma(\beta+1)\Gamma(\alpha-\beta)} \int_0^1 (1-r^2)^{\alpha-\beta-1} j_\beta(\lambda x r) r^{2\beta+1}\, dr.$$

This relation is called the Sonine's first integral for the normalized Bessel functions of first kind.

1.II.4. Product formula for the function j_α.

Let $\alpha \in \mathbb{R}$, such that $\alpha > -\frac{1}{2}$. From the relations (1.I.29) and (1.II.3) we obtain for all $x > 0$, $y > 0$ and $\lambda > 0$:

(1.II.21) $j_\alpha(\lambda x)\, j_\alpha(\lambda y) = \dfrac{\Gamma(\alpha+1)}{\sqrt{\pi}\,\Gamma(\alpha+\frac{1}{2})} \displaystyle\int_0^\pi j_\alpha(\lambda \sqrt{x^2+y^2-2xy\cos\theta})(\sin\theta)^{2\alpha}\, d\theta.$

We have also the same relation for all $x < 0$ and $y < 0$. As

$$\frac{\Gamma(\alpha+1)}{\sqrt{\pi}\,\Gamma(\alpha+\frac{1}{2})} \int_0^\pi (\sin\theta)^{2\alpha}\, d\theta = 1,$$

we deduce from (1.II.7) and the dominated convergence theorem that the relation (1.II.21) is available for all $x,y \in \mathbb{R}$ and $\lambda > 0$. On the other hand using (1.II.17) we see that the functions $\lambda \to j_\alpha(\lambda x) j_\alpha(\lambda y)$ and $\lambda \to \int_0^\pi j_\alpha(\sqrt{x^2+y^2-2xy\cos\theta})(\sin\theta)^{2\alpha}\, d\theta$ are even entire on $\mathbb{C}$. Then from the analytic extension theorem we obtain

(1.II.22) $$\forall\, x,y \in \mathbb{R},\ \forall\, \lambda \in \mathbb{C},\ j_\alpha(\lambda x)\, j_\alpha(\lambda y) = \frac{\Gamma(\alpha+1)}{\sqrt{\pi}\,\Gamma(\alpha+\frac{1}{2})}\int_0^\pi j_\alpha(\sqrt{x^2+y^2-2xy\cos\theta})(\sin\theta)^{2\alpha}\, d\theta.$$

This relation is called the product formula for the function j_α.

By making in this relation the change of variables

$$t = \sqrt{x^2+y^2-2xy\cos\theta}\,,\ \text{with}\ x>0,\ y>0,$$

we obtain

(1.II.23) $$\forall\, \lambda \in \mathbb{C}, j_\alpha(\lambda x)\ j_\alpha(\lambda x) = \int_0^\infty j_\alpha(\lambda t)\, W(x,y,t)\, d\mu_\alpha(t)\,,$$

with

(1.II.24) $$d\mu_\alpha(t) = \frac{t^{2\alpha+1}}{2^\alpha\, \Gamma(\alpha+1)}\, dt\,,$$

and

(1.II.25) $W(x, y, t) =$

$$\begin{cases} \dfrac{2^{1-\alpha}(\Gamma(\alpha+1))^2}{\sqrt{\pi}\,\Gamma(\alpha+\frac{1}{2})} \cdot \dfrac{[(x+y)^2-t^2]^{\alpha-\frac{1}{2}}[t^2-(x-y)^2]^{\alpha-\frac{1}{2}}}{(xyt)^{2\alpha}}, & \text{if } |x-y| < t < x+y, \\ 0 & , \text{ elswhere}. \end{cases}$$

This function satisfies the following properties

i) For all $t > 0$, $W(x, y, t) \geq 0$.

ii) We have for $x > 0$ and $y > 0$:

(1.II.26) $$\int_0^\infty W(x,y,t)\, d\mu_\alpha(t) = 1.$$

It suffices to take $\lambda = 0$ in the relation (1.II.23) and to use (1.II.7).

iii) We have

(1.II.27) $$\forall\, x, y, t > 0, \quad W(x,y,t) = W(y,x,t).$$

(1.II.28) $$\forall\, x, y, t > 0, \quad W(x, y, t) = W(x, t, y).$$

(1.II.29) $$\forall\, x, y, z, t, s > 0, \quad W(y, z, t)W(t, x, s) = W(x, z, t)\, W(t, y, s).$$

iv) We have

(1.II.30) $$\text{Supp}\, W(x, y, .) = [|x-y|, x - y].$$

v) In [1] we have the relation

(1.II.31) $$W(x, y, t) = \frac{2^{3\alpha-1}(\Gamma(\alpha+1))^2}{\sqrt{\pi}\,\Gamma(\alpha+\frac{1}{2})}\frac{[\Delta(x, y, t)]^{2\alpha-1}}{(xyt)^{2\alpha}},$$

where $\Delta(x, y, t)$ is the area of a triangle whose sides are x, y, t, if there is such a triangle, and otherwise $W(x, y, t) = 0$.

vi) The following formula is attributed to N.J. Sonine :

(1.II.32) $$\int_0^\infty j_\alpha(\lambda x) j_\alpha(\lambda y) j_\alpha(\lambda t)\, d\mu_\alpha(\lambda) = W(x, y, t).$$

(See [91] § 13. 46 (3) and [5] p. 36).

Remark.

Let $\alpha, \beta \in \mathbb{R}$, such that $\alpha > \beta > -\frac{1}{2}$. Using the relation (1.I.30) and applying the analytic extension theorem we obtain the following product formula for the normalized Bessel function of first kind which is more general than the formula (1.II.22). More precisely we have

(1.II.33) $$\forall\, x, y \in \mathbb{R}, \forall\, \lambda \in \mathbb{C},\ j_\alpha(\lambda x)\, j_\beta(\lambda y) = \int_0^1 \int_0^\pi j_\beta(\lambda\sqrt{x^2 + y^2 r^2 + 2xyr\cos\phi}\,)\, dm_{\alpha,\beta}(r, \phi).$$

where $dm_{\alpha,\beta}(r, \phi)$ is the measure given by the relation (1.I.31).

1.II.5. Useful formulas involving the function j_α.

i) For $\alpha \in]-\frac{1}{2}, \frac{1}{2}[$, we have

(1.II.34) $$\forall\, y \geq 0, \int_y^\infty (t^2 - y^2)^{-\alpha-\frac{1}{2}} \sin\lambda t\, dt = \frac{\sqrt{\pi}}{2^{2\alpha+1}} \frac{\Gamma(\frac{1}{2}-\alpha)}{\Gamma(\alpha+1)} \lambda^{2\alpha}\, j_\alpha(\lambda y).$$

ii) For $\mu \in]-1, +\infty[$, $\alpha > -\frac{1}{2}$ such that $\alpha - 2\mu > \frac{1}{2}$ with $\alpha \neq \mu$, we have

(1.II.35) $$\forall\, y \geq 0, \quad \int_y^\infty (t^2 - y^2)^\mu\, j_\alpha(\lambda t)\, t\, dt = \frac{2^{2\mu+1}\,\Gamma(\mu+1)\,\Gamma(\alpha+1)}{\Gamma(\alpha-\mu)}\, \lambda^{-2(\mu+1)}\, j_{\alpha-\mu-1}(\lambda y).$$

iii) For $\alpha \in]0, +\infty[$ we have

(1.II.36) $$\forall\, x > 0, \ \forall\, (\lambda, \mu) \in \mathbb{C}^2, \ j_\alpha(x\sqrt{\lambda^2 + \mu^2}) = \frac{2\Gamma(\alpha+1)x^{-2\alpha}}{\sqrt{\pi}\,\Gamma(\alpha + \frac{1}{2})} \times \int_0^x (x^2 - t^2)^{\alpha-\frac{1}{2}} \cos\mu t\, j_{\alpha-\frac{1}{2}}(\lambda\sqrt{x^2 - t^2})\, dt.$$

1.III. Problems.

Problem 1III.1.

A - Let $\nu \in \mathbb{C}$, such that $\operatorname{Re} \nu \geq 0$. We consider the Bessel differential equation

(E) $$y'' + \frac{1}{x}\, y' + \left(1 - \frac{\nu^2}{x^2}\right) y = 0, \quad x \in]0, +\infty[.$$

1°) Determine the solution of (E) of the form

$$y = x^\nu \sum_{n=0}^\infty a_n x^n,$$

with $a_0 = 2^{-\nu}(\Gamma(\nu+1))^{-1}$.

Show that the solution obtained is the function $J_\nu(x)$ given by the relation (1.I.1).

2°) i) Show that the function $J_{-\nu}(x)$ given by the relation (1.I.2), is also a solution of (E).

ii) Deduce the general solution of (E) for ν noninteger.

3°) We put

$$N_\nu(x) = \frac{\cos\nu\pi\, J_\nu(x) - J_{-\nu}(x)}{\sin\nu\pi}, \quad \text{for } \nu \in \mathbb{C},\ \nu \neq p,\ p \in \mathbb{N}.$$

$$N_p(x) = \lim_{\nu \to p} N_\nu(x).$$

Show that for $\nu \in \mathbb{C}$, such that $\operatorname{Re} \nu \geq 0$, the function $N_\nu(x)$ is a solution of (E).

4°) Using the fact that the function $N_p(x)$, $p \in \mathbb{N}$, possesses the asymtotic formula when $x \to 0^+$:

$$N_p(x) \sim \begin{cases} -x^{-p} , & \text{if } p > 0, \\ \mathrm{Log}(\frac{x}{2}) , & \text{if } p = 0 . \end{cases}$$

Determine the general solution of (E) for $\nu \in \mathbb{C}$, such that $\mathrm{Re}\, \nu \geq 0$.

B - Let $\nu \in \mathbb{C}$ such that $\mathrm{Re}\,(\nu + \frac{1}{2}) > 0$. We consider the function f_ν defined on $\mathbb{R}$ by

$$f_\nu(t) = \begin{cases} (1-t^2)^{\nu - \frac{1}{2}} , & \text{if } |t| < 1, \\ 0 , & \text{elswhere} . \end{cases}$$

1°) Show that the function f_ν is integrable on $\mathbb{R}$ with respect to the Lebesgue measure.

2°) We put

$$\forall\ x > 0, Z_\nu(x) = x^\nu\, \overline{\mathfrak{F}}(f)(x).$$

Where $\overline{\mathfrak{F}}$ is the classical Fourier transform on $\mathbb{R}$ given by

$$\overline{\mathfrak{F}}(g)(x) = \int_{\mathbb{R}} g(t)\, e^{ixt}\, dt.$$

Show that the function Z_ν is a solution of (E).

3°) Deduce that the function $J_\nu(x)$ possesses the Poisson integral representations (1.I.20) and (1.I.22).

(See [76]).

Problem 1.III.2.

Let $\nu \in \mathbb{C}$, such that $\mathrm{Re}\,(\nu + \frac{1}{2}) > 0$. We define the Struve function of order ν, denoted by $\mathbb{H}_\nu(x)$, by the relation

$$\forall\ x > 0, \mathbb{H}_\nu(x) = \frac{2(\frac{x}{2})^\nu}{\sqrt{\pi}\, \Gamma(\nu + \frac{1}{2})} \int_0^1 (1-t^2)^{\nu - \frac{1}{2}} \sin xt\, dt.$$

1°) Show that the function $\mathbb{H}_\nu(x)$ possesses also the integral representation

$$\forall\, x > 0,\ \mathbb{H}_\nu(x) = \frac{2(\frac{x}{2})^\nu}{\sqrt{\pi}\,\Gamma(\nu + \frac{1}{2})}\int_0^{\pi/2} \sin(x\cos\theta)\,(\sin\theta)^{2\nu}\,d\theta\ .$$

2°) Use the precedent question to obtain the series representation

$$\forall\, x \geq 0,\ \mathbb{H}_\nu(x) = \sum_{k=0}^{\infty} \frac{(-1)^k(\frac{x}{2})^{2k+\nu+1}}{\Gamma(k+\frac{3}{2})\Gamma(k+\nu+\frac{3}{2})}\ .$$

3°) Show that

$$\forall\, x > 0,\ \int_{-\pi/2}^{\pi/2} e^{-ix\cos\theta}\cos\theta d\theta = 2 - \pi[\mathbb{H}_1(x) + i\,J_1(x)].$$

4°) Verify that $\mathbb{H}_\nu(x)$ is a solution of the differential equation

$$y'' + \frac{1}{x}y' + (1 - \frac{\nu^2}{x^2})y = \frac{x^{\nu-1}}{2^{\nu+1}\sqrt{\pi}\,\Gamma(\nu + \frac{1}{2})},\quad x \in\,]0, +\infty[\ .$$

5°) Show that

$$\mathbb{H}_{\nu-1}(x) + \mathbb{H}_{\nu+1}(x) = \frac{2\nu}{x}\mathbb{H}_\nu(x) + \frac{x^\nu}{2^\nu\sqrt{\pi}\,\Gamma(\nu + \frac{3}{2})}\ .$$

$$\mathbb{H}_{\nu-1}(x) - \mathbb{H}_{\nu+1}(x) = 2\mathbb{H}'_\nu(x) - \frac{x^\nu}{2^\nu\sqrt{\pi}\,\Gamma(\nu + \frac{3}{2})}\ .$$

$$\mathbb{H}_0(x) = \frac{2}{\pi} - \mathbb{H}_1(x)\ .$$

$$\mathbb{H}_{1/2}(x) = \sqrt{\frac{2}{\pi x}}\,(1 - \cos x)\ .$$

$$\mathbb{H}_{-1/2}(x) = \sqrt{\frac{2}{\pi x}}\,\sin x\ .$$

(Use the second question to define $\mathbb{H}_{-\nu}(x)$).

6°) Show that

$$\mathbb{H}_{-(n+\frac{1}{2})}(x) = (-1)^n J_{n+\frac{1}{2}}(x),\quad n = 0, 1, 2,...$$

(See [3]).

Problem 1.III.3.

A - Let $\nu \in \mathbb{C}$. We define the Anger function by

$$\forall\, x \geq 0,\ \mathbb{J}_\nu(x) = \frac{1}{\pi}\int_0^\pi \cos(\nu\theta - x\sin\theta)d\theta\ .$$

Show that

$$\forall\, x \geq 0,\ \mathbb{J}_{\nu-1}(x) - \mathbb{J}_{\nu+1}(x) = 2\mathbb{J}'_\nu(x)$$

$$\forall\, x > 0,\ \mathbb{J}_{\nu-1}(x) + \mathbb{J}_{\nu+1}(x) = \frac{2\nu}{x}\mathbb{J}_\nu(x) - \frac{2}{\pi x}\sin\nu x\ .$$

B - Let $\nu \in \mathbb{C}^*$. We define the Weber function by

$$\forall\, x \geq 0,\ \mathbb{E}_\nu(x) = \frac{1}{\nu}\int_0^\pi \sin(\nu\theta - x\sin\theta)d\theta\ .$$

Show that

$$\forall\, x \geq 0,\ \mathbb{E}_{\nu-1}(x) - \mathbb{E}_{\nu+1}(x) = 2\mathbb{E}'_\nu(x)\ .$$

$$\forall\, x > 0,\ \mathbb{E}_{\nu-1}(x) + \mathbb{E}_{\nu+1}(x) = \frac{2\nu}{x}\mathbb{E}_\nu(x) - \frac{2}{\pi x}(1-\cos\nu x).$$

(See [3]).

Problem 1.III.4.

Using the generating function relation (1.I.7), prove the formulas

1°) $\forall\, x, \theta \in \mathbb{R},\ \cos(x\sin\theta) = J_0(x) + 2\sum_{n=1}^{\infty} J_{2n}(x)\cos 2n\theta\ .$

2°) $\forall\, x, \theta \in \mathbb{R},\ \sin(x\sin\theta) = 2\sum_{n=0}^{\infty} J_{2n+1}(x)\sin(2n+1)\theta\ .$

3°) $\forall\, x \in \mathbb{R},\ \cos x = J_0(x) + 2\sum_{n=1}^{\infty} (-1)^n J_{2n}(x).$

4°) $\forall\, x \in \mathbb{R},\ \sin x = 2\sum_{n=0}^{\infty} (-1)^n J_{2n+1}(x).$

Problem 1.III.5.

We consider the function $u(x,t)$ given by

$$u(x,t) = \exp[\frac{1}{2}x(t - \frac{1}{t})],\ t \in \mathbb{R}^*,\ x \in \mathbb{R}.$$

1°) Prove the relation

$$\forall\ x, y \in \mathbb{R},\ u(x+y,t) = u(x,t)\ u(y,t).$$

2°) From 1°) deduce the addition formula

$$\forall\ x, y \in \mathbb{R},\ J_n(x+y) = \sum_{k=-\infty}^{\infty} J_k(x)\ J_{n-k}(y)\ .$$

3°) From 2°) derive the relations

$$-\ \forall\ x \in \mathbb{R},\ J_0(2x) = [J_0(x)]^2 + 2\sum_{k=1}^{\infty} (-1)^k [J_k(x)]^2\ .$$

$$-\ \forall\ x \in \mathbb{R},\ [J_0(x)]^2 + 2\sum_{k=1}^{\infty} [J_k(x)]^2 = 1.$$

4°) Deduce from 3°) that

$$\forall\ x \in \mathbb{R},\ |J_0(x)| \leq 1$$

$$\forall\ x \in \mathbb{R},\ |J_n(x)| \leq \frac{\sqrt{2}}{2},\quad n = 1, 2, \ldots$$

Problem 1.III.6.

1°) Let $\nu \in \mathbb{R}$, such that $\nu > -\frac{1}{2}$. Replacing $J_\nu(x)$ by its series representation and using termwise integration, deduce the integral formula

$$\int_0^\infty e^{-ax}\, x^\nu\, J_\nu(bx)dx = \frac{(2b)^\nu\, \Gamma(\nu + \frac{1}{2})}{\sqrt{\pi}\,(a^2 + b^2)^{\nu+\frac{1}{2}}},\quad a, b > 0.$$

2°) Let $\nu \in \mathbb{R}$, such that $\nu > -\frac{1}{2}$. Replacing $J_\nu(x)$ by its integral representation (1.I.22), deduce the integral formula

$$\int_0^\infty x^{2m-\nu-1}\, J_\nu(x)\, dx = \frac{2^{2m-\nu-1}\,\Gamma(m)}{\Gamma(\nu - m - 1)},\ 0 < m < \frac{1}{2}\ .$$

Problem 1.III.7.

We provide $\mathbb{R}^n$, $n \geq 2$, with the euclidean structure defined by the scalar product

$$(x,y) = \sum_{j=1}^{n} x_j y_j ,$$

where $x = (x_1, ..., x_n)$, $y = (y_1,..., y_n) \in \mathbb{R}^n$.

The norm of $x \in \mathbb{R}^n$ is given by

$$\|x\|^2 = (x, x) = \sum_{j=1}^{n} x_j^2 .$$

We consider the space E of functions f defined on $\mathbb{R}^n$, of class C^2 and satisfying the equation

(a) $$\Delta f + f = 0 ,$$

where Δ is the Laplacian on $\mathbb{R}^n$:

$$\Delta = \sum_{j=1}^{n} \frac{\partial^2}{\partial x_j^2} .$$

We denote by E_0 the subspace of radial functions of E.

1°) Show that E_0 is of dimension 1, and formed by the functions φ given by

$$\forall\, x \in \mathbb{R}^n,\ \varphi(x) = k\, j_{\frac{n-2}{2}}(\|x\|), \quad k \in \mathbb{R},$$

where $j_{\frac{n-2}{2}}$ is the function defined by the relation (1.II.1).

2) i) Lest S^{n-1} be the unit sphere of $\mathbb{R}^n$, and σ the normalized measure on S^{n-1}. We consider the function Ψ on $\mathbb{R}^n$ defined by

$$\forall\, x \in \mathbb{R}^n,\ \Psi(x) = \int_{S^{n-1}} e^{i(u,x)}\, d\sigma(u).$$

Prove that Ψ is radial and satisfies the equation (a).

ii) Deduce that the function $j_{\frac{n-2}{2}}$ possesses the integral representation

(b) $$\forall\, x \in \mathbb{R}^n,\ j_{\frac{n-2}{2}}(\|x\|) = \int_{S^{n-1}} e^{i(u,x)}\, d\sigma(u).$$

3) Using the relation (b), find again the Poisson integral representations (1.II.12) and (1.II.13) of the function $j_{\frac{n-2}{2}}(r)$:

$$\forall\, r \in \mathbb{R},\ j_{\frac{n-2}{2}}(r) = \frac{\Gamma(\frac{n}{2})}{\sqrt{\pi}\,\Gamma(\frac{n-1}{2})} \int_{-1}^{1} (1-t^2)^{\frac{n-3}{2}} e^{irt}\, dt.$$

$$\forall\, r \in \mathbb{R},\ j_{\frac{n-2}{2}}(r) = \frac{\Gamma(\frac{n}{2})}{\sqrt{\pi}\,\Gamma(\frac{n-1}{2})} \int_0^\pi \cos(r\cos\theta)(\sin\theta)^{n-2}d\theta.$$

4) i) Let $G = SO(n)$ be the orthogonal group and dg the normalized Haar measure on G. We consider the function F defined on $\mathbb{R}^n$ by

$$F(x) = \int_G f(gx)\,dg.$$

Show that the function F is radial and satisfies the equation (a).

ii) Let f be a function in E. Deduce the relation

(c) $$\forall\, x \in \mathbb{R}^n,\ \forall\, r \geq 0,\ \int_{S^{n-1}} f(x+ru)d\sigma(u) = j_{\frac{n-2}{2}}(r)f(x).$$

iii) By applying the relation (c) to the function $f(x) = j_{\frac{n-2}{2}}(\|x\|)$, find again the product formula (1.II.22) for the function $j_{\frac{n-2}{2}}$:

$$\forall\, r, \rho \in \mathbb{R},\ \ j_{\frac{n-2}{2}}(r)\, j_{\frac{n-2}{2}}(\rho) = \frac{\Gamma(\frac{n}{2})}{\sqrt{\pi}\,\Gamma(\frac{n-1}{2})} \int_0^\pi j_{\frac{n-2}{2}}(\sqrt{r^2+\rho^2-2r\rho\cos\theta})(\sin\theta)^{n-2}d\theta.$$

(See [29]).

Problem 1.III.8.

1°) We consider the function U defined by

$$\forall\, x \geq 0,\ U(x) = \int_0^\pi \cos(x-t)j_0(t)dt.$$

Show that U satisfies the differential equation

$$U''(x) + U(x) = -\frac{x}{2}j_1(x).$$

2°) Deduce that

$$U(x) = A\sin x + B\cos x + xj_0(x),\ A, B \in \mathbb{R}.$$

3) Prove the formula

$$\forall\, x \geq 0,\ \int_0^\pi \cos(x-t)\, j_0(t)dt = xj_0(x).$$

Problem 1.III.9.

We consider the partial differential operators

$$\begin{cases} \Delta_1 = \dfrac{\partial}{\partial x}, \\ \Delta_2 = \dfrac{\partial^2}{\partial r^2} + \dfrac{2\alpha+1}{r}\dfrac{\partial}{\partial r} - \dfrac{\partial^2}{\partial x^2}, \end{cases}$$

with $(r, x) \in]0, +\infty[\times \mathbb{R}$, and $\alpha \in \mathbb{R}$, $\alpha \geq 0$.

1°) Show that the system

$$\begin{cases} \Delta_1 u(r, x) = i\lambda u(r,x) \quad, \quad \lambda \in \mathbb{C}, \\ \Delta_2 u(r, x) = -\mu^2 u(r,x), \quad \mu \in \mathbb{C}, \\ u(0, 0) = 1, \quad \dfrac{\partial}{\partial r} u(0, x) = 0, \quad \text{for all } x \in \mathbb{R}. \end{cases}$$

admits a unique solution $\varphi_{\lambda, \mu}$ of class C^∞ on $[0, \infty[\times \mathbb{R}$ given by

$$\varphi_{\lambda,\mu}(r, x) = e^{i\lambda x} j_\alpha(r\sqrt{\lambda^2 + \mu^2}).$$

2°) Prove that for all $(r, x) \in [0, +\infty[\times \mathbb{R}$, $|\varphi_{\lambda,\mu}(r, x)| \leq 1$, if and only if (λ, μ) belongs to the subset Γ of $\mathbb{C}^2$ given by

$$\Gamma = \mathbb{R} \times [0, \infty[\cup \{(\lambda, i\mu) / \lambda, \mu \in \mathbb{R}, |\mu| \leq |\lambda|\}.$$

3°) Show that the function $\varphi_{\lambda,\mu}$, $(\lambda, \mu) \in \mathbb{C}^2$, possesses the following integral representations

- If $\alpha > 0$:

$$\varphi_{\lambda,\mu}(r, x) = \begin{cases} \dfrac{2\alpha}{\pi} r^{-2\alpha} \displaystyle\int_0^\pi \int_{-\sqrt{r^2-t^2}}^{\sqrt{r^2-t^2}} (r^2 - t^2 - y^2)^{\alpha-1} \cos\mu t \, e^{i\lambda(x+y)} dt dy, & \text{if } r > 0 \text{ and } x \in \mathbb{R}, \\ e^{i\lambda x} & , \text{ if } r = 0 \text{ and } x \in \mathbb{R}. \end{cases}$$

- If $\alpha = 0$:

$$\varphi_{\lambda,\mu}(r, x) = \begin{cases} \dfrac{1}{\pi} \displaystyle\int_0^\pi (r^2 - t^2)^{-\frac{1}{2}} \cos\mu t \, [e^{i\lambda(x+\sqrt{r^2-t^2})} + e^{i\lambda(x-\sqrt{r^2-t^2})}] dt, & \text{if } r > 0 \text{ and } x \in \mathbb{R}, \\ e^{i\lambda x} & , \text{ if } r = 0 \text{ and } x \in \mathbb{R}. \end{cases}$$

4°) Establish the following product formula for the function $\varphi_{\lambda,\mu}$, $(\lambda, \mu) \in \mathbb{C}^2$:

$\forall$ (r, x), (s, y) $\in [0, \infty[\times \mathbb{R}$, $\varphi_{\lambda,\mu}(r,x)\, \varphi_{\lambda,\mu}(s, y) =$

$$\frac{\Gamma(\alpha+1)}{\sqrt{\pi}\,\Gamma(\alpha+\frac{1}{2})}\int_0^{\pi} \varphi_{\lambda,\mu}(\sqrt{r^2+s^2-2rs\cos\theta}\,,\; x+y)(\sin\theta)^{2\alpha}\, d\theta.$$

Problem 1.III.10.

Derive the integral relations

1°) $\forall x \in \mathbb{R},\ \int_0^{\pi/2} j_0(x\cos\phi)\cos\phi\, d\phi = j_{1/2}(x)$.

2°) $\forall x \in \mathbb{R},\ \int_0^{\pi} j_0(2x\cos\phi)\ d\phi = \pi\, [j_0(x)]^2$.

3°) $\forall x \in \mathbb{R},\ \int_0^{\pi/2} j_1(x\cos\phi)\cos\phi\, d\phi = [j_{1/2}(\frac{x}{2})]^2.$

4°) $\forall x \in \mathbb{R}^*,\ \frac{1}{2}\int_0^{\pi/2} [j_1(x\sin\phi)]^2 \sin\phi\, d\phi = \frac{1-j_1(2x)}{x^2}$.

5°) $\forall x, y \in \mathbb{R},\ \int_0^{\pi/2} j_0(x\cos\phi) j_0(y\sin\phi)\sin 2\phi\, d\phi = j_1(\sqrt{x^2+y^2})$.

6°) $\int_0^{\infty} e^{-st} j_n(at)\, t^{2n+1}\, dt = \frac{(2n+1)!}{a^n}\, \frac{s}{(s^2+a^2)^{n+\frac{3}{2}}}$, $a, s > 0,\ n \in \mathbb{N}$.

7°) $\int_0^{\infty} e^{-ax^2} j_\nu(bx) x^{2\nu+1}\, dx = \frac{\Gamma(\nu+1)}{2a^{\nu+1}}\, e^{-b^2/4a}$, $\nu > -1$, $a, b > 0$.

2 Riemann-Liouville and Weyl Integral Transforms

Introduction.

Using the Poisson integral representation of the normalized Bessel function $j_\alpha(\lambda x)$ studied in Chapter 1, J. Delsarte has defined in [21] the Poisson transform $\mathcal{R}_\alpha$ by

$$\forall\, x > 0,\ \mathcal{R}_\alpha(f)(x) = \frac{2\Gamma(\alpha+1)x^{-2\alpha}}{\sqrt{\pi}\,\Gamma(\alpha+\frac{1}{2})}\int_0^x (x^2-y^2)^{\alpha-\frac{1}{2}} f(y)dy.$$

Where f is a continuous function on $[0, +\infty[$.

J.L.Lions has considered in [57] the operator $\mathcal{R}_\alpha$. He has called it Delsarte operator, and has proved that it is a transmutation operator of the Bessel operator ℓ_α given by

$$\ell_\alpha = \frac{d^2}{dx^2} + \frac{2\alpha+1}{x}\frac{d}{dx},$$

into D^2, where $D = \dfrac{d}{dx}$.

The transform $\mathcal{R}_\alpha$ has been called in [83] Riemann-Liouville integral transform. We use this appellation in this chapter. In [83] we have also considered the dual operator W_α of $\mathcal{R}_\alpha$, i.e. the operator satisfying the relation

$$\int_0^\infty \mathcal{R}_\alpha(f)(x)\, g(x)\, x^{2\alpha+1}\, dx = \int_0^\infty W_\alpha(g)(y)\, f(y)\, dy.$$

Where f is continuous on $[0, +\infty[$, and g is continuous on $[0, +\infty[$ with compact support.

The operator W_α is given by the relation

$$\forall\, y \geq 0,\ W_\alpha(g)(y) = \frac{2\Gamma(\alpha+1)}{\sqrt{\pi}\,\Gamma(\alpha+\frac{1}{2})}\int_y^\infty (x^2-y^2)^{\alpha-\frac{1}{2}} f(x)x\, dx.$$

This operator has been called in [83] Weyl integral transform. It is also a transmutation operator of ℓ_α into D^2.

The study of transmutation operators associated with a singular differential operator generalizing the Bessel operator is given in [22] [59] [84] [85].

The operators $\mathcal{R}_\alpha$ and W_α are related to the Radon transform and its dual (see [45] [61] [66] [68] and Problem 2.VI.1).

In this chapter we study the operators $\mathcal{R}_\alpha$ and W_α on different spaces of functions and distributions.

We obtain also the following results

- We determine spaces of functions and distributions on which the operators $\mathcal{R}_\alpha$ and W_α are bijectives.

- We give the expressions of the inverse operators $\mathcal{R}_\alpha^{-1}$ and W_α^{-1}.

The operators $\mathcal{R}_\alpha$ and W_α are useful to study many questions, in particular J.L.Lions has used the Weyl transform to obtain the solution of the radiation problem associated with the Bessel operator (see Problem 2.VI.4 and [58]), I.A. Kiprijanov and V.V. Katrahov [49] have used the Riemann-Liouville integral transform $\mathcal{R}_\alpha$ to define and study a one dimensional singular pseudo-differential operators associated with the Bessel operators.

Generalizations of these questions are given in [85] [18].

Also in this chapter we define and study the Sonine integral transform $S_{\alpha,\beta}$ and its dual ${}^tS_{\alpha,\beta}$, and we prove that these transforms are related to the Riemann-Liouville and Weyl integral transforms $\mathcal{R}_\alpha$ and W_α and their inverse, and we deduce that they are transmutation operators of ℓ_α into ℓ_β.

2.I. The Riemann-Liouville integral transform.

In this chapter we consider $\alpha \in \mathbb{R}$ such that $\alpha > -\frac{1}{2}$.

2.I.1. Definition and properties.

Let f be a continuous function on $[0, +\infty[$. The transform $f \to \mathcal{R}_\alpha(f)$ defined by

(2.I.1) $$\forall\ x \geq 0,\ \mathcal{R}_\alpha(f)(x) = \frac{2\Gamma(\alpha+1)}{\sqrt{\pi}\,\Gamma(\alpha+\frac{1}{2})}\int_0^1 (1-t^2)^{\alpha-\frac{1}{2}} f(tx)dt$$

is called the Riemann-Liouville integral transform. It is also called the Poisson transform (see [21] p. 267).

Remark.

Let f be an even continuous function on $\mathbb{R}$. We define the function $x \dashrightarrow \mathcal{R}_\alpha(f)(x)$ on $]-\infty, 0]$ by parity. Then in this case the relation (2.I.1) remains true for all $x \in \mathbb{R}$.

We can obtain the operator $\mathcal{R}_\alpha$ by the following manner : We consider the partial differential equation

(2.I.2) $$\begin{cases} \phi_{xx} + \dfrac{2\alpha+1}{x}\phi_x - \phi_{yy} = 0, & x > 0,\ y > 0, \\ \phi(0, y) = f(y),\ \phi_x(0, y) = 0, & \text{for all } y \geq 0, \end{cases}$$

where f is an even function on $\mathbb{R}$ which is of classe C^2.

To resolve this equation, we put

(2.I.3) $$x + y = s\sqrt{2}, \quad y - x = t\sqrt{2}$$

and

(2.I.4) $$u(s,t) = \phi\left(\frac{s-t}{\sqrt{2}}, \frac{s+t}{\sqrt{2}}\right).$$

The function u is a solution of the equation

(2.I.5) $$\begin{cases} u_{st} - \dfrac{\alpha+\frac{1}{2}}{s-t}u_s + \dfrac{\alpha+\frac{1}{2}}{s-t}u_t = 0, \\ u(s,s) = f(s\sqrt{2}), \\ (u_s - u_t)_{/s=t} = 0. \end{cases}$$

This equation possesses the solution

(2.I.6) $$u(s,t) = \frac{\Gamma(2\alpha+1)}{(\Gamma(\alpha+\frac{1}{2}))^2}\int_0^1 f((s+(t-s)\rho)\sqrt{2})(1-\rho)^{\alpha-\frac{1}{2}}\rho^{\alpha-\frac{1}{2}}\,d\rho.$$

(See [19]).

Thus

$$\phi(x, 0) = u(\frac{x}{\sqrt{2}}, -\frac{x}{\sqrt{2}})$$
$$= \frac{\Gamma(2\alpha+1)}{(\Gamma(\alpha+\frac{1}{2}))^2}\int_0^1 f(x-2x\rho)(1-\rho)^{\alpha-\frac{1}{2}}\rho^{\alpha-\frac{1}{2}}\,d\rho.$$

If we make the change of variables

$$r = 1 - 2\rho$$

we obtain

$$\phi(x,0) = \frac{\Gamma(2\alpha+1)}{2^{2\alpha}(\Gamma(\alpha+\frac{1}{2}))^2}\int_{-1}^1 (1-r^2)^{\alpha-\frac{1}{2}} f(rx)dr$$
$$= \frac{\Gamma(2\alpha+1)}{2^{2\alpha-1}(\Gamma(\alpha+\frac{1}{2}))^2}\int_0^1 (1-r^2)^{\alpha-\frac{1}{2}} f(rx)dr.$$

But

$$\frac{\Gamma(2\alpha+1)}{2^{2\alpha-1}(\Gamma(\alpha+\frac{1}{2}))^2} = \frac{2\Gamma(\alpha+1)}{\sqrt{\pi}\,\Gamma(\alpha+\frac{1}{2})}.$$

Thus using the relation (2.I.1) we obtain

(2.I.7) $$\phi(x, 0) = \mathcal{R}_\alpha(f)(x).$$

(See [57] p. 58-60).

Remarks.

i) The operator $\mathcal{R}_{\frac{n-2}{2}}$, $n \in \mathbb{N}$, $n \geq 2$, is related to the spherical average $\mathcal{M}_r(h)(x)$ of a continuous function h on $\mathbb{R}^n$, on the sphere of center x and radius r. More precisely we have

(2.I.8) $$\mathcal{M}_r(h)(x) = \frac{1}{\omega_n}\int_{S^{n-1}} h(x + r\omega)\, d\sigma(\omega).$$

where S^{n-1} is the unit sphere of $\mathbb{R}^n$, σ the surface measure on S^{n-1} normalized by

$$\frac{1}{\omega_n}\int_{S^{n-1}} d\sigma(\omega) = 1$$

and

$$\omega_n = \frac{2\pi^{\frac{n}{2}}}{\Gamma(\frac{n}{2})}.$$

In particular if

$$h(x_1, x_2, ..., x_n) = f(x_1)$$

where f is an even continuous function on $\mathbb{R}$, then $\mathcal{M}_r(h)(x)$ takes the form

$$\mathcal{M}_r(h)(x) = \frac{\Gamma(\frac{n}{2})}{\sqrt{\pi}\,\Gamma(\frac{n-1}{2})}\int_{-1}^{1} (1-t^2)^{\frac{n-3}{2}} f(x_1+rt)\, dt.$$

We put $x_1 = 0$. Thus

$$\mathcal{M}_r(h)(0, x_2,..., x_n)) = \frac{2\Gamma(\frac{n}{2})}{\sqrt{\pi}\,\Gamma(\frac{n-1}{2})}\int_{0}^{1} (1-t^2)^{\frac{n-3}{2}} f(rt)dt.$$

From the relation (2.I.1) we see that

(2.I.9) $$\mathcal{M}_r(h)(0, x_2,..., x_n)) = \mathcal{R}_{\frac{n-2}{2}}(f)(r).$$

(See [16] p. 701-703).

ii) The operator $\mathcal{R}_{\frac{n-2}{2}}$ is also related to the wave and Darboux equations. In the following we present this property.

Let f be a function of classe C^2 on $\mathbb{R}^n$, $n \geq 2$. We consider the wave equation

(2.I.10) $$\begin{cases} \dfrac{\partial^2 u}{\partial t^2}(x, t) = \Delta u(x, t)\ , \quad t \in]0, +\infty[, \quad x \in \mathbb{R}^n, \\ u(x, 0) = f(x)\ ; \quad \dfrac{\partial u}{\partial t}(x, 0) = 0, \quad \text{for all } x \in \mathbb{R}^n, \end{cases}$$

where $\Delta = \sum_{j=1}^{n} \dfrac{\partial^2}{\partial x_j^2}$ is the Laplacian on $\mathbb{R}^n$,

and the Darboux equation

(2.I.11) $$\begin{cases} \dfrac{\partial^2 v}{\partial t^2}(x, t) + \dfrac{n-1}{t}\dfrac{\partial v}{\partial t}(x, t) = \Delta v(x, t), \quad t \in]0, +\infty[, \quad x \in \mathbb{R}^n, \\ v(x, 0) = f(x)\ ; \quad \dfrac{\partial v}{\partial t}(x, 0) = 0, \quad \text{for all } x \in \mathbb{R}^n. \end{cases}$$

From [16] p. 699-703, the solutions u and v of theses equations are connected by the relation

(2.I.12) $$v(x,t) = \mathcal{R}_{\frac{n-2}{2}}(u(x,\,.))(t).$$

From the relations (2.I.1) and (1.II.11) we deduce the following properties

i) For f continuous on $[0, +\infty[$ and bounded, the function $\mathcal{R}_\alpha(f)$ is continuous on $[0, +\infty[$ and we have

(2.I.13) $$\bullet\ \mathcal{R}_\alpha(f)(0) = f(0)\,.$$

(2.I.14) $$\bullet\ \|\mathcal{R}_\alpha(f)\|_u \leq \|f\|_u\,,$$

where

$$\|f\|_u = \sup_{x\geq 0} |f(x)|\,.$$

ii) We have

(2.I.15) $$\forall\, x \geq 0,\quad \forall\, \lambda \in \mathbb{C},\ \mathcal{R}_\alpha(\cos\lambda.)(x) = j_\alpha(\lambda x).$$

Theorem 2.I.1. For f of class C^2 on $[0, +\infty[$ such that $f'(0) = 0$, we have

(2.I.16) $$\ell_\alpha(\mathcal{R}_\alpha(f)) = \mathcal{R}_\alpha\left(\frac{d^2}{dt^2}f\right),$$

where ℓ_α is the Bessel operator given by the relation (1.II.9).

Proof.

By derivation under the integral sign, the function $\mathcal{R}_\alpha(f)$ is of class C^2 on $[0, +\infty[$ such that

$$(\mathcal{R}(f))'(0) = 0$$

and we have

(2.I.17) $$\forall\ x \geq 0,\ (\mathcal{R}_\alpha(f))'(x) = \frac{2\Gamma(\alpha+1)}{\sqrt{\pi}\,\Gamma(\alpha+\frac{1}{2})}\int_0^1 (1-t^2)^{\alpha-\frac{1}{2}}\,tf'(tx)\,dt.$$

(2.I.18) $$\forall\ x \geq 0,\ (\mathcal{R}_\alpha(f))''(x) = \frac{2\Gamma(\alpha+1)}{\sqrt{\pi}\,\Gamma(\alpha+\frac{1}{2})}\int_0^1 (1-t^2)^{\alpha-\frac{1}{2}}\,t^2f''(tx)\,dt.$$

By making an integration by parts in the integral of the relation (2.I.17) we obtain

(2.I.19) $$\forall\ x \geq 0,\ (\mathcal{R}_\alpha(f))'(x) = \frac{2\Gamma(\alpha+1)x}{\sqrt{\pi}\,\Gamma(\alpha+\frac{1}{2})(2\alpha+1)}\int_0^1 (1-t^2)^{\alpha+\frac{1}{2}}\,f''(tx)\,dt.$$

On the other hand the relation (2.I.18) can also be written in the form

(2.I.20) $$\forall\ x \geq 0,\ (\mathcal{R}_\alpha(f))''(x) = -\frac{2\Gamma(\alpha+1)}{\sqrt{\pi}\,\Gamma(\alpha+\frac{1}{2})}\int_0^1 (1-t^2)^{\alpha-\frac{1}{2}} f''(tx)\,dt$$
$$+\frac{2\Gamma(\alpha+1)}{\sqrt{\pi}\,\Gamma(\alpha+\frac{1}{2})}\int_0^1 (1-t^2)^{\alpha-\frac{1}{2}} f''(tx)\,dt.$$

From the relations (2.I.19) and (2.I.20) we deduce

(2.I.21) $$\forall\ x \geq 0,\ \ell_\alpha(f))(x) = \frac{2\Gamma(\alpha+1)}{\sqrt{\pi}\,\Gamma(\alpha+\frac{1}{2})}\int_0^1 (1-t^2)^{\alpha-\frac{1}{2}} f''(tx)\,dt.$$

The relation (2.I.1) gives the desired result.

Remark.

The relation (2.I.16) has been given in [57] p. 65-66.

2.I.2. Inversion of the operator $\mathcal{R}_\alpha$.

Notations. We denote by

i) $\mathcal{E}_*(\mathbb{R})$ the space of even C^∞-functions on $\mathbb{R}$. We provide it with the topology of uniform convergence on every compact of the functions and their derivatives. This topology is defined by the semi-norms

(2.I.22) $$p_{m,a}(f) = \sup_{\substack{0\leq k\leq n\\ x\in[-a,a]\\ a\geq 0}} \left|\frac{d^k}{dx^k} f(x)\right|.$$

Equiped with this topology the space $\mathcal{E}_*(\mathbb{R})$ is a Fréchet space.

ii) E_k, $k \in \mathbb{N}$, the space of odd C^∞-functions on $\mathbb{R}$, such that $f^{(p)}(0) = 0$, p = 0, 1, 2, ..., 2k.

iii) We put

(2.I.23) $$\frac{d}{dx^2} = \frac{1}{2x}\frac{d}{dx}.$$

Proposition 2.I.1. The operator $\mathcal{R}_\alpha$ is continuous from $\mathcal{E}_*(\mathbb{R})$ into itself.

Proof.

Let f be in $\mathcal{E}_*(\mathbb{R})$. The function x ----> $\mathcal{R}_\alpha(f)(x)$ is an even function on $\mathbb{R}$. By derivation under the integral sign we deduce that for all $m \in \mathbb{N}$, the function $\mathcal{R}_\alpha(f)$ is of class C^m on $[0, +\infty[$ and we have

(2.I.24) $$\forall\, x \geq 0,\ \frac{d^m}{dx^m}\mathcal{R}_\alpha(f)(x) = \frac{2\Gamma(\alpha+1)}{\sqrt{\pi}\,\Gamma(\alpha+\frac{1}{2})}\int_0^1 (1-t^2)^{\alpha-\frac{1}{2}} t^m f^{(m)}(tx)\,dt.$$

Then the function x ---> $\mathcal{R}_\alpha(f)(x)$ belongs to $\mathcal{E}_*(\mathbb{R})$.

On the other hand we deduce from the relation (2.I.24) that for all $m \in \mathbb{N}$ and a ≥ 0, we have

(2.I.25) $$p_{m,a}(\mathcal{R}_\alpha(f)) \leq p_{m,a}(f).$$

Thus the operator $\mathcal{R}_\alpha$ is continuous from $\mathcal{E}_*(\mathbb{R})$ into itself.

In the following we need the operator Q_k, $k \in \mathbb{N}$, defined on $\mathcal{E}_*(\mathbb{R})$ by

(2.I.26) $$\forall\, x \geq 0,\ Q_k(f)(x) = \frac{2}{k!}\int_0^x (x^2 - y^2)^k f(y)dy$$

By a chang of variables we deduce that

(2.I.27) $$\forall\, x \geq 0,\ Q_k(f)(x) = \frac{\sqrt{\pi}}{\Gamma(k+\frac{3}{2})} x^{2k+1}\mathcal{R}_{k+\frac{1}{2}}(f)(x).$$

Proposition 2.I.2. The operator Q_k, $k \in \mathbb{N}$, is bijective from $\mathcal{E}_*(\mathbb{R})$ onto E_k. The inverse operator Q_k^{-1} is given by

(2.I.28) $$\forall\, x \geq 0,\ Q_k^{-1}(g)(x) = x\left(\frac{d}{dx^2}\right)^{k+1}(g)(x).$$

Proof.

Let f be in $\mathcal{E}_*(\mathbb{R})$. From the relations (2.I.27) and (2.I.1) we deduce that the function x ----> $Q_k(f)(x)$ belongs to E_k.

We consider the operator A_k, $k \in \mathbb{N}$, defined on E_k by

$$\forall\, x \geq 0,\ A_k(g)(x) = x\left(\frac{d}{dx^2}\right)^{k+1}(g)(x),$$

it is clear that the function x ------> $A_k(g)(x)$ belongs to $\mathcal{E}_*(\mathbb{R})$.

For all f in $\mathcal{E}_*(\mathbb{R})$ we obtain by derivation

$$\forall\, x \geq 0,\ A_k(Q_k(f))(x) = f(x).$$

On the other hand, for all g in E_k we deduce by integration by parts that

$$\forall\, x \geq 0,\ Q_k(A_k(g))(x) = g(x).$$

Then the operator Q_k is bijective from $\mathcal{E}_*(\mathbb{R})$ onto E_k and we have

$$\forall\, x \geq 0,\ Q_k^{-1}(g)(x) = A_k(g)(x).$$

Theorem 2.I.2. The operator $\mathcal{R}_\alpha$ is a topological isomorphism from $\mathcal{E}_*(\mathbb{R})$ onto itself. The inverse operator $\mathcal{R}_\alpha^{-1}$ is given by

i) If $\alpha = k+r,\ k \in \mathbb{N},\ -\frac{1}{2} < r < \frac{1}{2}$.

$$\forall\, x \geq 0,\ \mathcal{R}_\alpha^{-1}(f)(x) = C_1 x \left(\frac{d}{dx^2}\right)^{k+1} \left[x^{2k+1} \int_0^1 \frac{t^{2\alpha+1}}{(1-t^2)^{r+\frac{1}{2}}} f(tx)\, dt\right], \tag{2.I.29}$$

where

$$C_1 = \frac{2\sqrt{\pi}}{\Gamma(\alpha+1)\Gamma(\frac{1}{2} - r)}. \tag{2.I.30}$$

ii) If $\alpha = k + \frac{1}{2},\ k \in \mathbb{N}$.

$$\forall\, x \geq 0,\ \mathcal{R}_\alpha^{-1}(f)(x) = C_2 x \left(\frac{d}{dx^2}\right)^{k+1} \left[x^{2k+1} f(x)\right], \tag{2.I.31}$$

where

$$C_2 = 2^{2k+1} \frac{\sqrt{\pi}}{\Gamma(k+\frac{3}{2})}. \tag{2.I.32}$$

Proof.

We denote by B_α the operator defined on $\mathcal{E}_*(\mathbb{R})$ by the second members of the relations (2.I.29) and (2.I.31). It is clear that for all f in $\mathcal{E}_*(\mathbb{R})$ the function $x \dashrightarrow B_\alpha(f)(x)$ belongs to $\mathcal{E}_*(\mathbb{R})$.

Let f be in $\mathcal{E}_*(\mathbb{R})$. Calculate $B_\alpha \circ \mathcal{R}_\alpha(f)$.

- Case 1 : $-\frac{1}{2} < \alpha < \frac{1}{2}$.

By the change of variables $y = tx$ in (2.I.1) and (2.I.29) the operators $\mathcal{R}_\alpha$ and B_α can be written in the form

$$\forall\, x > 0,\ \mathcal{R}_\alpha(f)(x) = \frac{2\Gamma(\alpha+1)x^{-2\alpha}}{\sqrt{\pi}\,\Gamma(\alpha+\frac{1}{2})} \int_0^x (x^2 - y^2)^{\alpha - \frac{1}{2}} f(y)dy \tag{2.I.33}$$

(2.I.34) $$\forall\, x > 0,\ B_\alpha(f)(x) = \frac{C_1}{2}\frac{d}{dx}\int_0^x f(y)\frac{y^{2\alpha+1}}{(x^2-y^2)^{\alpha+\frac{1}{2}}}dy.$$

For all $x > 0$, we have

$$I(x) = \int_0^x \frac{y^{2\alpha+1}}{(x^2-y^2)^{\alpha+\frac{1}{2}}}\mathcal{R}_\alpha(f)(y)dy = \frac{2\Gamma(\alpha+1)}{\sqrt{\pi}\,\Gamma(\alpha+\frac{1}{2})}\times$$
$$\int_0^x \frac{y}{(x^2-y^2)^{\alpha+\frac{1}{2}}}\left(\int_0^y (y^2-u^2)^{\alpha-\frac{1}{2}} f(u)du\right)dy.$$

From Fubini theorem we deduce

$$I(x) = \frac{2\Gamma(\alpha+1)}{\sqrt{\pi}\,\Gamma(\alpha+\frac{1}{2})}\int_0^x \left(\int_u^x (x^2-y^2)^{-\alpha-\frac{1}{2}}(y^2-u^2)^{\alpha-\frac{1}{2}}ydy\right) f(u)du.$$

But by the change of variables

$$v = \frac{y^2-u^2}{x^2-u^2},$$

we obtain

(2.I.35) $$\int_u^x (x^2-y^2)^{-\alpha-\frac{1}{2}}(y^2-u^2)^{\alpha-\frac{1}{2}}y\,dy = \frac{1}{2}\,\Gamma(\frac{1}{2}-\alpha)\ \Gamma(\frac{1}{2}+\alpha).$$

Thus

$$I(x) = \frac{\Gamma(\alpha+1)\,\Gamma(\frac{1}{2}-\alpha)}{\sqrt{\pi}}\int_0^x f(u)du.$$

Using the fait that

(2.I.36) $$C_1 = \frac{2\sqrt{\pi}}{\Gamma(\alpha+1)\,\Gamma(\frac{1}{2}-\alpha)},$$

we obtain

$$\forall\, x > 0,\ B_\alpha \circ \mathcal{R}_\alpha(f)(x) = \frac{C_1}{2}\frac{d}{dx}I(x) = f(x).$$

Then

(2.I.37) $$\forall\, x \geq 0,\ B_\alpha \circ \mathcal{R}_\alpha(f)(x) = f(x).$$

Calculate now $\mathcal{R}_\alpha \circ B(f)$. We have

$$\forall\, y \geq 0,\quad f(y) = f(0) + \int_0^y f'(x)dx.$$

By the change of variables $x = yu$, we obtain

$$\forall\, y \geq 0,\quad f(y) = f(0) + y\int_0^1 f'(yu)du,$$

we write this relation in the form

(2.I.38) $$\forall\, y \geq 0,\quad f(y) = f(0) + g(y)$$

with

$$g(y) = y\int_0^1 f'(yu)du\ .$$

Thus

$$\forall\, x > 0,\ B_\alpha(f)(x) = \frac{C_1}{2}\frac{d}{dx}\int_0^x f(0)\,(x^2 - y^2)^{-\alpha-\frac{1}{2}}\, y^{2\alpha+1}\,dy + \frac{C_1}{2}\frac{d}{dx}\int_0^x g(y)(x^2 - y^2)^{-\alpha-\frac{1}{2}}\, y^{2\alpha+1}\,dy.$$

The change of variables $y = xv$ in the first integrale gives

$$\frac{d}{dx}\int_0^x (x^2 - y^2)^{-\alpha-\frac{1}{2}}\, y^{2\alpha+1}\,dy = \int_0^1 (1-u^2)^{-\alpha-\frac{1}{2}}\, u^{2\alpha+1}\,du = \frac{\Gamma(\frac{1}{2}-\alpha)\,\Gamma(\frac{1}{2}+\alpha)}{\sqrt{\pi}} = \frac{C_1}{2}.$$

Thus

$$\forall\, x > 0,\ B_\alpha(f)(x) = f(0) + \frac{C_1}{2}\frac{d}{dx}\int_0^x (g(y)\,y^{2\alpha})\,(x^2 - y^2)^{-\alpha-\frac{1}{2}}\,y\,dy.$$

By making an integration by parts we obtain

$$\forall\, x > 0,\ B_\alpha(f)(x) = f(0) + \frac{C_1}{2(1-2\alpha)}\frac{d}{dx}\int_0^x (x^2 - y^2)^{\frac{1}{2}-\alpha}\Big(\frac{d}{dy}g(y)y^{2\alpha}\Big)dy.$$

Thus by derivation

$$\forall\, x > 0,\ B_\alpha(f)(x) = f(0) + \frac{C_1}{2}x\int_0^x (x^2 - y^2)^{-\alpha-\frac{1}{2}}\Big(\frac{d}{dy}g(y)y^{2\alpha}\Big)dy.$$

Using this relation and (2.I.33) we have

$$\forall\, x > 0,\ \mathcal{R}_\alpha \circ B_\alpha(f)(x) = f(0) + \frac{C_1\,\Gamma(\alpha+1)}{\sqrt{\pi}\,\Gamma(\alpha+\frac{1}{2})}x^{-2\alpha}\int_0^x (x^2 - y^2)^{\alpha-\frac{1}{2}} \times \Big(\int_0^y (y^2 - t^2)^{-\alpha-\frac{1}{2}}\Big(\frac{d}{dt}g(t)t^{2\alpha}\Big)dt\Big)y\,dy.$$

From Fubini's theorem we deduce

$$\forall\, x > 0,\ \mathcal{R}_\alpha \circ B_\alpha(f)(x) = f(0) + \frac{C_1\,\Gamma(\alpha+1)}{\sqrt{\pi}\,\Gamma(\alpha+\frac{1}{2})}x^{-2\alpha}\int_0^x \Big(\int_t^x (x^2 - y^2)^{\alpha-\frac{1}{2}}(y^2 - t^2)^{-\alpha-\frac{1}{2}}y\,dy\Big)\Big(\frac{d}{dt}g(t)t^{2\alpha}\Big)dt.$$

But by changing α in $(-\alpha)$ in the relation (2.I.35) we have

$$\int_t^x (x^2 - y^2)^{\alpha-\frac{1}{2}} (y^2 - t^2)^{-\alpha-\frac{1}{2}} y\,dy = \frac{1}{2}\,\Gamma(\frac{1}{2} + \alpha)\,\Gamma(\frac{1}{2} - \alpha).$$

Thus

$$\forall\, x > 0,\ \mathcal{R}_\alpha \circ B_\alpha(f)(x) = f(0) + \frac{C_1\,\Gamma(\alpha+1)\,\Gamma(\frac{1}{2} - \alpha)}{\sqrt{\pi}} x^{-2\alpha} \int_0^x \frac{d}{dt}(g(t)t^{2\alpha})dt.$$

Using the relations (2.I.36) and (2.I.38) we obtain

$$\forall\, x > 0,\ \mathcal{R}_\alpha \circ B_\alpha(f)(x) = f(0) + g(x) = f(x).$$

Thus

(2.I.39) $$\forall\, x \geq 0,\ \mathcal{R}_\alpha \circ B_\alpha(f)(x) = f(x).$$

- Case 2 : $\alpha = k + r,\ k \in \mathbb{N}, k \geq 1, -\frac{1}{2} < r < \frac{1}{2}.$

By the change of variables $y = tx$ in (2.I.1) we have

(2.I.40) $$\forall\, x > 0,\ \mathcal{R}_\alpha (f)(x) = \frac{2\Gamma(\alpha+1)}{\sqrt{\pi}\,\Gamma(\alpha + \frac{1}{2})} x^{-2\alpha} \int_0^x (x^2 - y^2)^{\alpha-\frac{1}{2}} f(y)dy.$$

Integrations by parts give

$\forall\, x > 0,\ \mathcal{R}_\alpha (f)(x) =$

$$\frac{\Gamma(2\alpha)\,\Gamma(\alpha-k)\,\Gamma(\alpha+1)\,x^{-2\alpha}}{2^{2k}\sqrt{\pi}\,\Gamma(2\alpha-2k)\,\Gamma(\alpha)\Gamma(\frac{1}{2} + \alpha)} \int_0^x (x^2 - y^2)^{r-\frac{1}{2}} \frac{d}{dy} Q_k(f)(y)dy,$$

where Q_k is the operator given by the relation (2.I.26).

A proof analogous to the precedent , Proposition 2.I.2 and the fact that

$$\Gamma(2z)) = \pi^{-1/2}\, 2^{2z-1}\, \Gamma(z)\, \Gamma(\frac{1}{2}+z)\ ,$$

we obtain

(2.I.41) $$\forall\, x \geq 0,\ B_\alpha \circ \mathcal{R}_\alpha (f)(x) = \mathcal{R}_\alpha \circ B_\alpha(f)(x).$$

- Case 3 : $\alpha = k + \frac{1}{2},\ k \in \mathbb{N}.$

From the relation (2.I.27) we have

(2.I.42) $$\forall\, x > 0\ \mathcal{R}_{k+\frac{1}{2}}(f)(x) = \frac{1}{C_2} x^{-2k-1}\, Q_k(f)(x),$$

where C_2 is the constant given by the relation (2.I.32).

Using Proposition 2.I.2 we obtain

(2.I.43) $$\forall\, x \geq 0,\ B_\alpha \circ \mathcal{R}_\alpha(f)(x) = \mathcal{R}_\alpha \circ B_\alpha(f)(x).$$

From the relations (2.I.37), (2.I.39), (2.I.41) and (2.I.43) we deduce that the operator $\mathcal{R}_\alpha$ is bijective from $\mathcal{E}_*(\mathbb{R})$ onto itself. The inverse operator is B_α.

As the space $\mathcal{E}_*(\mathbb{R})$ is a Fréchet space we deduce from Proposition 2.I.1 that the operator $\mathcal{R}_\alpha^{-1}$ is continuous from $\mathcal{E}_*(\mathbb{R})$ onto itself. Then the operator $\mathcal{R}_\alpha$ is a topological isomorphism from $\mathcal{E}_*(\mathbb{R})$ onto itself.

Theorem 2.I.3. The operator $\mathcal{R}_\alpha$ is the unique topological isomorphism from $\mathcal{E}_*(\mathbb{R})$ onto itself such that

(2.I.44) $$\begin{cases} \forall\, x \geq 0,\ \ell_\alpha(\mathcal{R}_\alpha(f))(x) = \mathcal{R}_\alpha(\frac{d^2}{dt^2} f)(x), \\ \mathcal{R}_\alpha(f)(0) = f(0). \end{cases}$$

Proof.

From Theorems 2.I.1 and 2.I.2 and the relation (2.I.13) we deduce that the operator $\mathcal{R}_\alpha$ is a topological isomorphism from $\mathcal{E}_*(\mathbb{R})$ onto itself satisfying (2.I.44). We prove the unicity of $\mathcal{R}_\alpha$ as follows : From the relation (2.I.15) the functions $j_\alpha(\lambda x) = \mathcal{R}_\alpha(\cos \lambda .)(x)$ are uniquely defined for all $\lambda \in \mathbb{R}$. As the functions x ---> $\cos\lambda x$, for all $\lambda \in \mathbb{R}$, are dense in $\mathcal{E}_*(\mathbb{R})$, we deduce that $\mathcal{R}_\alpha$ is uniquely defined.

Remarks.

i) The operator $\mathcal{R}_\alpha$ is a transmutation operator of ℓ_α into $\frac{d^2}{dt^2}$ on $\mathcal{E}_*(\mathbb{R})$. (See [84]).

ii) Applying a process very close to that of M. Riesz for the extension of Riemann-Liouville integrals, J.L.Lions has studied in [57] p. 75-80, the analytic extension of the operator $\alpha \to \mathcal{R}_\alpha$, and has proved that for $\alpha \in \mathbb{C}$, $\alpha \neq -1, -2, \ldots$, the transform $f \to \mathcal{R}_\alpha(f)$ is a topological isomorphism from $\mathcal{E}_*(\mathbb{R})$ onto itself satisfying (2.I.44).

2.I.3. The Riemann-Liouville integral transform on the spaces $L^p([0,+\infty[, dx)$, $1 \leq p \leq +\infty$.

Notations. We denote by

- $C([0,+\infty[)$ the space of continuous functions on $[0,+\infty[$.

- $L^p([0,+\infty[, dx)$, $1 \leq p \leq +\infty$, the space of measurable functions on $[0,+\infty[$, such that

(2.I.45) $$\|f\|_p = \left(\int_0^\infty |f(x)|^p dx\right)^{1/p} < +\infty, \ 1 \leq p < +\infty.$$

(2.I.46) $$\|f\|_\infty = \operatorname*{ess\,sup}_{x\geq 0} |f(x)| < +\infty.$$

We define the Riemann-Liouville integral transform $\mathcal{R}_\alpha$ on the spaces $L^p([0,+\infty[, dx)$, $1 \leq p \leq +\infty$, by

(2.I.47) $$\mathcal{R}_\alpha(f)(x) = \frac{2\Gamma(\alpha+1)}{\sqrt{\pi}\,\Gamma(\alpha+\frac{1}{2})}\int_0^1 (1-t^2)^{\alpha-\frac{1}{2}} f(tx)dt, \text{ a.e. } x \geq 0.$$

Theorem 2.I.4. For all f in $L^p([0,+\infty[, dx)$, $1 \leq p \leq +\infty$, the function $\mathcal{R}_\alpha(f)$ belongs to $L^p([0,+\infty[, dx)$, $1 \leq p \leq +\infty$, and we have

(2.I.48) $$\|\mathcal{R}_\alpha(f)\|_p \leq k_p \|f\|_p,$$

with

(2.I.49) $$k_p = \begin{cases} 1, \text{ if } p = 1, +\infty, \\ \dfrac{2\Gamma(\alpha+1)}{\sqrt{\pi}\,\Gamma(\alpha+\frac{1}{2})}\left(\dfrac{\sqrt{\pi}}{2}\dfrac{\Gamma(\frac{p-1}{p}(\alpha-\frac{1}{2})+1)}{\Gamma(\frac{p-1}{p}(\alpha-\frac{1}{2})+\frac{3}{2})}\right)^{\frac{p-1}{p}}, \ p \in \,]1,+\infty[. \end{cases}$$

Proof.

- If $p = 1, +\infty$.

We deduce from the relation (2.I.1), Fubini-Tonelli's theorem and the fact that

$$\frac{2\Gamma(\alpha+1)}{\sqrt{\pi}\,\Gamma(\alpha+\frac{1}{2})}\int_0^1 (1-t^2)^{\alpha-\frac{1}{2}} dt = 1$$

that

$$\|\mathcal{R}_\alpha(f)\|_1 \leq \|f\|_1,$$

and

$$\|\mathcal{R}_\alpha(f)\|_\infty \leq \|f\|_\infty .$$

- If $1 < p < +\infty$.

From Hölder's inequality we have

$$\int_0^1 (1-t^2)^{\alpha-\frac{1}{2}} |f(tx)| dt \leq \left(\int_0^1 (1-t^2)^{q(\alpha-\frac{1}{2})} dt\right)^{1/q} \left(\int_0^1 |f(tx)|^p dt\right)^{1/p},$$

where q is the dual exponent of p, i.e. $\frac{1}{p} + \frac{1}{q} = 1.$

Thus using the relation (2.I.1) we have

$$\int_0^\infty |\mathcal{R}_\alpha(f)(x)|^p dx \leq \left(\frac{2\Gamma(\alpha+1)}{\sqrt{\pi}\,\Gamma(\alpha+\frac{1}{2})}\right)^p \left(\int_0^1 (1-t^2)^{q(\alpha-\frac{1}{2})} dt\right)^{p/q} \int_0^\infty \int_0^1 |f(tx)|^p dt\, dx.$$

But from Fubini-Tonelli's theorem we have

$$\int_0^\infty \int_0^1 |f(tx)|^p dt\, dx = \|f\|_p^p.$$

On the other hand we have

$$\int_0^1 (1-t^2)^{q(\alpha-\frac{1}{2})} dt = \frac{\sqrt{\pi}}{2} \frac{\Gamma(q(\alpha-\frac{1}{2})+1)}{\Gamma(q(\alpha-\frac{1}{2})+\frac{3}{2})}.$$

Thus

$$\|\mathcal{R}_\alpha(f)\|_p \leq \frac{2\Gamma(\alpha+1)}{\sqrt{\pi}\,\Gamma(\alpha+\frac{1}{2})} \left(\frac{\sqrt{\pi}}{2} \frac{\Gamma(\frac{p-1}{p}(\alpha-\frac{1}{2})+1)}{\Gamma(\frac{p-1}{p}(\alpha-\frac{1}{2})+\frac{3}{2})}\right)^{\frac{p-1}{p}} \|f\|_p .$$

Theorem 2.I.5. The operator $\mathcal{R}_\alpha$ is injective from $L^p([0,+\infty[, dx) \cap C([0,+\infty[)$, $1 \leq p \leq +\infty$, into itself.

Proof.

From Theorem 2.I.4, for all f in $L^p([0,+\infty[, dx) \cap C([0,+\infty[)$, $1 \leq p \leq +\infty$, the function $x \longrightarrow \mathcal{R}_\alpha(f)(x)$ belong to $L^p([0,+\infty[, dx) \cap C([0,+\infty[)$, $1 \leq p \leq +\infty$.

Let f be in $L^p([0,+\infty[, dx) \cap C([0,+\infty[)$, $1 \leq p \leq +\infty$, such that

(2.I.50) $$\forall\, x \geq 0,\ \mathcal{R}_\alpha(f)(x) = 0.$$

i) Case 1 : $-\frac{1}{2} < \alpha < \frac{1}{2}$.

From the relation (2.I.33) we obtain

$$\forall\, x > 0,\ \int_0^x (x^2 - y^2)^{\alpha-\frac{1}{2}} f(y)dy = 0.$$

Thus

$$\forall\, x > 0,\ \int_0^x \frac{y}{(x^2 - y^2)^{\alpha+\frac{1}{2}}} \Big(\int_0^y (y^2 - u^2)^{\alpha-\frac{1}{2}} f(u)du\Big)\, dy = 0.$$

Using Fubini-Tonelli's theorem and the relation (2.I.35) we have

$$\int_0^x \int_0^y (x^2 - y^2)^{-\alpha-\frac{1}{2}} (y^2 - u^2)^{\alpha-\frac{1}{2}}\, y\, |f(u)| dy\, du$$

$$= \int_0^x \Big(\int_u^x (x^2 - y^2)^{-\alpha-\frac{1}{2}} (y^2 - u^2)^{\alpha-\frac{1}{2}}\, y\, dy\Big) |f(u)| du$$

$$= \frac{1}{2}\Gamma\Big(\frac{1}{2} - \alpha\Big)\, \Gamma\Big(\frac{1}{2} + \alpha\Big) \int_0^x |f(u)|\, du < +\infty.$$

From Fubini's theorem we obtain

$$\forall\, x > 0,\ \int_0^x \Big(\int_u^x (x^2 - y^2)^{-\alpha-\frac{1}{2}} (y^2 - u^2)^{\alpha-\frac{1}{2}}\, y\, dy\Big) f(u)\, du = 0,$$

the relation (2.I.35) implies

$$\forall\, x > 0,\ \int_0^x f(u)\, du = 0.$$

Thus

(2.I.51) $$\forall\, x \geq 0,\ f(x) = 0.$$

ii) Case 2 : $\alpha = k + r,\ k \in \mathbb{N},\ k \geq 1,\ -\frac{1}{2} < r < \frac{1}{2}.$

From the relation (2.I.40) we have

$$\forall\, x > 0,\ \int_0^x (x^2 - y^2)^{k+r-\frac{1}{2}} f(y)\, dy = 0.$$

By integration by parts we obtain

$$\forall\, x > 0,\ \int_0^x (x^2 - y^2)^{r-\frac{1}{2}} \frac{d}{dy} Q_k(f)(y)\, dy = 0,$$

where Q_k is the operator defined by the relation (2.I.26) for f in $L^p([0, +\infty[, dx) \cap C([0,+\infty[)$, $1 \leq p \leq +\infty$.

The same proof than for the i) gives

$$\forall\, x > 0,\ \frac{d}{dx} Q_k(f)(x) = 0.$$

Thus

(2.I.52) $$\forall\ x \geq 0,\ f(x) = 0.$$

iii) Case 3 : $\alpha = k + \dfrac{1}{2},\ k \in \mathbb{N}.$

From the relation (2.I.42) we obtain

$$\forall\ x > 0,\ Q_k f(x) = 0.$$

Then

(2.I.53) $$\forall\ x \geq 0,\quad f(x) = 0.$$

The relations (2.I.51), (2.I.52) and (2.I.53) imply that the operator $\mathcal{R}_\alpha$ is injective on $L^p([0, +\infty[, dx) \cap C([0,+\infty[),\ 1 \leq p \leq +\infty.$

2.II. The Weyl integral transform.

2.II.1. Definition and properties.

Let f be a continuous function on $[0, +\infty[$ with compact support and g a continuous function on $[0, +\infty[$. From Fubini's theorem, the relation (2.I.1) and a change of variables, we deduce the relation

$$\int_0^\infty \mathcal{R}_\alpha(g)(x)f(x)\, x^{2\alpha+1}\, dx = \int_0^\infty \Big[\frac{2\Gamma(\alpha+1)x^{-2\alpha}}{\sqrt{\pi}\,\Gamma(\alpha+\frac{1}{2})}\int_0^x (x^2 - y^2)^{\alpha-\frac{1}{2}} g(y)dy\Big] \times f(x)\, x^{2\alpha+1}\, dx.$$

$$= \int_0^\infty \Big[\frac{2\Gamma(\alpha+1)}{\sqrt{\pi}\,\Gamma(\alpha+\frac{1}{2})}\int_y^\infty (x^2 - y^2)^{\alpha-\frac{1}{2}} f(x)xdx\Big]\, g(y)\, dy.$$

We put

(2.II.1) $$\forall\ y \geq 0,\ W_\alpha(f)(y) = \frac{2\Gamma(\alpha+1)}{\sqrt{\pi}\,\Gamma(\alpha+\frac{1}{2})}\int_y^\infty (x^2 - y^2)^{\alpha-\frac{1}{2}} f(x)xdx.$$

The transform $f \dashrightarrow W_\alpha(f)$ is called the Weyl integral transform. It is the dual of the Riemann-Liouville integral transform $\mathcal{R}_\alpha$ in the following sens

(2.II.2) $$\int_0^\infty \mathcal{R}_\alpha(g)(x)\, f(x)\, x^{2\alpha+1}\, dx = \int_0^\infty W_\alpha(f)(y)g(y)dy.$$

Proposition 2.II.1.

i) Let f be a continuous function on $[0, +\infty[$ with support in the interval $[0,a]$, $a \geq 0$. Then the function $W_\alpha(f)$ is continuous on $[0, +\infty[$ with support in $[0,a]$ and we have

(2.II.3) $$\|W_\alpha(f)\|_u \leq (2a^2)^{\alpha+\frac{1}{2}} \|f\|_u,$$

where

(2.II..4) $$\|f\|_u = \sup_{x\geq 0} |f(x)|.$$

ii) We have for $\alpha \in]-\frac{1}{2}, \frac{1}{2}[$:

(2.II.5) $$\forall\ y \geq 0, \forall\ \lambda \in \mathbb{C},\ W_\alpha(\sin(\lambda.))(y) = \frac{\Gamma(1-\alpha)}{\sqrt{\pi}\,\Gamma(\alpha+\frac{1}{2})}\lambda^{2\alpha} j_\alpha(\lambda y).$$

Proof.

i) Let f be a continuous function on $[0, +\infty[$ with compact support such that $\mathrm{Supp} f \subset [0, a]$, $a \geq 0$. From the relation (2.II.1) it is clear that

$$\forall\ y \notin [0,a],\ W_\alpha(f)(y) = 0,$$

thus

$$\mathrm{supp}\ W_\alpha(f) \subset [0, a],$$

and we have

(2.II.6) $$\forall\ y \in [0,a],\ W_\alpha(f)(y) = \frac{2\Gamma(\alpha+1)}{\sqrt{\pi}\,\Gamma(\alpha+\frac{1}{2})}\int_y^a (x^2 - y^2)^{\alpha-\frac{1}{2}} f(x) x\, dx.$$

The change of variables $t = x - y$, gives

$$\forall\ y \in [0,a],\ W_\alpha(f)(y) = \frac{2\Gamma(\alpha+1)}{\sqrt{\pi}\,\Gamma(\alpha+\frac{1}{2})}\int_0^\infty t^{\alpha-\frac{1}{2}}(t+2y)^{\alpha-\frac{1}{2}}(t+y)f(t+y)\, \mathbb{1}_{]0,a-y[}(t)\, dt,$$

where $\mathbb{1}_{]0,a-y[}$ is the characteristic function of the interval $]0, a-y[$.

We have

- For all $t \in]0, +\infty[$, the function $y \dashrightarrow t^{\alpha-\frac{1}{2}}(t+2y)^{\alpha-\frac{1}{2}}(t+y)f(t+y)\mathbb{1}_{]0,a-y[}(t)$ is continuous on $[0, a]$.

- For all $y \in [0, a]$ we have

$$|t^{\alpha-\frac{1}{2}}(t+2y)^{\alpha-\frac{1}{2}}(t+y)f(t+y)\,1_{]0,a-y[}(t)|$$
$$\leq \begin{cases} \|f\|_u\, 3^{\alpha+\frac{1}{2}}\, a^{2\alpha}\, 1_{]0,a[}(t), & \text{if } \alpha \in [\frac{1}{2}, +\infty[. \\ \|f\|_u\, (2a)^{\alpha+\frac{1}{2}}\, t^{\alpha-\frac{1}{2}}\, 1_{]0,a[}(t), & \text{if } \alpha \in]-\frac{1}{2}, \frac{1}{2}[. \end{cases}$$

It is clear that the function in the second member of this inequality is integrable on $]0, +\infty[$ with respect to the Lebesgue measure, then from the dominated convergence theorem we deduce that the function $y \dashrightarrow W_\alpha(f)(y)$ is continuous on $[0, a]$.

We obtain the inequality (2.II.3) from the relation (2.II.6).

ii) The relations (2.II.1) and (1.II.32) give the result.

Proposition 2.II.1. Let f be a function of class C^2 on $[0, +\infty[$ such that $f'(0) = 0$, and with compact support. Then the function $W_\alpha(f)$ is of class C^1 on $[0, +\infty[$, with compact support, such that $W_\alpha(f))'(0) = 0$, and we have

(2.II.7) $$\forall\, y \geq 0,\ (W_\alpha(f))'(y) = yW_\alpha\Big(\frac{1}{x}\frac{d}{dx}f\Big)(y).$$

Proof.

By an integration by parts in the relation (2.II.1) we have

$$\forall\, y \geq 0,\ W_\alpha(f)(y) = -\frac{\Gamma(\alpha+1)}{\sqrt{\pi}\,\Gamma(\alpha+\frac{3}{2})}\int_y^\infty (x^2-y^2)^{\alpha+\frac{1}{2}} f'(x)\,dx.$$

We apply the derivation theorem to the integral of this relation and we obtain

$$\forall\, y \geq 0,\ (W_\alpha(f))'(y) = \frac{2\Gamma(\alpha+1)y}{\sqrt{\pi}\,\Gamma(\alpha+\frac{1}{2})}\int_y^\infty (x^2-y^2)^{\alpha-\frac{1}{2}} f'(x)\,dx,$$

thus

$$\forall\, y \geq 0,\ (W_\alpha(f))'(y) = y\,W_\alpha\Big(\frac{1}{x}\frac{d}{dx}f\Big)(y),$$

which is the relation (2.II.7).

This relation and Proposition 2.II.1 show that the function $y \dashrightarrow W_\alpha(f)(y)$ is of class C^1 on $[0, +\infty[$, such that

$$(W_\alpha(f))'(0) = 0,$$

and has compact support.

2.II.2. Inversion of the operator W_α.

Notations. We denote by

- $\mathcal{D}_{*,a}(\mathbb{R})$ the space of even C^∞-functions on $\mathbb{R}$ with support in $[-a,a]$, $a \geq 0$. This space is provided with the topology of uniform convergence of functions and their derivatives. This topology is defined by the seminorms

$$p_n(f) = \sup_{\substack{x\in[-a,a]\\ 0\leq k\leq n}} \left|\frac{d^k}{dx^k} f(x)\right| .$$

- $\mathcal{D}_*(\mathbb{R})$ the space of even C^∞-functions on $\mathbb{R}$ with compact support.

$$\mathcal{D}_*(\mathbb{R}) = \bigcup_{a\geq 0} \mathcal{D}_{*,a}(\mathbb{R}).$$

This space shall be given the inductive limit topology.

We say that a sequence $\{f_n\}_{n\in\mathbb{N}}$ of $\mathcal{D}_*(\mathbb{R})$ converges to f in $\mathcal{D}_*(\mathbb{R})$ when $n \longrightarrow +\infty$, if

- There exists $a \geq 0$ such that

$$\forall\, n\in \mathbb{N},\ \operatorname{supp} f_n \subset [-a,a].$$

- f_n converges to f in $\mathcal{D}_{*,a}(\mathbb{R})$.

Equiped with this topology the space $\mathcal{D}_*(\mathbb{R})$ is a Fréchet space.

Lemma 2.II.1. Let f be a function in $\mathcal{E}_*(\mathbb{R})$. Then there exists a function g of class C^∞ on $[0, +\infty[$ such that

$$\forall\, x\in \mathbb{R},\ f(x) = g(x^2), \tag{2.II.8}$$

and for all $n\in\mathbb{N}$ and $a\in[0,+\infty[$, we have

$$\sup_{x\in[-a,a]} \left|\left(\frac{d}{dx^2}\right)^n f(x)\right| \leq \frac{n!}{(2n)!} |f^{(2n)}(0)| + b_n \sup_{x\in[-a,a]} |f^{(2n+1)}(x)| \tag{2.II.9}$$

with b_n a positive constant.

Proof.

The equality

$$\forall\, x\in \mathbb{R},\quad f(x) = g(x^2),$$

defines a function g of class C^∞ on $]0,+\infty[$. It remains to study the derivability of g in the point 0.

We have

$$g'(x^2) = \frac{1}{2x}\frac{d}{dx}(g(x^2)) = \frac{d}{dx^2}(g(x^2)), \tag{2.II.10}$$

and by induction

$$\left(\frac{1}{2x}\frac{d}{dx}\right)^n = \sum_{p=1}^{n} a_{n,p}\, x^{p-2n}\left(\frac{d}{dx}\right)^p, \tag{2.II.11}$$

where $x \in]0, +\infty[$, $n \in \mathbb{N}^*$, $a_{n,p} \in \mathbb{Z}$, $a_{n,n} = 1$.

Then

$$g^{(n)}(x^2) = \sum_{p=1}^{n} a_{n,p}\, x^{p-2n} f^{(p)}(x). \tag{2.II.12}$$

From Taylor's formula we have

$$f^{(p)}(x) = \sum_{q=p}^{2n} f^{(q)}(0)\frac{x^{q-p}}{(q-p)!} + x^{2n-p+1} r_{n,p}(x), \tag{2.II.13}$$

with

$$r_{n,p}(x) = \int_0^1 \frac{(1-s)^{2n-p}}{(2n-p)!} f^{(2n+1)}(xs)ds, \tag{2.II.14}$$

which is a C^∞-function on $\mathbb{R}$.

If we replace in (2.II.12) the function $f^{(p)}(x)$ by its expression (2.II.13), and we use the fact that f is even, we obtain

$$g^{(n)}(x^2) = \sum_{q=1}^{n} c_{n,q}\, x^{2(q-n)} f^{(2n)}(0) + x\, r_n(x), \tag{2.II.15}$$

where $c_{n,q}$, $1 \le q \le n$, are contants independent of f, and

$$r_n(x) = \sum_{p=1}^{n} a_{n,p}\, r_{n,p}(x). \tag{2.II.16}$$

If f is an even polynomial of degree less or equal to $2n$, then $r_n = 0$ and

$$g^{(n)}(x^2) = \sum_{q=1}^{n} c_{n,q}\, x^{2(q-n)} f^{(2n)}(0).$$

But g is a polynomial of degree less or equal to n. Then if we consider the terms of higest degree of f and g, we obtain

$$g^{(n)}(x^2) = \frac{n!}{(2n)!} f^{(2n)}(0).$$

By identification we deduce that

(2.II.17) $$C_{n,n} = \frac{n!}{(2n)!},\quad C_{n,q} = 0, \text{ for } 1 \leq q < n.$$

For f in $\mathcal{E}_*(\mathbb{R})$, we have from (2.II.13) :

(2.II.18) $$g^{(n)}(x^2) = \frac{n!}{(2n)!} f^{(2n)}(0) + x r_n(x).$$

Then $g^{(n)}(x^2)$ admits a limit when $x \to 0$.

We deduce that g is of class C^∞ on $[0, +\infty[$ and for all $a \in [0, +\infty[$ we obtain from (2.II.14) :

(2.II.19) $$\sup_{x\in[-a,a]} |g^{(n)}(x^2)| \leq \frac{n!}{(2n)!}|f^{(2n)}(0)| + b_n \sup_{x\in[-a,a]} |f^{(2n+1)}(x)|,$$

with b_n a positive constant.

Thus (2.II.10) implies

(2.II.20) $$\sup_{x\in[-a,a]} \left|\left(\frac{1}{2x}\frac{d}{dx}\right)^n f(x)\right| \leq \frac{n!}{(2n)!}|f^{(2n)}(0)| + b_n \sup_{x\in[-a,a]} |f^{(2n+1)}(x)|.$$

We obtain (2.II.9) by remarking that this inequality is also true for $n = 0$.

Proposition 2.II.3.

i) For all f in $\mathcal{D}_*(\mathbb{R})$ the function $W_\alpha(f)$ belongs to $\mathcal{D}_*(\mathbb{R})$.

ii) The operator W_α is continuous from $\mathcal{D}_*(\mathbb{R})$ into itself.

Proof.

i) Let f be in $\mathcal{D}_*(\mathbb{R})$. From Lemma 2.II.1 there exists a C^∞-function g on $[0, +\infty[$ with compact support such that

$$\forall x \in \mathbb{R},\ f(x) = g(x^2).$$

Then from (2.II.1) we have

$$\forall y \geq 0,\ W_\alpha(f)(y) = \frac{2\Gamma(\alpha+1)}{\sqrt{\pi}\,\Gamma(\alpha+\frac{1}{2})}\int_y^\infty (x^2 - y^2)^{\alpha-\frac{1}{2}} g(x^2)x\,dx.$$

By the change of variables $t = x^2$ we obtain

$$\forall y \geq 0,\ W_\alpha(f)(y) = \frac{2\Gamma(\alpha+1)}{\sqrt{\pi}\,\Gamma(\alpha+\frac{1}{2})}\int_{y^2}^\infty (t-y^2)^{\alpha-\frac{1}{2}} g(t)dt.$$

This relation permits to extend the function $y \longrightarrow W_\alpha(f)(y)$ to an even function on $\mathbb{R}$. Then the relation (2.II.1) remains true for all $x \in \mathbb{R}$.

From Propositions 2.II.1 and 2.II.2, we deduce that the function $y \longrightarrow W_\alpha(f)(y)$ is a C^∞-function on $[0, +\infty[$, with compact support. Then the function $W_\alpha(f)$ belongs to $\mathcal{D}_*(\mathbb{R})$.

ii) Let $a \in [0, +\infty[$. For all f in $\mathcal{D}_*(\mathbb{R})$ with support in the interval $[-a,a]$ and $n \in \mathbb{N}$, we deduce from the relation (2.II.7), that for all $q \in \mathbb{N}$, $0 \le q \le n$:

$$\sup_{y\in[-a,a]} \left|\frac{d^q}{dy^q} W_\alpha(f)(y)\right| \le \sum_{p=0}^{q} d_p \sup_{y\in[-a,a]} \left|W_\alpha\left(\left(\frac{1}{x}\frac{d}{dx}\right)^p f\right)(y)\right|,$$

where d_p, $0 \le p \le q$, are positive constants.

Using the relations (2.II.9) and (2.II.3), we deduce that there exists a positive constant k_q such that

$$\sup_{y\in[-a,a]} \left|\frac{d^q}{dy^y} W_\alpha(f)(y)\right| \le k_q \sup_{\substack{x\in[a,a]\\ 0\le\ell\le q}} \left|\frac{d^\ell}{dx^\ell} f(x)\right| .$$

Thus there exists $q_0 \in \mathbb{N}$, $0 \le q_0 \le n$, and a positive constant k_{q_0} such that

$$p_q(W_\alpha(f)) \le k_{q_0}\, p_{q_0}(f).$$

This inequality shows that the operator W_α is continuous from $\mathcal{D}_*(\mathbb{R})$ into itself.

Theorem 2.II.1. For f in $\mathcal{D}_*(\mathbb{R})$ we have

$$\forall\ y \ge 0,\ W_\alpha(\ell_\alpha(f))(y) = \frac{d^2}{dy^2} W_\alpha(f)(y), \tag{2.II.21}$$

where ℓ_α is the Bessel operator given by the relation (1.II.9).

Proof.

From the relation (2.II.2) we have for g in $\mathcal{E}_*(\mathbb{R})$ and f in $\mathcal{D}_*(\mathbb{R})$:

$$\int_0^\infty \mathcal{R}_\alpha(g)(x)\, \ell_\alpha(f))(x) x^{2\alpha+1}\, dx = \int_0^\infty W_\alpha(\ell_\alpha(f))(y)\, g(y)\, dy. \tag{2.II.22}$$

If we remark that

$$\ell_\alpha(f)(x) = \frac{1}{x^{2\alpha+1}} \frac{d}{dx}\left[x^{2\alpha+1} \frac{d}{dx} f(x)\right],$$

and we make two integrations by parts in the integral of the first member of (2.II.22), we obtain

(2.II.23) $$\int_0^\infty \ell_\alpha(\mathcal{R}_\alpha(g))(x)\, f(x)\, x^{2\alpha+1}\, dx = \int_0^\infty W_\alpha(\ell_\alpha(f))(y)\, g(y)\, dy.$$

By applying Theorem 2.I.1 to the integral of the first member, we have

(2.II.24) $$\int_0^\infty \mathcal{R}_\alpha(\frac{d^2}{dt^2} g)(x)\, f(x)\, x^{2\alpha+1}\, dx = \int_0^\infty W_\alpha(\ell_\alpha(f))(y)\, g(y)\, dy.$$

Using the relation (2.II.2), the integral of the first member of this relation can also be written in the form

(2.II.25) $$\int_0^\infty \mathcal{R}_\alpha(\frac{d^2}{dt^2} g)(x)\, f(x)\, x^{2\alpha+1}\, dx = \int_0^\infty W_\alpha(f)(y)\, \frac{d^2}{dy^2}\, g(y)\, dy.$$

By making now two integrations by parts in the integral of the second member, we obtain

(2.II.26) $$\int_0^\infty \mathcal{R}_\alpha(\frac{d^2}{dt^2} g)(x)\, f(x)\, x^{2\alpha+1}\, dx = \int_0^\infty \frac{d^2}{dy^2} W_\alpha(f)(y)\;\; g(y)\, dy.$$

Thus from the relations (2.II.24) and (2.II.26) we have

$$\int_0^\infty [W_\alpha(\ell_\alpha(f))(y) - \frac{d^2}{dy^2} W_\alpha(f)(y)]g(y)dy = 0.$$

If suppf $\subset$ [-a,a], $a \geq 0$, then from Proposition 2.II.1, Supp $W_\alpha(f)$) and supp $(\frac{d^2}{dy^2} W_\alpha(f))$ are contained in [-a,a], and thus for all g in $\mathcal{E}_*(\mathbb{R})$:

$$\int_0^a [W_\alpha(\ell_\alpha(f))(y) - \frac{d^2}{dy^2} W_\alpha(f)(y)]\, g(y)dy = 0.$$

Then

$$\forall\, y \geq 0,\ W_\alpha(\ell_\alpha(f))(y) = \frac{d^2}{dy^2} W_\alpha(f)(y).$$

Theorem 2.II.2. The Weyl integral transform is a topological isomorphism from $\mathcal{D}_*(\mathbb{R})$ onto itself. The inverse operator W_α^{-1} is given by

i) If $\alpha = k + r,\ k \in \mathbb{N},\ -\frac{1}{2} < r < \frac{1}{2}$.

(2.II.27) $$\forall\, y \geq 0,\ \ W_\alpha^{-1}(f)(y) = (-1)^{k+1} C_1 \int_y^\infty (x^2 - y^2)^{-r-\frac{1}{2}} (\frac{d}{dx^2})^{k+1} f(x) x\, dx.$$

ii) If $\alpha = k + \frac{1}{2},\ k \in \mathbb{N}$.

(2.II.28) $$\forall\, y \geq 0,\ W_\alpha^{-1}(f)(y) = (-1)^{k+1} C_2 \left(\frac{d}{dy^2}\right)^{k+1} f(y).$$

Where C_1 and C_2 are the constants given in Theorem 2.I.2, and $\frac{d}{dy^2} = \frac{1}{2y}\frac{d}{dy}$.

Remark.

For $\alpha = k + r$, $k \in \mathbb{N}$, $-\frac{1}{2} < r < \frac{1}{2}$, the operator W_α^{-1} can also be written in the form

(2.II.29) $$\forall\, y \geq 0,\ W_\alpha^{-1}(f)(y) = (-1)^{k+1} C_1 \frac{\sqrt{\pi}\,\Gamma(\frac{1}{2} - r)}{2\Gamma(1-r)} W_{-r}\left(\left(\frac{d}{dx^2}\right)^{k+1} f\right)(y).$$

Proof of Theorem 2.II.2.

We denote by V_α the operator defined on $\mathcal{D}_*(\mathbb{R})$ by the second members of the relations (2.II.27) and (2.II.28).

- If $\alpha = k + r$, $k \in \mathbb{N}$, $-\frac{1}{2} < r < \frac{1}{2}$, we deduce from (2.II.29) and Proposition 2.II.3. ii) that the operator V_α is continuous from $\mathcal{D}_*(\mathbb{R})$ into itself.

- If $\alpha = k + \frac{1}{2}$, $k \in \mathbb{N}$, the relation (2.II.9) implies that the operator V_α is continuous from $\mathcal{D}_*(\mathbb{R})$ into itself.

In the following we will prove that

$$\forall\, f \in \mathcal{D}_*(\mathbb{R}),\ W_\alpha \circ V_\alpha(f) = V_\alpha \circ W_\alpha(f) = f.$$

- Case 1 : $\alpha = k + r$, $k \in \mathbb{N}$, $-\frac{1}{2} < r < \frac{1}{2}$.

For g in $\mathcal{E}_*(\mathbb{R})$ and f in $\mathcal{D}_*(\mathbb{R})$ we have from the relation (2.II.27) :

$$\int_0^\infty V_\alpha(f)(y)\, g(y)\, y^{2\alpha+1} dy = (-1)^{k+1} C_1 \int_0^\infty \left[\int_y^\infty (x^2 - y^2)^{-r-\frac{1}{2}} \left(\frac{d}{dx^2}\right)^{k+1} f(x) x\, dx\right] \times g(y)\, y^{2\alpha+1} dy.$$

By applying Fubini's theorem to the second member, we obtain

$$\int_0^\infty V_\alpha(f)(y)\, g(y)\, y^{2\alpha+1} dy = (-1)^{k+1} C_1 \int_0^\infty \left[\int_0^x (x^2 - y^2)^{-r-\frac{1}{2}} g(y)\, y^{2\alpha+1} dy\right] \times \left(\frac{d}{dx^2}\right)^{k+1} f(x) x\, dx.$$

We make integrations by parts in the first integral of the second member and we deduce that

$$\int_0^\infty V_\alpha(f)(y)\, g(y)\, y^{2\alpha+1}\, dy = C_1 \int_0^\infty f(x) \left(\frac{d}{dx^2}\right)^{k+1} \left[\int_0^x (x^2 - y^2)^{-r-\frac{1}{2}} g(y)\, y^{2\alpha+1}\, dy\right] x\, dx.$$

From the relation (2.I.29) we have

$$\int_0^\infty V_\alpha(f)(y)\, g(y)\, y^{2\alpha+1}\, dy = \int_0^\infty f(x)\, \mathcal{R}_\alpha^{-1}(g)(x) dx. \tag{2.II.30}$$

- Case 2 : $\alpha = k + \frac{1}{2}$, $k \in \mathbb{N}$.

Using the relation (2.II.28) and integrations by parts, we show that for g in $\mathcal{E}_*(\mathbb{R})$ and f in $\mathcal{D}_*(\mathbb{R})$ the relation (2.II.30) is also true in this case.

From the relations (2.II.30) and (2.II.2) we deduce that for all $\alpha > -\frac{1}{2}$, we have the following two relations

$$\int_0^\infty V_\alpha \circ W_\alpha(f)(y)\, g(y)\, y^{2\alpha+1}\, dy = \int_0^\infty f(y)\, g(y)\, y^{2\alpha+1}\, dy .$$

$$\int_0^\infty W_\alpha \circ V_\alpha(f)(y)\, g(y)\, dy = \int_0^\infty f(x)\, g(x)\, dx.$$

But from Proposition 2.II.1, the functions $V_\alpha \circ W_\alpha(f)$ and $W_\alpha \circ V_\alpha(f)$ are with compact support. Then

$$V_\alpha \circ W_\alpha(f) = W_\alpha \circ V_\alpha(f) = f.$$

We deduce that the operator W_α is bijective from $\mathcal{D}_*(\mathbb{R})$ onto itself, and the inverse operator W_α^{-1} is V_α.

As the space $\mathcal{D}_*(\mathbb{R})$ is a Fréchet space, we deduce from Proposition 2.II.3 ii) that the operator W_α^{-1} is continuous from $\mathcal{D}_*(\mathbb{R})$ into itself. Then the operator W_α is a topological isomorphism from $\mathcal{D}_*(\mathbb{R})$ onto itself.

Remark.

Theorems 2.II.1 and 2.II.2 show that the operator W_α is a transmutation operator of ℓ_α into $\frac{d^2}{dy^2}$ on $\mathcal{D}_*(\mathbb{R})$.

2.III. The Weyl integral transfrom on the space $\mathcal{S}_*(\mathbb{R})$.

Notation. We denote by $\mathcal{S}_*(\mathbb{R})$ the space of even C^∞-functions on $\mathbb{R}$ which are rapidly decreasing together with all their derivatives, i.e.

(2.III.1) $$\forall\, p, k \in \mathbb{N}, \quad \sup_{x\geq 0} x^p |f^{(k)}(x)| < +\infty.$$

Examples.

i) Let $a \in]0, +\infty[$. The function e^{-ax^2} is in $\mathcal{S}_*(\mathbb{R})$.

ii) The functions of $\mathcal{D}_*(\mathbb{R})$ belong to $\mathcal{S}_*(\mathbb{R})$.

The space $\mathcal{S}_*(\mathbb{R})$ is provided with the topology defined by the seminorms

(2.III.2) $$P_{p,k}(f) = \sup_{\substack{0\leq \ell \leq k \\ x\geq 0}} (1+x^2)^p \left|\frac{d^\ell}{dx^\ell} f(x)\right|.$$

We say that a sequence $\{f_n\}_{n\in\mathbb{N}}$ of $\mathcal{S}_*(\mathbb{R})$ converges to f in $\mathcal{S}_*(\mathbb{R})$ when $n \dashrightarrow +\infty$, if and only if

(2.III.3) $$\forall\, p, k \in \mathbb{N},\ P_{p,k}(f_n - f) \xrightarrow[n\to+\infty]{} 0.$$

Properties

i) We consider the differential operator with constant coefficients

$$D_n = \sum_{k=0}^{n} a_k \frac{d^{2k}}{dx^{2k}}, \quad a_k \in \mathbb{C},$$

and the even polynomial

$$P(x) = \sum_{\ell=0}^{m} b_\ell\, x^{2\ell}, \ b_\ell \in \mathbb{C}.$$

Then for all f in $\mathcal{S}_*(\mathbb{R})$ the function $P(x)\, D_n(f)(x)$ belongs to $\mathcal{S}_*(\mathbb{R})$.

ii) For f and g in $\mathcal{S}_*(\mathbb{R})$, the function $f.g$ belongs to $\mathcal{S}_*(\mathbb{R})$.

Proposition 2.III.1.

i) The space $\mathcal{D}_*(\mathbb{R})$ is dense in $\mathcal{S}_*(\mathbb{R})$.

ii) The injection of $\mathcal{D}_*(\mathbb{R})$ into $\mathcal{S}_*(\mathbb{R})$ is continuous.

Proof.

i) We have

$$\mathcal{D}_*(\mathbb{R}) \subset \mathcal{S}_*(\mathbb{R}).$$

Let f be in $\mathcal{S}_*(\mathbb{R})$. We consider the sequence $\{\varphi_n\}_{n\in\mathbb{N}^*}$ of $\mathcal{D}_*(\mathbb{R})$ defined by

$$\forall\, x \in \mathbb{R},\ \varphi_n(x) = \varphi(\frac{x}{n}),$$

where φ is a function in $\mathcal{S}_*(\mathbb{R})$ such that

$$\begin{cases} \varphi(x) = 1, & \text{if } |x| \le 1, \\ \varphi(x) = 0, & \text{if } |x| \ge 2, \\ 0 \le \varphi(x) \le 1, & \text{elsewhere.} \end{cases}$$

The support of φ_n is contained in the interval [-2n, 2n].

The sequence $\{f_n\}_{n\in\mathbb{N}^*}$ of $\mathcal{S}_*(\mathbb{R})$ given by

$$f_n = f.\varphi_n ,$$

converges to f in $\mathcal{S}_*(\mathbb{R})$ when n tends to infinity.

ii) Let $\{f_n\}_{n\in\mathbb{N}}$ be a sequence of $\mathcal{D}_*(\mathbb{R})$ which converges to zero in $\mathcal{D}_*(\mathbb{R})$ when n tends to infinity, i.e.

- There exists $a \in [0, +\infty[$, such that

$$\forall\, n \in \mathbb{N},\ \operatorname{supp} f_n \subset [-a, a].$$

- For all $k \in \mathbb{N}$, the sequence $\{f_n^{(k)}\}_{n\in\mathbb{N}}$ converges to zero uniformly when n tends to infinity.

But we have

$$\forall\, p, k \in \mathbb{N},\ \sup_{x\ge 0} (1+x^2)^p |f_n^{(k)}(x)| = \sup_{x\in[-a,a]} (1+x^2)^p |f_n^{(k)}(x)|$$
$$\le (1+a^2)^p \sup_{x\ge 0} |f_n^{(k)}(x)|.$$

Thus the sequence $\{f_n\}_{n\in\mathbb{N}}$ converges to zero in $\mathcal{S}_*(\mathbb{R})$ when n tends to infinity.

This completes the proof.

Definition 2.III.1. We define the Weyl integral transform W_α on $\mathcal{S}_*(\mathbb{R})$ by

$$\text{(2.III.4)} \qquad \forall\, y \ge 0,\ W_\alpha(f)(y) = \frac{2\Gamma(\alpha+1)}{\sqrt{\pi}\,\Gamma(\alpha+\frac{1}{2})} \int_y^\infty (x^2 - y^2)^{\alpha-\frac{1}{2}} f(x)x\,dx.$$

Proposition 2.III.2.

i) For all f in $\mathcal{S}_*(\mathbb{R})$, the function $W_\alpha(f)$ belongs to $\mathcal{S}_*(\mathbb{R})$.

ii) The operator W_α is continuous from $\mathcal{S}_*(\mathbb{R})$ into itself.

Proof.

i) The same proof as for Proposition 2.II.3 shows that the function $y \dashrightarrow W_\alpha(f)(y)$ is even on $\mathbb{R}$ and of class C^∞.

On the other hand the change of variables $x = ty$ in the relation (2.III.4) gives

$$\text{(2.III.5)}\qquad \forall\, y > 0,\ W_\alpha(f)(y) = \frac{2\Gamma(\alpha+1)}{\sqrt{\pi}\,\Gamma(\alpha+\frac{1}{2})}\, y^{2\alpha+1}\int_1^\infty (t^2-1)^{\alpha-\frac{1}{2}} f(ty)t\,dt.$$

For all $p, q \in \mathbb{N}$, we obtain from the derivation theorem

$$\forall\, y > 0,\ y^p(W_\alpha(f))^{(q)}(y) = \frac{2\Gamma(\alpha+1)}{\sqrt{\pi}\,\Gamma(\alpha+\frac{1}{2})}\sum_{\ell=0}^{q}\binom{q}{\ell}(2\alpha+1)(2\alpha)\dots(2\alpha-q+\ell+2)\times$$
$$\int_1^\infty (t^2-1)^{\alpha-\frac{1}{2}} y^{2\alpha-q+\ell+p+1} f^{(\ell)}(ty)\, t^{\ell+1}dt.$$

But for all $y \geq 1$:

$$\int_1^\infty (t^2-1)^{\alpha-\frac{1}{2}} y^{2\alpha-q+\ell+p+1}|f^{(\ell)}(ty)|t^{\ell+1}\,dt$$
$$\leq \int_1^\infty (t^2-1)^{\alpha-\frac{1}{2}} (yt)^{2[\alpha]+p+2q+3}|f^{(\ell)}(ty)|\,\frac{dt}{t^{2[\alpha]+p+q+2}},$$

where $[\alpha]$ is the entire part of α.

Thus there exists a positive constant $k_{p,q}$ such that

$$\text{(2.III.6)}\qquad \forall\, y \geq 1,\ y^p\,|(W_\alpha(f))^{(q)}(y)| \leq k_{p,q}\sup_{\substack{0\leq \ell\leq q\\ y\geq 1\\ t\geq 1}} (yt)^{2[\alpha]+p+2q+3}|f^{(\ell)}(yt)|,$$
$$\leq k_{p,q}\, P_{2[\alpha]+p+2q+3\,,\,q}(f).$$

using the continuity of the function $(W_\alpha(f))^{(q)}$ on $[0, +\infty[$ we obtain

$$\sup_{y\geq 0} y^p|(W_\alpha(f))^{(q)}(y)| < +\infty.$$

Then the function $W_\alpha(f)$ belongs to $\mathcal{S}_*(\mathbb{R})$.

ii) From the relation (2.III.6) we deduce that for all $p, k \in \mathbb{N}$, there exist $p_0, k_0 \in \mathbb{N}$, and a positive constant C such that

(2.III.7) $$\forall\ f \in \mathcal{S}_*(\mathbb{R}),\ P_{p,k}(W_\alpha(f)) \leq C\ \ P_{p_0\,,\,k_0}(f).$$

Then the operator W_α is continuous from $\mathcal{S}_*(\mathbb{R})$ into itself.

Lemme 2.III.1.

i) For all f in $\mathcal{S}_*(\mathbb{R})$ the function $x \dashrightarrow \frac{d}{dx^2}f(x)$ belongs to $\mathcal{S}_*(\mathbb{R})$.

ii) For $k \in \mathbb{N}$, the operator $(\frac{d}{dx^2})^{k+1}$ is continuous from $\mathcal{S}_*(\mathbb{R})$ into itself.

Proof.

We deduce the results from the relation

(2.III.8) $$\forall\ x \geq 0,\ f'(x) = -\ x \int_1^\infty f''(tx)dt$$

Proposition 2.III.3. The operator W_α^{-1} defined on $\mathcal{S}_*(\mathbb{R})$ by

- If $\alpha = k + r,\ k \in \mathbb{N},\ -\frac{1}{2} < r < \frac{1}{2}$.

(2.III.9) $$\forall\ y \geq 0,\ W_\alpha^{-1}(f)(y) = (-1)^{k+1}\,C_1 \int_y^\infty (x^2 - y^2)^{-r-\frac{1}{2}}\,(\frac{d}{dx^2})^{k+1} f(x)xdx.$$

- If $\alpha = k + \frac{1}{2},\ k \in \mathbb{N}$.

(2.III.10) $$\forall\ y \geq 0,\ W_\alpha^{-1}(f)(y) = (-1)^{k+1}\,C_2(\frac{d}{dy^2})^{k+1}\,f(y).$$

where C_1 and C_2 are the constants given in Theorem 2.I.2.

satisfies the properties

i) For all f in $\mathcal{S}_*(\mathbb{R})$ the function $y \dashrightarrow W_\alpha^{-1}(f)(y)$ belongs to $\mathcal{S}_*(\mathbb{R})$.

ii) The operator W_α^{-1} is continuous from $\mathcal{S}_*(\mathbb{R})$ into itself.

Proof.

- If $\alpha = k + r,\ k \in \mathbb{N},\ -\frac{1}{2} < r < \frac{1}{2}$.

The relation (2.III.9) can also be written in the form

(2.III.11) $$\forall\ y \geq 0,\ W_\alpha^{-1}(f)(y) = (-1)^{k+1}\,C_1\ \frac{\sqrt{\pi}\,\Gamma(\frac{1}{2} - r)}{\Gamma(1 - r)}\,W_{-r}((\frac{d}{dx^2})^{k+1}\,f)(y).$$

We obtain the i) and ii) from this relation, Lemma 2.III.1 and Proposition 2.III.2.

- If $\alpha = k + \frac{1}{2}$, $k \in \mathbb{N}$.

The i) and ii) are given by Lemma 2.III.1.

Theorem 2.III.1. The Weyl integral transform W_α is a topological isomorphism from $\mathcal{S}_*(\mathbb{R})$ onto itself. The inverse operator is given by W_α^{-1} .

Proof.

From Theorem 2.II.2 we have

$$\forall f \in \mathcal{D}_*(\mathbb{R}),\ W_\alpha \circ W_\alpha^{-1}(f) = W_\alpha^{-1} \circ W_\alpha(f) = f,$$

we deduce the result from Propositions 2.III.1, 2.III.2 and 2.III.3.

Remarks.

i) For f in $\mathcal{S}_*(\mathbb{R})$ we have

$$\forall y \geq 0,\ W_\alpha(\ell_\alpha(f))(y) = \frac{d^2}{dy^2} W_\alpha(f)(y), \tag{2.III.12}$$

where ℓ_α is the Bessel operator given by the relation (1.II.9).

ii) From the relation (2.III.12) and Theorem 2.III.1 we deduce that the operator W_α is a transmutation operator of ℓ_α into $\frac{d^2}{dy^2}$ on $\mathcal{D}_*(\mathbb{R})$.

2.IV. The Weyl transform on the space $\mathcal{E}'_*(\mathbb{R})$.

Notations. We denote by

- $\mathcal{D}'_*(\mathbb{R})$ the space of even distributions on $\mathbb{R}$. It is the topological dual of $\mathcal{D}_*(\mathbb{R})$.

- $\mathcal{E}'_*(\mathbb{R})$ the space of even distributions on $\mathbb{R}$ which are of compact support. It is the topological dual of $\mathcal{E}_*(\mathbb{R})$.

We denote also by W_α the Weyl transform defined on $\mathcal{E}'_*(\mathbb{R})$ by

$$\langle W_\alpha(T), g\rangle = \langle T, \mathcal{R}_\alpha(g)\rangle,\ g \in \mathcal{E}_*(\mathbb{R}) . \tag{2.IV.1}$$

Where $\mathcal{R}_\alpha$ is the Riemann-Liouvelle integral transform defined by the relation (2.I.1).

Proposition 2.IV.1. Let f be in $\mathcal{D}_*(\mathbb{R})$ and $T_{x^{2\alpha+1}f}$ the distribution of $\mathcal{E}'_*(\mathbb{R})$ given by the function $x^{2\alpha+1}f$. Then the distribution $W_\alpha(T_{x^{2\alpha+1}f})$ of $\mathcal{E}'_*(\mathbb{R})$ is given by the function $W_\alpha(f)$ defined by the relation (2.II.1).

Proof.

We have for g in $\mathcal{E}_*(\mathbb{R})$:

$$\begin{aligned}<W_\alpha(T_{x^{2\alpha+1}f}), g> &= <T_{x^{2\alpha+1}f}\ , \mathcal{R}_\alpha(g)> \\ &= \int_0^\infty \mathcal{R}_\alpha(g)(x)f(x)x^{2\alpha+1}\, dx.\end{aligned}$$

Using the relation (2.II.2) we obtain

$$\begin{aligned}<W_\alpha(T_{x^{2\alpha+1}f}), g> &= \int_0^\infty W_\alpha(f)(y)\, g(y)dy \\ &= <T_{W_\alpha(f)}, g>.\end{aligned}$$

Which gives the result.

Theorem 2.IV.1. The Weyl transform W_α is a topological isomorphism from $\mathcal{E}'_*(\mathbb{R})$ onto itself. The inverse operator is given by

(2.IV.2) $$<W_\alpha^{-1}(T), g> = <T, \mathcal{R}_\alpha^{-1}(g)>, \ g \in \mathcal{E}_*(\mathbb{R}).$$

Proof.

We deduce the result from Theorem 2.I.2.

2.V. The Sonine integral transform and its dual.

In this section we consider $\alpha, \beta \in \mathbb{R}$ such that $\alpha > \beta > -\frac{1}{2}$.

2.V.1. The Sonine integral transform.

Let f be a continuous function on $[0, +\infty[$. We define the Sonine integral transform f ---> $S_{\alpha,\beta}(f)$ by

(2.V.1) $$\forall\ x \geq 0,\ S_{\alpha,\beta}(f)(x) = \frac{2\Gamma(\alpha+1)}{\Gamma(\beta+1)\Gamma(\alpha-\beta)}\int_0^1 (1-r^2)^{\alpha-\beta-1} f(rx)r^{2\beta+1}\, dr.$$

Using this relation and (1.II.20) we obtain the following properties

i) For f continuous on $[0, +\infty[$ and bounded, the function $S_{\alpha,\beta}(f)$ is continous on $[0, +\infty[$, and we have

(2.V.2) $\quad \bullet\ S_{\alpha,\beta}(f)(0) = f(0).$

(2.V.3) $\quad \bullet\ \|S_{\alpha,\beta}(f)\|_u \leq \|f\|_u$,

where

(2.V.4)
$$\|f\|_u = \operatorname{Sup}_{x\geq 0} |f(x)|.$$

ii) We have

(2.V.5)
$$\forall\ x \geq 0,\ \forall\, \lambda \in \mathbb{C},\ S_{\alpha,\beta}(j_\beta(\lambda .))(x) = j_\alpha(\lambda x).$$

Theorem 2.V.1. For f of class C^2 on $[0, +\infty[$ such that $f'(0) = 0$, we have

(2.V.6)
$$\forall\ x \geq 0,\ \ell_\alpha(S_{\alpha,\beta}\ (f))(x) = S_{\alpha,\beta}(\ell_\beta(f))(x).$$

where ℓ_α and ℓ_β are the Bessel operators given by

(2.V.7)
$$\ell_\alpha = \frac{d^2}{dx^2} + \frac{2\alpha+1}{x}\frac{d}{dx}.$$

(2.V.8)
$$\ell_\beta = \frac{d^2}{dx^2} + \frac{2\beta+1}{x}\frac{d}{dx}.$$

Proof.

By derivation under the integral sign, the function $S_{\alpha,\beta}(f)$ is of class C^2 on $[0, +\infty[$ such that

$$(S_{\alpha,\beta}(f))'(0) = 0,$$

and we have

(2.V.9)
$$\forall\ x \geq 0,\ (S_{\alpha,\beta}(f))'(x) = \frac{2\Gamma(\alpha+1)}{\Gamma(\beta+1)\Gamma(\alpha-\beta)}\int_0^1 (1-r^2)^{\alpha-\beta-1}\, f'(rx)\, r^{2\beta+2}\, dr.$$

By the change of variables $t = rx$, we have

(2.V.10)
$$\forall\ x \geq 0,\ (S_{\alpha,\beta}(f))'(x) = \frac{2\Gamma(\alpha+1)x^{-2\alpha-1}}{\Gamma(\beta+1)\Gamma(\alpha-\beta)}\int_0^x (x^2 - t^2)^{\alpha-\beta-1}\, f'(t)\, t^{2\beta+2}\, dt.$$

Using an integration by parts and the fact that

(2.V.11)
$$\ell_\beta(f)(t) = \frac{1}{t^{2\beta+1}}\frac{d}{dt}[t^{2\beta+1}\frac{d}{dt}f(t)],$$

we obtain

(2.V.12)
$$\forall\ x \geq 0,\ (S_{\alpha,\beta}(f))'(x) = \frac{\Gamma(\alpha+1)x^{-2\alpha-1}}{\Gamma(\beta+1)\Gamma(\alpha-\beta+1)}\int_0^x (x^2 - t^2)^{\alpha-\beta}\, \ell_\beta(f)(t) t^{2\beta+1}\, dt.$$

By derivation we have

(2.V.13) $$\forall\, x \geq 0,\ (S_{\alpha,\beta}(f))''(x) = - \frac{(2\alpha+1)\Gamma(\alpha+1)x^{-2\alpha-2}}{\Gamma(\beta+1)\Gamma(\alpha-\beta+1)} \int_0^x (x^2 - t^2)^{\alpha-\beta}\, \ell_\beta(f)(t)t^{2\beta+1}\, dt + \frac{2\Gamma(\alpha+1)x^{-2\alpha}}{\Gamma(\beta+1)\Gamma(\alpha-\beta)} \int_0^x (x^2 - t^2)^{\alpha-\beta-1}\, \ell_\beta(f)(t)t^{2\beta+1}\, dt.$$

From this relation and (2.V.11) we have

(2.V.14) $$\forall\, x \geq 0,\ \ell_\alpha(S_{\alpha,\beta}(f))(x) = \frac{2\Gamma(\alpha+1)x^{-2\alpha}}{\Gamma(\beta+1)\Gamma(\alpha-\beta)} \int_0^x (x^2 - t^2)^{\alpha-\beta-1} \ell_\beta(f)(t)t^{2\beta+1}\, dt.$$

A change of variables and the relation (2.V.1) implies the result.

The same proof as for Proposition 2.I.1 gives the following proposition.

Proposition 2.V.1

i) For all f in $\mathcal{E}_*(\mathbb{R})$ the function $x \dashrightarrow S_{\alpha,\beta}(f)(x)$ belongs to $\mathcal{E}_*(\mathbb{R})$.

ii) Th operator $S_{\alpha,\beta}$ is continuous from $\mathcal{E}_*(\mathbb{R})$ into itself.

Theorem 2.V.2.

i) For all f in $\mathcal{E}_*(\mathbb{R})$ we have

(2.V.15) $$S_{\alpha,\beta}(f) = \mathcal{R}_\alpha \circ \mathcal{R}_\beta^{-1}(f).$$

ii) The operator $S_{\alpha,\beta}$ is a topological isomorphism from $\mathcal{E}_*(\mathbb{R})$ onto itself. The inverse operator is given by

(2.V.16) $$\forall\, f \in \mathcal{E}_*(\mathbb{R}),\ S_{\alpha,\beta}^{-1}(f) = \mathcal{R}_\beta(f) \circ \mathcal{R}_\alpha^{-1}(f).$$

Proof.

i) We put for f in $\mathcal{E}_*(\mathbb{R})$:

$$P_{\alpha,\beta}(f) = S_{\alpha,\beta}(f) - \mathcal{R}_\alpha \circ \mathcal{R}_\beta^{-1}(f).$$

From Theorems 2.I.1, 2.V.1, and the relations (2.I.13), (2.V.2), (2.I.15), (2.V.5), we have

$$\begin{cases} \ell_\beta P_{\alpha,\beta}(j_\alpha(\lambda\cdot))(x) = -\lambda^2 P_{\alpha,\beta}(j_\alpha(\lambda\cdot))(x), & x > 0,\ \lambda \in \mathbb{R}, \\ P_{\alpha,\beta}(j_\alpha(\lambda\cdot))(0) = 0. \end{cases}$$

The unicity of the solution of this equation gives

$$P_{\alpha,\beta}(j_\alpha(\lambda\cdot))(x) = 0.$$

But from Proposition 2.I.1 and Theorem 2.I.2, the operator $P_{\alpha,\beta}$ is continuous from $\mathcal{E}_*(\mathbb{R})$ into itself. Thus the density of the family of functions $\{j_\alpha(\lambda x)\}_{\lambda\in\mathbb{R}}$ in $\mathcal{E}_*(\mathbb{R})$ implies

$$P_{\alpha,\beta}(f) = 0, \text{ for all } f \in \mathcal{E}_*(\mathbb{R}),$$

which gives the relation (2.V.15).

ii) We deduce the desired results from the i) and Theorem 2.I.2.

Remarks.

i) The operator $S_{\alpha,\beta}$ is the unique topological isomorphism from $\mathcal{E}_*(\mathbb{R})$ onto itself such that

$$\begin{cases} \ell_\alpha(S_{\alpha,\beta}(f)) = S_{\alpha,\beta}(\ell_\beta(f)), \\ S_{\alpha,\beta}(f)(0) = 0. \end{cases} \tag{2.V.17}$$

ii) The operator $S_{\alpha,\beta}$ is a transmutation operator from ℓ_α into ℓ_β on $\mathcal{E}_*(\mathbb{R})$.

2.V.2. The Sonine integral transform on the spaces $L^p([0, +\infty[, dx)$, $1 \leq p \leq +\infty$.

We define the Sonine integral transform $S_{\alpha,\beta}$ on the spaces $L^p([0, +\infty[, dx)$, $1 \leq p \leq +\infty$, by

$$S_{\alpha,\beta}(f)(x) = \frac{2\Gamma(\alpha+1)}{\Gamma(\beta+1)\Gamma(\alpha-\beta)} \int_0^1 (1-r^2)^{\alpha-\beta-1} f(rx)\, r^{2\beta+1}\, dr, \quad \text{a.e. } x \geq 0. \tag{2.V.18}$$

Using the same proofs as for Theorems 2.I.4 and 2.I.5 we obtain the following results.

Theorem 2.V.3. For all f in $L^p([0, +\infty[, dx)$, $1 \leq p \leq +\infty$, the function $S_{\alpha,\beta}(f)$ belongs to $L^p([0, +\infty[, dx)$, $1 \leq p \leq +\infty$, and we have

$$\|S_{\alpha,\beta}(f)\|_p \leq k_p^o \|f\|_p, \tag{2.V.19}$$

with

(2.V.20) $k_p^o =$

$$\begin{cases} 1, & \text{if } p = 1, +\infty, \\ \dfrac{2\Gamma(\alpha+1)}{\Gamma(\beta+1)\Gamma(\alpha-\beta)}\left(\dfrac{1}{2}\,\dfrac{\Gamma(\frac{p-1}{p}(\alpha-\beta-1)+1)\Gamma(\frac{p-1}{p}(\beta+\frac{1}{2})+\frac{1}{2})}{\Gamma(\frac{p-1}{p}(\alpha-\frac{1}{2})+\frac{3}{2})}\right)^{\frac{p-1}{p}}, & \text{if } p \in]1,+\infty[. \end{cases}$$

Theorem 2.V.4. The operator $S_{\alpha,\beta}$ is injective from $L^P([0, +\infty[, dx) \cap C([0, +\infty[)$, $1 \leq p \leq +\infty$, into itself.

2.V.3. The dual Sonine integral transform.

For f continuous on $[0, +\infty[$, with compact support, we put

$$\forall\, y \geq 0,\ {}^tS_{\alpha,\beta}(f)(y) = \frac{\sqrt{\pi}\,\Gamma(\alpha+1)}{\Gamma(\beta+1)\Gamma(\alpha-\beta+\frac{1}{2})} W_{\alpha-\beta-\frac{1}{2}}(f)(y). \tag{2.V.21}$$

The transform $f \dashrightarrow {}^tS_{\alpha,\beta}(f)$ is called the dual Sonine integral transform. It is the dual of the Sonine integral transform $S_{\alpha,\beta}$ in the following sens

$$\int_0^\infty S_{\alpha,\beta}(g)(x)\, f(x)\, x^{2\alpha+1}dx = \int_0^\infty {}^tS_{\alpha,\beta}(f)(y)g(y)y^{2\beta+1}\, dy \tag{2.V.22}$$

where g is a continous function on $[0, +\infty[$.

Using Theorems 2.II.1, 2.II.2, 2.III.1 and the relation (2.III.12), we deduce the following properties of the operator ${}^tS_{\alpha,\beta}$:

- The operator ${}^tS_{\alpha,\beta}$ is a topological isomorphism from $\mathcal{D}_*(\mathbb{R})$ (resp. $\mathcal{S}_*(\mathbb{R})$) onto itself.

- For all f in $\mathcal{D}_*(\mathbb{R})$ (resp. $\mathcal{S}_*(\mathbb{R})$) we have the permutation relation

$$\forall\, y \geq 0,\ {}^tS_{\alpha,\beta}(\ell_{\alpha-\beta-\frac{1}{2}}(f))(y) = \frac{d^2}{dy^2} S_{\alpha,\beta}(f)(y), \tag{2.V.23}$$

where $\ell_{\alpha-\beta-\frac{1}{2}}$ is the Bessel operator given by

$$\ell_{\alpha-\beta-\frac{1}{2}} = \frac{d^2}{dx^2} + \frac{2(\alpha-\beta)}{x}\frac{d}{dx}. \tag{2.V.24}$$

These properties show that the operator ${}^tS_{\alpha,\beta}$ is a transmutation operator from $\ell_{\alpha-\beta-\frac{1}{2}}$ into $\frac{d^2}{dy^2}$ on $\mathcal{D}_*(\mathbb{R})$ and $\mathcal{S}_*(\mathbb{R})$.

Theorem 2.V.5.

i) For all f in $\mathcal{D}_*(\mathbb{R})$ (resp $\mathcal{S}_*(\mathbb{R})$) we have

$$ {}^tS_{\alpha,\beta}(f) = W_\beta^{-1} \circ W_\alpha(f). \tag{2.V.25}$$

ii) For f in $\mathcal{D}_*(\mathbb{R})$ (resp $\mathcal{S}_*(\mathbb{R})$) we have the permutation relation

$$ {}^tS_{\alpha,\beta}(\ell_\alpha(f)) = \ell_\beta({}^tS_{\alpha,\beta}(f)). \tag{2.V.26}$$

Proof.

i) From Theorem 2.V.2 i) and the relations (2.II.2), (2.V.22) we obtain for f in $\mathcal{D}_*(\mathbb{R})$ and g in $\mathcal{E}_*(\mathbb{R})$:

$$\int_0^\infty {}^tS_{\alpha,\beta}(f)(y)\, g(y)\, y^{2\beta+1}\, dy = \int_0^\infty W_\beta^{-1} \circ W_\alpha(f)(y)\, g(y)\, y^{2\beta+1} dy.$$

As the functions ${}^tS_{\alpha,\beta}(f)$ and $W_\beta^{-1} \circ W_\alpha(f)$ are with compact support, then

$$ {}^tS_{\alpha,\beta}(f) = W_\beta^{-1} \circ W_\alpha(f).$$

Using the continuity of the operators ${}^tS_{\alpha,\beta}$, W_α and W_β^{-1} and the density of $\mathcal{D}_*(\mathbb{R})$ in $\mathcal{S}_*(\mathbb{R})$, we deduce that this relation is also true for f in $\mathcal{S}_*(\mathbb{R})$.

ii) We obtain the result from the i) and the relation (2.III.12).

Remark.

Theorem 2.V.5 shows that the operator ${}^tS_{\alpha,\beta}$ is also a transmutation operator from ℓ_α into ℓ_β on $\mathcal{D}_*(\mathbb{R})$ and $\mathcal{S}_*(\mathbb{R})$.

2.V.4. The Sonine transform on the space $\mathcal{E}'_*(\mathbb{R})$.

We denote also by ${}^tS_{\alpha,\beta}$ the Sonine transform defined on $\mathcal{E}'_*(\mathbb{R})$ by

$$ \langle {}^tS_{\alpha,\beta}(T), g\rangle = \langle T, S_{\alpha,\beta}(g)\rangle, \quad g \in \mathcal{E}_*(\mathbb{R}). \tag{2.V.27}$$

where $S_{\alpha,\beta}$ is the Sonine integral transform defined by the relation (2.V.1).

For all f in $\mathcal{D}_*(\mathbb{R})$ we have the relation.

$$ {}^tS_{\alpha,\beta}(T_{x^{2\alpha+1}f}) = T_{y^{2\beta+1}\,{}^tS_{\alpha,\beta}(f)}\,. \tag{2.V.28}$$

where $f \dashrightarrow {}^tS_{\alpha,\beta}(f)$ is the dual Sonine integral transform given by the relation (2.V.21).

From Theorem 2.V.2.ii) we deduce the following result.

Theorem 2.V.6. The Sonine transform ${}^t S_{\alpha,\beta}$ is a topological isomorphism from $\mathcal{E}'_*(\mathbb{R})$ onto itself. The inverse operator ${}^t S^{-1}_{\alpha,\beta}$ is given by

(2.V.29) $$< {}^t S^{-1}_{\alpha,\beta}(T), g> = <T, S^{-1}_{\alpha,\beta}(g)>, \quad g \in \mathcal{E}_*(\mathbb{R}).$$

2.VI.Problems.

Problem 2.IV.1.

The Radon transform $R(f)$ of a function f in $L^1(\mathbb{R}^n)$ (the space of integrable functions on $\mathbb{R}^n$, $n \geq 2$, with respect to the Lebesgue measure) by

$$R(f)(\theta, s) = R_\theta(f)(s) = \int_{(x,\theta)=s} f(x)dm(x), \quad (\theta, s) \in S^{n-1} \times \mathbb{R}.$$

Where S^{n-1} is the unit sphere in $\mathbb{R}^n$, $(.,.)$ denotes the scalar product of $\mathbb{R}^n$ and dm is the Lebesgue measure on the hyperplan $\{x \in \mathbb{R}^n / (x, \theta) = s\}$.

The dual Radon transform ${}^tR$ is defined for an essentially bounded function g on $S^{n-1} \times \mathbb{R}$ by

$${}^tR(g)(x) = \int_{S^{n-1}} g(\theta, (x, \theta))d\theta,$$

where $d\theta$ is the surface measure on S^{n-1}.

1°) Let $f(x) = f_0(\|x\|)$, with $\|x\|^2 = (x, x)$, a radial function in $L^1(\mathbb{R}^n)$. Show that $R(f)$ is independent of $\theta \in S^{n-1}$ and we have

$$R_\theta(f) = \frac{\pi^{n/2}}{\Gamma(n/2)} W_{\frac{n-2}{2}}(f_0),$$

where $W_{\frac{n-2}{2}}$ is the Weyl integral transform.

2°) Let g be in $L^\infty(\mathbb{R})$ (the space of essentially bounded functions on $\mathbb{R}$ with respect to the Lebesgue measure). Prove that ${}^tR(g)$ is a radial function on $\mathbb{R}^n$ and we have

$${}^tR(g)(x) = \frac{2\pi^{n/2}}{\Gamma(n/2)} \mathcal{R}_{\frac{n-2}{2}}(f)(\|x\|),$$

where $\mathcal{R}_{\frac{n-2}{2}}$ is the Riemann-Liouville integral transform.

3°) We denote by ${}^tR_\theta$ the operator defined for each $\theta \in S^{n-1}$, on $L^\infty(\mathbb{R})$ by

$${}^tR_\theta(g)(x) = g((x, \theta)).$$

Show that for $f \in L^1(\mathbb{R}^n)$ and $g \in L^\infty(\mathbb{R})$, we have

$$ {}^tR_\theta\,(g * R_\theta(f)) = ({}^tR_\theta\,(g)) * f. $$

Where $*$ is the usual convolution product on $\mathbb{R}^n$.

(See [66] and [45] [61]).

Problem 2.VI.2.

We denote by Ω the open half line $]0, +\infty[$ and by

- $\mathcal{D}_+(\Omega)$ the space of C^∞-functions on Ω, which are zero near the origin.

- $\mathcal{D}'_-(\Omega)$ the space of distributions on Ω with support limited to the right. It is the topological dual of $\mathcal{D}_+(\Omega)$.

- $\mathcal{D}_-(\mathbb{R})$ the space of C^∞-functions on $\mathbb{R}$, with support limited to the right.

- $\mathcal{D}'_+(\mathbb{R})$ the space of distributions on $\mathbb{R}$, with support limited to the left. It is the topological dual of $\mathcal{D}_-(\mathbb{R})$.

1°) Prove that the Riemann-Liouville integral transform $\mathcal{R}_\alpha$ defined by the relation (2.I.1) is a topological isomorphism from $\mathcal{D}_+(\mathbb{R})$ onto itself, satisfying the relations

$$ \begin{cases} \ell_\alpha(\mathcal{R}_\alpha(f)) = \mathcal{R}_\alpha(D^2 f), & f \in \mathcal{D}_+(\Omega), \\ \mathcal{R}_\alpha^{-1}(\ell_\alpha(f)) = D^2\mathcal{R}_\alpha^{-1}(f), & f \in \mathcal{D}_+(\Omega). \end{cases} $$

Where ℓ_α is the Bessel operator given by

$$ \ell_\alpha = \frac{d^2}{dx^2} + \frac{2\alpha+1}{x}\frac{d}{dx}, $$

and

$$ D = \frac{d}{dx}. $$

2°) We denote also by W_α the Weyl transform defined on $\mathcal{D}'_-(\Omega)$ by

$$ \langle W_\alpha(T), g\rangle = \langle T, \mathcal{R}_\alpha(g)\rangle, \; g \in \mathcal{D}_+(\Omega). $$

Show that W_α is a topological isomorphism from $\mathcal{D}'_-(\Omega)$ onto itself satisfying the relations

$$\begin{cases} W_\alpha(x^{2\alpha+1}\ell_\alpha(T)) = D^2 W_\alpha(x^{2\alpha+1}T), \ T \in \mathcal{D}'_-(\Omega). \\ \ell_\alpha(x^{-2\alpha-1}W_\alpha^{-1}(T)) = x^{-2\alpha-1}W_\alpha^{-1}(D^2T), \ T \in \mathcal{D}'_-(\Omega). \end{cases}$$

Where $\ell_\alpha(T)$ is the distribution in $\mathcal{D}'_-(\Omega)$ given by

$$<\ell_\alpha(T), g> = <T, \ell_\alpha(g)>, \ g \in \mathcal{D}_+(\Omega).$$

3°) Let f be a C^∞-function on Ω with support limited to the right and $T_{x^{2\alpha+1}f}$ the distribution of $\mathcal{D}'_-(\Omega)$ given by the function $x^{2\alpha+1}f$. Prove that the distribution $W_\alpha(T_{x^{2\alpha+1}f})$ of $\mathcal{D}'_-(\Omega)$ is given by the function $W_\alpha(f)$ defined by the relation (2.II.1).

4°) Let f be a C^∞-function on Ω with support limited to the right and T_f the distribution of $\mathcal{D}'_-(\Omega)$ given by the function f. Show that the distribution $W_\alpha^{-1}(T_f)$ of $\mathcal{D}'_-(\Omega)$ is given by the function denoted $y^{2\alpha+1}W_\alpha^{-1}(f)(y)$, where $W_\alpha^{-1}(f)$ is the function defined by the relations (2.II.27) and (2.II.28).

(See [58] and [85]).

Problem 2.VI.3.

Let $a \in]0, +\infty[$. We denote by $Z_{\alpha,a}$ the distribution of $\mathcal{D}'_+(\mathbb{R})$ associated with the Bessel operator ℓ_α, defined for $f \in \mathcal{D}_-(\mathbb{R})$ by

- If $\alpha = k + r$, $k \in \mathbb{N}$, $-\frac{1}{2} < r < \frac{1}{2}$.

$$<Z_{\alpha,a}, f> = (-1)^{k+1} C_1 \int_a^{+\infty} (x^2-a^2)^{-r-\frac{1}{2}} \left(\frac{d}{dx^2}\right)^{k+1} f(x)x\,dx.$$

- If $\alpha = k + \frac{1}{2}$, $k \in \mathbb{N}$.

$$<Z_{\alpha,a}, f> = (-1)^{k+1} C_2 \left(\frac{d}{dy^2}\right)^{k+1} f(y)_{/y=a}.$$

Where

$$C_1 = \frac{2\sqrt{\pi}}{\Gamma(\alpha+1)\Gamma(\frac{1}{2}-r)}, \text{ and } C_2 = \frac{\sqrt{\pi}}{\Gamma(k+\frac{3}{2})}.$$

The definition of the spaces $\mathcal{D}_-(\mathbb{R})$ and $\mathcal{D}'_+(\mathbb{R})$ is given in Problem 2.VI.2.

1°) Prove that the function $a \to Z_{\alpha,a}$ is infinitely differentiable on $]0, +\infty[$ with values in $\mathcal{D}'_+(\mathbb{R})$.

2°) i) Show that for $\alpha \in]-\frac{1}{2},0[$, the net $\{Z_{\alpha,a}\}_{a>0}$ admits a limit Z_α in $\mathcal{D}'_+(\mathbb{R})$, when a tends to zero, given by

$$Z_\alpha = -\frac{(2\alpha+1)\sqrt{\pi}}{\Gamma(\alpha+1)\Gamma(\frac{1}{2}-\alpha)} \operatorname{Pf}(\frac{Y(x)}{x^{2\alpha+2}}).$$

Where Y is the Heaviside function and Pf the finite part (see [76][77]).

ii) Prove that the distibution Z_α is inversible in $\mathcal{D}'_+(\mathbb{R})$ and we have

$$Z_\alpha^{-1} = \frac{2\Gamma(\alpha+1)}{\sqrt{\pi}\,\Gamma(\alpha+\frac{1}{2})} \operatorname{Pf}(Y(x)x^{2\alpha}).$$

3°) Prove that for $\alpha = 0$, the net $\{Z_{\alpha,a} - (\log a)\delta'\}_{a>0}$ admits a limit in $\mathcal{D}'_+(\mathbb{R})$, when a tends to zero, given by

$$\lim_{a\to 0^+} [Z_{\alpha,a} - (\operatorname{Log} a)\delta'] = (\log 2)\,\delta' - \operatorname{Pf}(\frac{Y(x)}{x^2}).$$

Where δ' is the derivative of the Dirac measure δ at the point zero.

4°) We suppose that $\alpha \in [\frac{1}{2}, +\infty[$. Show that when a tends to zero, the net $\{a^{2\alpha} Z_{\alpha,a}\}_{a>0}$ admits in $\mathcal{D}'_+(\mathbb{R})$ a limit given by

$$\lim_{a\to 0^+} a^{2\alpha} Z_{\alpha,a} = -\frac{1}{2\alpha}\delta'.$$

(See [58] and [85]).

Problem 2.VI.4.

We consider the following subsets of $\mathbb{R}^2$:

$$P = \{(x, y) \in \mathbb{R}^2 \ / \ y > 0\}.$$

$$Q = \{(x, y) \in P \ / \ x \geq y\}.$$

We denote by $\mathcal{D}'(P)$ the space of distributions in P. If $u \in \mathcal{D}'(P)$ we shall write also $(x, y) \to u(x, y)$.

We consider a distribution $u(x, y) \in \mathcal{D}'(P)$ which satisfies

$$(2.\text{VI}.1) \qquad (\frac{\partial^2}{\partial x^2} - \ell_{\alpha,y})\, u(x, y) = 0, \quad \alpha \in \mathbb{R},$$

where $\ell_{\alpha,y}$ is the operator ℓ_α defined in Problem 2.VI.2 acting on the y variable, and

(2.VI.2) u has its support in Q.

The radiation problem associated with the Bessel operator, consists to look for $u \in \mathcal{D}'(P)$, satisfying (2.VI.1), (2.VI.2) and

$$u(.,y) \longrightarrow f, \text{ in } \mathcal{D}'_+(\mathbb{R}), \text{ as } y \longrightarrow 0^+,$$

where f is given in $\mathcal{D}'_+(\mathbb{R})$, $f(x) = 0$, for $x < 0$.

The definition of the space $\mathcal{D}'_+(\mathbb{R})$ is given in Problem 2.VI.2.

1°) Let u(x, y) be a distribution in $\mathcal{D}'(P)$ satisfying (2.VI.1) and (2.VI.2).

i) Show that for $\alpha \in]-\frac{1}{2}, 0[$, we have

$$u(x, y) = Z_{\alpha,y} * f(x),$$

where $f \in \mathcal{D}'_+(\mathbb{R})$, $f(t) = 0$, for $t < 0$, $Z_{\alpha,y}$ the distribution defined in Problem 2.VI.3, and $*$ is the usual convolution product of distributions on $\mathbb{R}$.

(Use the distribution $v(x, y) = (I \otimes W_\alpha)\, x^{2\alpha+1} u(x, y)$, where I is the identity operator and W_α the Weyl transform given in Problem 2.VI.2).

ii) Prove that for $\alpha \in]-\infty, -\frac{1}{2}]$, we have

$$u(x, y) = (y^{-2\alpha}\, \tilde{Z}_{\alpha,y}) * f(x),$$

where $f \in \mathcal{D}'_+(\mathbb{R})$, $f(t) = 0$, for $t < 0$, and $\tilde{Z}_{\alpha,y}$ the distribution associated with the Bessel operator

$$\tilde{\ell}_\alpha = \frac{d^2}{dy^2} + \frac{1-2\alpha}{y}\frac{d}{dy},$$

which we can define as we have made for the distribution $Z_{\alpha,y}$ associated with the Bessel operator ℓ_α.

2°) We consider a distribution u(x, y) in $\mathcal{D}'(P)$ satisfying (2.VI.1) and (2.VI.2). Using the precedent question deduce the following results.

i) u(x, y) is equal to a function $y \to u(.,y)$ infinitely differentiable on $]0, +\infty[$ with values in $\mathcal{D}'_+(\mathbb{R})$.

ii) For $\alpha \in]-\infty, 0[$, u(.,y) converges in $\mathcal{D}'_+(\mathbb{R})$, when y tends to zero, to u(x, 0) given by

$$u(x, 0) = \begin{cases} Z_\alpha * f(x) , & \text{if } \alpha \in]-\frac{1}{2}, 0[, \\ -\frac{1}{2\alpha} \delta' * f(x), & \text{if } \alpha \in]-\infty, -\frac{1}{2}] , \end{cases}$$

with $u(x, 0) = 0$, for $x < 0$.

Where Z_α is the distribution defined in Problem 2.VI.3, and $*$ is the usual convolution product of distributions on $\mathbb{R}$.

iii) For $\alpha \in [0, +\infty[$, when y tends to zero, $u(.,y)$ is not bounded in $\mathcal{D}'_+(\mathbb{R})$, except if $u = 0$.

3°) Prove that the radiation problem associated with the Bessel operator admits for $\alpha \in]-\infty, 0[$ a unique solution $u(x, y)$ which depends continuously from f.

(See [58] and [85]).

3 Convolution Product and Fourier-Cosine Transform of Functions, measures and Distributions

Introduction

We consider in this chapter spaces of even functions on $\mathbb{R}$. We define on these spaces the translation operators σ_x, $x \geq 0$, the convolution product $*_0$ and the Fourier-cosine transform $\mathfrak{F}_0$, respectively by

$$\forall\, y \geq 0,\ \sigma_x(f)(y) = \frac{1}{2}[f(x+y) + f(x-y)].$$

$$\forall\, x \geq 0,\ f *_0 g(x) = \int_0^\infty \sigma_x(f)(y)\, g(y)\, dy.$$

$$\forall\, \lambda \geq 0,\ \mathfrak{F}_0(f)(\lambda) = \int_0^\infty f(x) \cos\lambda x dx.$$

We define also the convolution product $*_0$ and the Fourier-cosine transform on the space of complex bounded measures on $[0, +\infty[$ and on even distributions on $\mathbb{R}$.

Next we give without proofs the main results of the Harmonic Analysis obtained by using σ_x, $*_0$ and $\mathfrak{F}_0$. We deduce these results from the classical Harmonic Analysis on $\mathbb{R}$ (see [24] [27] [43] [53] [74] [76] [77] [79]).

The most of the results of this chapter are used in the following chapters, and we have given them to facilitate the proofs of the main results of these chapters.

3.I. Convolution product of functions and distributions.

Notations. We denote by

- $C_*(\mathbb{R})$ (resp. $C_{*,c}(\mathbb{R})$) the space of even continuous functions on $\mathbb{R}$ (resp. with compact support).

We provide these spaces with the usual topology.

- $C_{*,0}(\mathbb{R})$ the space of even continuous functions on $\mathbb{R}$ which tend to zero at the infinity. We equip it with the topology of uniform convergence. This topology is defined by the norm

$$\|f\|_u = \sup_{x\geq 0} |f(x)|.$$

Provided with this topology $C_{*,0}(\mathbb{R})$ is a Banach space.

3.I.1. The translation operator .

Let f be in $C_*(\mathbb{R})$. We define the translation operators σ_x, $x \geq 0$, by

(3.I.1) $$\forall\ y \geq 0, \sigma_x(f)(y) = \frac{1}{2}[f(x+y) + f(x-y)].$$

Remark.

For f continuous on $[0, +\infty[$, we define σ_x, $x \geq 0$, by

(3.I.2) $$\forall\ y \geq 0, \sigma_x(f)(y) = \frac{1}{2}[f(x+y) + f(|x-y|)].$$

The operators σ_x, $x \geq 0$, satisfy the following properties

i) For all f in $C_*(\mathbb{R})$ we have

(3.I.3) $$\forall\ y \geq 0, \sigma_0(f)(y) = f(y) .$$

(3.I.4) $$\forall\ y \geq 0, \sigma_x(f)(y) = \sigma_y(f)(x).$$

ii) We have the product formula

(3.I.5) $$\forall\ x \geq 0,\ \lambda \in \mathbb{C},\ \sigma_x(\cos(\lambda.))(y) = \cos\lambda x \cos\lambda y.$$

Proposition 3.I.1.

i) Let f be in $C_*(\mathbb{R})$. Then for all $x \geq 0$, the function y ----> $\sigma_x(f)(y)$ belongs to $C_*(\mathbb{R})$.

ii) Let f be in $C_{*,c}(\mathbb{R})$ such that supp f $\subset [-a, a]$, $a \geq 0$. Then for all $x \geq 0$, the function y ----> $\sigma_x(f)(y)$ belongs to $C_{*,c}(\mathbb{R})$ and we have

(3.I.6) $$\text{supp}\ \sigma_x(f) \subset [-a -x, a + x].$$

Proposition 3.I.2.

i) Let f be in $\mathcal{E}_*(\mathbb{R})$. Then for all $x \geq 0$, the function y ----> $\sigma_x(f)(y)$ belongs to $\mathcal{E}_*(\mathbb{R})$.

ii) For all $x \geq 0$, the transform f ---> $\sigma_x(f)$ is continuous from $\mathcal{E}_*(\mathbb{R})$. into itself.

iii) The mapping $x \dashrightarrow \sigma_x$ is even on $\mathbb{R}$ and of class C^∞.

iv) We have

(3.I.7) $$D_x^2 \sigma_x = \sigma_x D^2, \text{ where } D = \frac{d}{dx}.$$

Proposition 3.I.3.

i) Let f be in $L^p([0, +\infty[, dx)$, $1 \le p \le +\infty$. Then for all $x \ge 0$, the function $y \dashrightarrow \sigma_x(f)(y)$ belongs to $L^p([0, +\infty[, dx)$, $1 \le p \le +\infty$, and we have

(3.I.8) $$\|\sigma_x(f)\|_p \le \|f\|_p .$$

ii) For all f in $L^p([0, +\infty[, dx)$, $1 \le p \le +\infty$, we have

(3.I.9) $$\lim_{x \to 0} \|\sigma_x(f) - f\|_p = 0.$$

Proposition 3.I.4. Let f be in $C_{*,0}(\mathbb{R})$. Then for all $x \ge 0$, the function $y \dashrightarrow \sigma_x(f)(y)$ belongs to $C_{*,0}(\mathbb{R})$, and we have

(3.I.10) $$\|\sigma_x(f)\|_u \le \|f\|_u ,$$

and

(3.I.11) $$\lim_{x \to 0} \|\sigma_x(f) - f\|_u = 0.$$

3.I.2. Convolution product of functions.

Definition 3.I.1. The convolution product of f and g in $C_{*,c}(\mathbb{R})$ is the function $f *_0 g$ defined by

(3.I.12) $$\forall\, x \ge 0,\ f *_0 g(x) = \int_0^\infty \sigma_x(f)(y)\, g(y)\, dy.$$

This convolution product possesses the following properties.

i) Let f and g be in $C_{*,c}(\mathbb{R})$ such that $\operatorname{supp} f \subset [-a, a]$, $a \ge 0$, and $\operatorname{supp} g \subset [-b, b]$, $b \ge 0$. Then the function $f *_0 g$ belongs to $C_{*,c}(\mathbb{R})$, and we have

(3.I.13) $$\operatorname{supp} f *_0 g \subset [-a - b, a + b].$$

ii) For all f, g, h in $C_{*,c}(\mathbb{R})$ we have

(3.I.14) $$\bullet\ f *_0 g = g *_0 g.$$

(3.I.15) $$\bullet\ (f *_0 g) *_0 h = f *_0 (g *_0 h).$$

Proposition 3.I.5. Let f be in $L^p([0, +\infty[, dx)$, $1 \le p \le +\infty$, and g be in $L^q([0, +\infty[, dx)$, $1 \le q \le +\infty$. Then the function $f *_0 g$ belongs to $L^r([0, +\infty[, dx)$, $1 \le r \le +\infty$, such that $\frac{1}{r} = \frac{1}{p} + \frac{1}{q} - 1$, and we have

(3.I.16) $$\|f *_0 g\|_r \le \|f\|_p \|g\|_q .$$

Proposition 3.I.6.

i) Let f be in $L^p([0, +\infty[, dx)$, $1 < p < +\infty$, and g be in $L^q([0, +\infty[, dx)$, $1 < q < +\infty$, such that $\frac{1}{p} + \frac{1}{q} = 1$. Then the function $f *_0 g$ is continuous on $[0, +\infty[$, tends to zero when $x \to +\infty$, and we have

(3.I.17) $$\sup_{x \ge 0} |f *_0 g(x)| \le \|f\|_p \|g\|_q .$$

ii) Let f be in $L^1([0, +\infty[, dx)$, and g be in $L^\infty([0, +\infty[, dx)$. Then the function $f *_0 g$ is bounded and continuous on $[0, +\infty[$, and we have

(3.I.18) $$\sup_{x \ge 0} |f *_0 g(x)| \le \|f\|_1 \|g\|_\infty .$$

iii) Let f be in $L^1([0, +\infty[, dx)$ and g be in $C_{*,0}(\mathbb{R})$. Then the function $f *_0 g$ belongs to $C_{*,0}(\mathbb{R})$, and we have

(3.I.19) $$\|f *_0 g\|_u \le \|f\|_1 \|g\|_u .$$

Proposition 3.I.7. Let f be in $L^p([0, +\infty[, dx)$, $1 \le p \le +\infty$, and g in $\mathcal{D}_*(\mathbb{R})$. Then the function $f *_0 g$ belongs to $\mathcal{E}_*(\mathbb{R})$ and we have for all $k \in \mathbb{N}$:

(3.I.20) $$D^{2k}(f *_0 g) = f * (D^{2k} g), \text{ where } D = \frac{d}{dx} .$$

Corollary 3.I.2. For f, g in $\mathcal{D}_*(\mathbb{R})$, the function $f *_0 g$ belongs to $\mathcal{D}_*(\mathbb{R})$.

Proposition 3.I.8. Let φ be a positive function in $L^1([0, +\infty[, dx)$ such that $\|\varphi\|_1 = 1$. We put for $n \in \mathbb{N}^*$:

(3.I.21) $$\varphi_n(x) = n\varphi(nx).$$

Then

i) for all f in $L^p([0, +\infty[, dx)$, $1 \le p < +\infty$, we have

(3.I.22) $$\lim_{n \to +\infty} \|f *_0 \varphi_n - f\|_p = 0.$$

ii) For all f in $C_{*,0}(\mathbb{R})$, we have

(3.I.23) $$\lim_{n\to+\infty} \|f *_0 \varphi_n - f\|_u = 0.$$

Corollary 3.I.3. The space $\mathcal{D}_*(\mathbb{R})$ is dense in $L^p([0, +\infty[, dx)$, $1 \le p < +\infty$.

3.I.3. Convolution product of measures.

We consider the space $C_c([0, +\infty[)$ of continuous functions on $[0, +\infty[$, with compact support. It will be topologized as the inductive limit of the spaces

$$C_a([0, +\infty[) = \{f \in C_c([0, +\infty[) \mid \text{supp } f \subset [0, a]\}, \quad a \ge 0\}.$$

each of which carries the uniform norm.

A complex Radon measure μ on $[0, +\infty[$ is a continuous linear functional on $C_c([0, +\infty[)$. Thus for every $a \ge 0$, there exists a constant k_a such that

$$|\mu(f)| \le k_a \|f\|_u \ , \quad \text{for all } f \in C_a([0, +\infty[).$$

Where

$$\|f\|_u = \sup_{x\ge 0} |f(x)|.$$

The space of Radon measures on $[0, +\infty[$ will be denoted by $M([0, +\infty[)$.

For every μ in $M([0, +\infty[)$ write

$$\|\mu\| = \sup\{\mu(f) \mid f \in C_c([0, +\infty[), \|f\|_u \le 1\}.$$

A measure μ is said to be bounded if $\|\mu\| < +\infty$. In addition μ is called a probability measure if $\mu \ge 0$ and $\|\mu\| = 1$.

The space $M([0, +\infty[)$ has various distinguished subsets, namely $M^b([0, +\infty[)$, $M_+([0, +\infty[)$ and $M^1([0, +\infty[)$, the space of bounded Radon measures, the subsets of positive measures and probability measures respectively.

$M([0, +\infty[)$ can be made into a locally convex space by giving it the $\sigma(M([0, +\infty[), C_c([0, +\infty[))$ topology of simple convergence on $C_c([0, +\infty[))$. This is called the vague topology which can be described by the seminorms $\mu \to \sup\{|\mu(f_i)| / 1 \le i \le n\}$ where $f_i \in C_c([0, +\infty[)$, $i = 1, 2, ..., n$ and $n \in \mathbb{N}$.

The vector spaces $M^b([0, +\infty[)$ and $C_b([0, +\infty[)$ (the space of continuous and bounded functions on $[0, +\infty[)$ form a dual pair. The topology $\sigma(M^b([0, +\infty[),$

$C_b([0, +\infty[)$ will be called the weak topology on $M^b([0, +\infty[)$ (we say also Bernoulli topology).

Definition 3.I.2. The convolution product of μ and ν in $M^b([0, +\infty[)$ is the measure $\mu *_0 \nu$ defined by

(3.I.24) $$\mu *_0 \nu(f) = \int_0^\infty \int_0^\infty \sigma_x(f)(y)\, d\mu(x)\, d\nu(y),\ f \in C_c([0, +\infty[).$$

Where σ_x, $x \geq 0$, are the translation operators given by the relation (3.I.2).

This convolution product possesses the following properties

i) For μ, ν in $M^b([0, +\infty[)$ the measure $\mu *_0 \nu$ belongs to $M^b([0, +\infty[)$ and we have

(3.I.25) $$\|\mu *_0 \nu\| \leq \|\mu\|\, \|\nu\|.$$

ii) For all μ, ν, ω in $M^b([0, +\infty[)$ we have

(3.I.26) $$\bullet\ \mu *_0 \nu = \nu *_0 \mu.$$

(3.I.27) $$\bullet\ \mu *_0 (\nu *_0 \omega) = (\mu *_0 \nu) *_0 \omega.$$

iii) For all $x \geq 0$ and $y \geq 0$, we have

(3.I.28) $$\delta_x *_0 \delta_y(f) = \sigma_x(f)(y),\quad f \in C_c([0, +\infty[).$$

Where δ_x is the Dirac measure at the point x.

iv) Let μ in $M^b([0, +\infty[)$ and $\nu = f\, dx$ with f in $L^1([0, +\infty[, dx)$. We have

(3.I.29) $$\mu *_0 \nu = \mu *_0 (f\, dx) = (\mu *_0 f)\, dx.$$

Where $\mu *_0 f$ is the function given by

(3.I.30) $$\forall\, x \geq 0,\ \mu *_0 f(x) = \int_0^\infty \sigma_x(f)(y) d\mu(y),$$

called convolution product of the measure μ and the function f.

v) If $\mu = g dx$ and $\nu = h dx$, with g, h in $L^1([0, +\infty[, dx)$. Then

(3.I.31) $$\mu *_0 \nu = (g dx) *_0 (h\, dx) = (g *_0 h)\, dx.$$

Where $g *_0 h$ is the convolution product of g and h given by the relation (3.I.12).

3.I.4. Convolution product of distributions.

Notations. We denote by $\mathcal{S}'_*(\mathbb{R})$ the space of even tempered distributions on $\mathbb{R}$. It is the topological dual of $\mathcal{S}_*(\mathbb{R})$.

Definition 3.I.3.

i) Let T be in $\mathcal{D}'_*(\mathbb{R})$ (resp. $\mathcal{S}'_*(\mathbb{R})$) and Ψ be in $\mathcal{D}_*(\mathbb{R})$.(resp. $\mathcal{S}_*(\mathbb{R})$). The convolution product of T and Ψ is the function $T *_0 \Psi$ defined by

(3.I.32) $$\forall\, x \geq 0,\ T *_0 \Psi\,(x) = \langle T, \sigma_x(\Psi)\rangle.$$

ii) Let T be in $\mathcal{D}'_*(\mathbb{R})$ and S be in $\mathcal{E}'_*(\mathbb{R})$. The convolution product of T and S is the distribution $T *_0 S$ of $\mathcal{D}'_*(\mathbb{R})$ defined by

(3.I.33) $$\langle T *_0 S, \varphi\rangle = \langle T_x, \langle S_y, \sigma_x(\varphi)(y)\rangle\rangle,\ \varphi \in \mathcal{D}_*(\mathbb{R}).$$

Proposition 3.I.9. For T in $\mathcal{D}'_*(\mathbb{R})$ and Ψ in $\mathcal{D}_*(\mathbb{R})$, the function $T *_0 \Psi$ belongs to $\mathcal{E}_*(\mathbb{R})$, and we have for all $k \in \mathbb{N}$:

(3.I.34) $$D^{2k}(T *_0 \Psi) = (D^{2k}T) *_0 \Psi = T *_0 (D^{2k}\Psi).$$

where

(3.I.35) $$D = \frac{d}{dx}.$$

Proposition 3.I.10.

i) For S in $\mathcal{S}'_*(\mathbb{R})$ and Ψ in $\mathcal{S}_*(\mathbb{R})$, the function $S *_0 \Psi$ belongs to $\mathcal{E}_*(\mathbb{R})$.

ii) Let S be in $\mathcal{E}'_*(\mathbb{R})$ and Ψ be in $\mathcal{S}_*(\mathbb{R})$. Then the function $S *_0 \Psi$ belongs to $\mathcal{S}_*(\mathbb{R})$.

iii) For S in $\mathcal{S}'_*(\mathbb{R})$ and T in $\mathcal{E}'_*(\mathbb{R})$, the distribution $S *_0 T$ belongs to $\mathcal{S}'_*(\mathbb{R})$.

3.II. The Fourier-cosine transform.

3.II.1. The Fourier-cosine transform on $L^1([0, +\infty[, dx)$.

Definition 3.II.1. Let f be a function in $L^1([0, +\infty[, dx)$. The Fourier-cosine transform of f is the function $\mathcal{F}_0(f)$ defined by

(3.II.1) $$\forall\, \lambda \geq 0,\ \mathcal{F}_0(f)(\lambda) = \int_0^\infty f(x) \cos\lambda x\, dx.$$

Examples.

Let $a \in]0, +\infty[$. We denote by $\mathbb{1}_{[-a,a]}$ the characteristic function of the interval $[-a,a]$.

$f(x)$	$\mathcal{F}_0(f)(\lambda)$
$\frac{1}{a}\mathbb{1}_{[-a,a]}(x)$	$j_{\frac{1}{2}}(\lambda a)$
$\frac{2}{a}(1-\frac{\lvert x\rvert}{a})\mathbb{1}_{[-a,a]}(x)$	$(j_{\frac{1}{2}}(\frac{\lambda a}{2}))^2$
$e^{-a\lvert x\rvert}$	$\frac{a}{a^2+\lambda^2}$
e^{-ax^2}	$\frac{1}{2}\sqrt{\frac{\pi}{a}}\,e^{-\lambda^2/4a}$

Properties.

i) Let f be in $L^1([0,+\infty[, dx)$. Then the function $\mathcal{F}_0(f)$ belongs to $C_{*,0}(\mathbb{R})$, and we have

(3.II.2) $$\|\mathcal{F}_0(f)\|_u \leq \|f\|_1.$$

ii) Let f be in $L^1([0,+\infty[, dx)$. Then for all $x \geq 0$, we have

(3.II.3) $$\forall\, \lambda \geq 0,\ \mathcal{F}_0(\sigma_x(f))(\lambda) = (\cos\lambda x)\,\mathcal{F}_0(f)(\lambda).$$

iii) Let f, g be in $L^1([0,+\infty[, dx)$. Then

(3.II.4) $$\forall\, \lambda \geq 0,\ \mathcal{F}_0(f *_0 g)(\lambda) = \mathcal{F}_0(f)(\lambda).\,\mathcal{F}_0(g)(\lambda).$$

Remark.

From the properties i) and ii) we deduce that the Fourier-cosine transform is a homomorphism from the algebra $L^1([0,+\infty[, dx)$ into the algebra $C_{*,0}(\mathbb{R})$.

Proposition 3.II.1. Let f be in $L^1([0,+\infty[, dx)$ such that there exists $p \in \mathbb{N}^*$, for which the function $x^{2p}f(x)$ belongs to $L^1([0,+\infty[, dx)$. Then the function $\mathcal{F}_0(f)$ is of class C^{2p} on $[0,+\infty[$, and its 2k-derivative, with $1 \leq k \leq p$, is given by

(3.II.5) $$\forall\, \lambda \geq 0,\ \frac{d^{2k}}{d\lambda^{2k}}\mathcal{F}_0(f)(\lambda) = (-1)^k\,\mathcal{F}_0(x^{2k}f)(\lambda).$$

Proposition 3.II.2. Let f be in $L^1([0, +\infty[, dx)$ such that its derivatives $\frac{d^\ell}{dx^\ell} f(x)$, $1 \leq \ell \leq 2p$, exist for all $x \geq 0$, and are in $L^1([0, +\infty[, dx)$. Then for all $1 \leq k \leq p$, we have

$$\forall \lambda \geq 0, \mathfrak{F}_0\left(\frac{d^{2k}}{dx^{2k}} f\right)(\lambda) = (-1)^k \lambda^{2k} \mathfrak{F}_0(f)(\lambda). \tag{3.II.6}$$

Theorem 3.II.1. Let f be in $L^1([0, +\infty[, dx)$ such that $\mathfrak{F}_0(f)$ belongs to $L^1([0, +\infty[, \frac{2}{\pi} d\lambda)$ (the space of integrable functions on $[0, +\infty[$ with respect to the measure $\frac{2}{\pi} d\lambda$). Then we have the following inversion formula for the Fourier-cosine transform $\mathfrak{F}_0$:

$$f(x) = \frac{2}{\pi} \int_0^\infty \mathfrak{F}_0(f)(\lambda) \cos\lambda x \, d\lambda, \quad \text{a.e. } x \geq 0. \tag{3.II.7}$$

Notation. We denote by $\overline{\mathfrak{F}}_0$ the transform defined on $L^1([0, +\infty[, \frac{2}{\pi} d\lambda)$ by

$$\forall x \geq 0, \overline{\mathfrak{F}}_0(f)(x) = \frac{2}{\pi} \int_0^\infty f(\lambda) \cos\lambda x \, d\lambda. \tag{3.II.8}$$

Corollary 3.I.1. The Fourier-cosine transform $\mathfrak{F}_0$ is injective from $L^1([0, +\infty[, dx)$ into $C_{*,0}(\mathbb{R})$.

3.II.2. The Fourier-cosine transform on $\mathcal{S}_*(\mathbb{R})$ and $\mathcal{D}_*(\mathbb{R})$.

Notations. We denote by

- $\mathbb{H}_{*,a}(\mathbb{C})$ the space of even entire functions Ψ on $\mathbb{C}$, rapidly decreasing and of exponential type a, that is

$$\forall m \in \mathbb{N}, \rho_m(\Psi) = \sup_{\lambda \in \mathbb{C}} (1 + |\lambda|^2)^m |\Psi(\lambda)| e^{-a|\mathrm{Im}\lambda|} < +\infty.$$

This space is provided with the topology defined by the seminorms ρ_m.

- $\mathbb{H}_*(\mathbb{C})$ the space of even entire functions on $\mathbb{C}$, rapidly decreasing and of exponential type

$$\mathbb{H}_*(\mathbb{C}) = \bigcup_{a \geq 0} \mathbb{H}_{*,a}(\mathbb{C}),$$

this space shall be given the inductive limit topology.

We say that a sequence $\{\Psi_n\}_{n\in\mathbb{N}}$ of $\mathbb{H}_*(\mathbb{C})$ converges to Ψ in $\mathbb{H}_*(\mathbb{C})$ when $n \longrightarrow +\infty$, if

- There exists $a \geq 0$ such that

$$\forall\, n \in \mathbb{N}, \Psi_n \in \mathbb{H}_{*,a}(\mathbb{C}).$$

- Ψ_n converges to Ψ in $\mathbb{H}_{*,a}(\mathbb{C})$ when $n \to +\infty$.

Equiped with this topology the space $\mathbb{H}_*(\mathbb{C})$ is a Fréchet space.

Theorem 3.II.2. The Fourier-cosine transform $\mathfrak{F}_0$ is a topological isomorphism from $\mathcal{S}_*(\mathbb{R})$ onto itself. The inverse transform is given by

(3.II.9) $$\forall\, x \geq 0\ \mathfrak{F}_0^{-1}(f)(x) = \frac{2}{\pi}\int_0^{\infty} f(\lambda) \cos\lambda x \, d\lambda.$$

Corollary 3.II.1. The space $\mathcal{S}_*(\mathbb{R})$ is an algebra for the convolution product, an algebra for the multiplication, and the Fourier-cosine transform $\mathfrak{F}_0$ is a topological isomorphism from the first onto the second.

Proposition 3.II.1 Let f, g be in $\mathcal{S}_*(\mathbb{R})$. Then we have the following relations

(3.II.10) $$\int_0^{\infty} f(x)\overline{g(x)}\,dx = \frac{2}{\pi}\int_0^{\infty} \mathfrak{F}_0(f)(\lambda)\,\overline{\mathfrak{F}_0(g)(\lambda)}\, d\lambda .$$

(3.II.11) $$\int_0^{\infty} |f(x)|^2 dx = \frac{2}{\pi}\int_0^{\infty} |\mathfrak{F}_0(f)(\lambda)|^2\, d\lambda.$$

Proposition 3.II.3. For f, g in $\mathcal{S}_*(\mathbb{R})$ we have

(3.II.12) $$\forall\, \lambda \geq 0,\ \mathfrak{F}_0(f.g)(\lambda) = \frac{2}{\pi}\mathfrak{F}_0(f)(\lambda) *_0 \mathfrak{F}_0(g)(\lambda).$$

Theorem 3.II.3. (Paley-Wiener theorem) The Fourier-cosine transform $\mathfrak{F}_0$ is a topological isomorphism from $\mathcal{D}_*(\mathbb{R})$ onto $\mathbb{H}_*(\mathbb{C})$.

3.II.3. The Fourier-cosine transform on $L^2([0, +\infty[, dx)$.

Theorem 3.II.4. (Plancherel theorem). The Fourier-cosine transform $\mathfrak{F}_0$ extends to an isometric isomorphism from $L^2([0, +\infty[, dx)$ onto $L^2([0, +\infty[, \frac{2}{\pi} d\lambda)$ (the space of square integrable functions on $[0, +\infty[$, with respect to the measure $\frac{2}{\pi} d\lambda$). The inverse transform is given by

(3.II.13) $$\mathfrak{F}^{-1}(f)(x) = \frac{2}{\pi}\int_0^{\infty} f(\lambda)\cos\lambda x \, d\lambda.$$

The convergence of this integral is in $L^2([0, +\infty[, dx)$.

Proposition 3.II.4. For all f, g in $L^2([0, +\infty[, dx)$, we have

(3.II.14) $$\forall x \geq 0, \ f *_0 g(x) = \overline{\mathfrak{F}}_0 (\mathfrak{F}_0(f). \mathfrak{F}_0(g))(x).$$

Corollary 3.II.1. We have

(3.II.15) $$L^2([0, +\infty[, dx) *_0 L^2([0, +\infty[, dx) = \overline{\mathfrak{F}}_0 (L^1([0, +\infty[, \frac{2}{\pi} d\lambda)).$$

Proposition 3.II.5. Let f, g be in $L^2([0, +\infty[, dx)$. Then the function $f*_0g$ belongs to $L^2([0, +\infty[, dx)$ if and only if the function $\mathfrak{F}_0(f). \mathfrak{F}_0(g)$ belongs to $L^2([0, +\infty[, dx)$, and we have

(3.II.16) $$\mathfrak{F}_0(f*_0g) = \mathfrak{F}_0(f). \mathfrak{F}_0(g).$$

Proposition 3.II.6. For all f, g in $L^2([0, +\infty[, dx)$, we have

(3.II.17) $$\int_0^\infty |f *_0 g(x)|^2 dx = \frac{2}{\pi} \int_0^\infty |\mathfrak{F}_0(f)(\lambda)|^2 |\mathfrak{F}_0(g)(\lambda)|^2 d\lambda,$$

the two sides are finite or infinite.

3.II.4. The Fourier-cosine transform on $M^b([0, +\infty[)$.

Definition 3.II.2. The Fourier-cosine transform of μ in $M^b([0, +\infty[)$ is the function $\mathfrak{F}_0(\mu)$ defined by

(3.II.18) $$\forall \lambda \geq 0, \mathfrak{F}_0(\mu)(\lambda) = \int_0^\infty \cos\lambda x \, d\mu(x).$$

This transform possesses the following properties.

i) Let μ in $M^b([0, +\infty[)$. Then the function $\mathfrak{F}_0(\mu)$ is continuous on $[0, +\infty[$ and we have

(3.II.19) $$\forall \lambda \geq 0, |\mathfrak{F}_0(\mu)(\lambda)| \leq \|\mu\|.$$

ii) For all μ in $M^b([0, +\infty[)$, we have

(3.II.20) $$\lim_{\lambda \to +\infty} \mathfrak{F}_0(\mu)(\lambda) = \mu(\{0\}).$$

iii) If $\mu = f\,dx$, with f in $L^1([0, +\infty[, dx)$, then $\mathfrak{F}_0(\mu)$ is the Fourier-cosine transform of f given by the relation (3.II.1).

iv) For all μ, ν in $M^b([0, +\infty[)$, we have

(3.II.21) $$\forall \lambda \geq 0, \mathcal{F}_0(\mu *_0 \nu)(\lambda) = \mathcal{F}_0(\mu)(\lambda).\ \mathcal{F}_0(\nu)(\lambda).$$

v) The transform $\mathcal{F}_0$ is injective on $M^b([0, +\infty[)$.

3.II.5. The Fourier-cosine transform on $\mathcal{E}'_*(\mathbb{R})$ and $\mathcal{S}'_*(\mathbb{R})$.

Notations. We denote by

- $\mathcal{H}_{*,a}(\mathbb{C})$ the space of even entire functions φ on $\mathbb{C}$, slowly increasing and of exponential type a, that is

$$\exists\ m \in \mathbb{N},\ q_m(\varphi) = \sup_{\lambda \in \mathbb{C}} (1+|\lambda|^2)^{-m} |\varphi(\lambda)|\, e^{-a|\mathrm{Im}\lambda|} < +\infty.$$

We provide this space with the topology defined by the seminorms q_m.

- $\mathcal{H}_*(\mathbb{C})$ the space of even entire functions on $\mathbb{C}$, slowly increasing and of exponential type

$$\mathcal{H}_*(\mathbb{C}) = \bigcup_{a \geq 0} \mathcal{H}_{*,a}(\mathbb{C}).$$

This space is equipped with the inductive limit topology.

For this topology the space $\mathcal{H}_*(\mathbb{C})$ is a Fréchet space.

Definition 3.II.3.

i) The Fourier-cosine transform of S in $\mathcal{S}'_*(\mathbb{R})$ is the distribution $\mathcal{F}_0(S)$ defined by

(3.II.22) $$\langle \mathcal{F}_0(S), \varphi \rangle = \langle S, \mathcal{F}_0(\varphi) \rangle,\quad \varphi \in \mathcal{S}_*(\mathbb{R}).$$

ii) The Fourier-cosine transform of T in $\mathcal{E}'_*(\mathbb{R})$ is the function $\mathcal{F}_0(T)$ defined by

(3.II.23) $$\forall \lambda \geq 0,\ \mathcal{F}_0(T)(\lambda) = \langle T, \cos(\lambda .) \rangle.$$

Remark.

For T in $\mathcal{E}'_*(\mathbb{R})$, the function $\mathcal{F}_0(T)(\lambda)$ is well defined for $\lambda \in \mathbb{C}$.

Properties.

i) For all S in $\mathcal{S}'_*(\mathbb{R})$ the distribution $\mathcal{F}_0(S)$ belongs to $\mathcal{S}'_*(\mathbb{R})$.

ii) For all T in $\mathcal{E}'_*(\mathbb{R})$. The function $\mathcal{F}_0(T)$ is even and entire on $\mathbb{C}$.

iii) Let S be in $\mathcal{S}'_*(\mathbb{R})$ and T in $\mathcal{E}'_*(\mathbb{R})$. Then we have

$$\mathcal{F}_0(T *_0 S) = \mathcal{F}_0(T)\, \mathcal{F}_0(S). \tag{3.II.24}$$

Theorem 3.II.5. The Fourier-cosine transform $\mathcal{F}_0$ is a topological isomorphism from

- $\mathcal{E}'_*(\mathbb{R})$ onto $\mathcal{H}_*(\mathbb{C})$.
- $\mathcal{S}'_*(\mathbb{R})$ onto itself.

Definition 3.II.4. Let S be in $\mathcal{S}'_*(\mathbb{R})$. The singular support of S, denoted sing supp S, is the set of points in $\mathbb{R}$ having no open neighborhood to which the restriction of S is a C^∞-function.

Theorem 3.II.6. Let S be in $\mathcal{E}'_*(\mathbb{R})$ and $a \in [0, +\infty[$. Then we have

$$\text{singsupp } S \subset [-a, a], \tag{3.II.25}$$

if and only if there exists a positive integer N and a sequence of positive constants $\{C_m\}_{m\in\mathbb{N}^*}$ such that for $m = 1, 2, \ldots$

$$|\mathcal{F}_0(S)(\lambda)| \leq C_m(1+|\lambda|^2)^N e^{a|\text{Im}\lambda|}, \quad \text{if } |\text{Im}\lambda| \leq m \operatorname{Log}(1 + |\lambda|^2),\ \lambda \in \mathbb{C}. \tag{3.II.26}$$

(See [48] p. 186).

3.III. Problems.

Problem 3.III.1.

Let f be a measurable function on $[0, +\infty[$ such that

$$\int_0^1 |f(x)|dx < +\infty, \text{ and } \int_1^\infty \frac{|f(x)|}{x^2}\, dx < +\infty.$$

1°) Show that for all $a \in]0, +\infty[$, we have

$$\int_0^\infty \sin ax \left(\int_0^\infty f(y)\, e^{-xy}\, dy\right)dx = a \int_0^\infty \frac{f(y)}{a^2 + y^2} dy.$$

2°) Prove the relation

$$\int_0^\infty \frac{\sin ax}{\sqrt{a^2 + x^2}} dx = a\int_0^\infty \frac{j_0(y)}{a^2 + y^2} dy.$$

Problem 3.III.2.

1°) Let f be a nonnegative continous function on $[0, +\infty[$ with compact support. We put

$$\forall\, x > 0,\ F(x) = \frac{1}{x}\int_0^x f(t)dt.$$

Show that

i) $xF' = f - F$.

ii) $\|F\|_p \leq \frac{p}{p-1}\|f\|_p$, $1 < p < +\infty$.

iii) the inequality ii) is also true when f is not nonnegative.

2°) Let f be in $L^p([0, +\infty[, dx)$, $1 < p < +\infty$. Show that F is well defined and the inequality ii) is always true.

3°) Show that if f is a nonnegative function in $L^1([0, +\infty[, dx)$, then F does not belong to $L^1([0, +\infty[, dx)$.

Problem 3.III.3.

1°) Calculate the integral

$$\int_0^\infty \frac{dt}{\sqrt{t}\,(1+t)}.$$

2°) Let f, g be in $L^2([0, +\infty[, dx)$. Show that for all $t > 0$, we have

$$\left|\int_0^\infty f(x)\,\overline{g(tx)}\,dx\right| \leq \frac{1}{\sqrt{t}}\|f\|_2\,\|g\|_2.$$

3°) Prove that

$$\left|\int_0^\infty\int_0^\infty \frac{f(x)g(y)}{x+y}dxdy\right| \leq \pi\,\|f\|_2\,\|g\|_2\,.$$

4°) Show that the function T(g) given by

$$T(g)(x) = \int_0^\infty \frac{g(y)}{x+y}dy,$$

is defined for almost all $x \geq 0$, belongs to $L^2([0, +\infty[, dx)$ and we have

$$\|T(g)\|_2 \leq \pi\,\|g\|_2\,.$$

Problem 3.III.4.

1°) We put for $\sigma \in \,]0, +\infty[$:

$$\forall\, x \geq 0\,,\ G_\sigma(x) = \frac{1}{\sigma}\sqrt{\frac{\pi}{2}}\, e^{-x^2/2\sigma^2}\,,$$

show that

$$G_\sigma *_0 G_\tau = G_{\sqrt{\sigma^2+\tau^2}}\,,\quad \sigma, \tau \in \,]0, +\infty[.$$

2°) We consider the function f_α, $\alpha \in \,]0, +\infty[$, given by

$$\forall\, x \geq 0,\ f_\sigma(x) = e^{-\alpha|x|},$$

calculate

$$f_\alpha *_0 f_\beta\,,\quad \alpha, \beta \in \,]0, +\infty[.$$

3°) We put for $a \in \,]0, +\infty[$:

$$\forall\, x \geq 0,\ \mathcal{G}_a(x) = \frac{a}{a^2 + x^2}\,.$$

Show that

$$\mathcal{G}_a *_0 \mathcal{G}_b = \mathcal{G}_{a+b}\,,\quad a\,,\, b \in \,]0, +\infty[.$$

Problem 5.III.5

We consider the function F defined by

$$\forall\, \lambda \geq 0,\ F(\lambda) = \int_0^\infty e^{-t^2} \cos\lambda t\, dt.$$

1°) Show that F is even on $\mathbb{R}$ and of class C^1 on $[0, +\infty[$, and satisfies the differential equation

$$F'(\lambda) - \frac{\lambda}{2} F(\lambda) = 0,\ \lambda \geq 0.$$

2°) Verify directly that

$$F(0) = \frac{\sqrt{\pi}}{2}.$$

3°) Solve the differential equation in 1°) subject to the initial condition in 2°) to deduce that

$$\forall\, \lambda \geq 0,\ \mathcal{F}_0(e^{-t^2})(\lambda) = \frac{\sqrt{\pi}}{2} e^{-\frac{\lambda^2}{4}}.$$

Problem 3.III.6.

Let f be in $L^1([0, +\infty[, dx)$ continuous at the point $x \in [0, +\infty[$. We want to show that

(a) $$f(x) = \lim_{\substack{\varepsilon \to 0 \\ \varepsilon > 0}} \frac{2}{\pi} \int_0^\infty \mathfrak{F}_0(f)(\lambda)\, e^{-\varepsilon\lambda} \cos\lambda x \, d\lambda.$$

1°) Study the existence of the integral

$$\int_0^\infty \mathfrak{F}_0(f)(\lambda) e^{-\varepsilon\lambda} \cos\lambda x \, d\lambda.$$

2°) Show that we can bring the study of the relation (a) to the case $x = 0$.

3°) Prove the relation

$$\int_0^\infty \mathfrak{F}_0(f)(\lambda)\, e^{-\varepsilon\lambda}\, d\lambda = \int_0^\infty f(y) \frac{\varepsilon}{\varepsilon^2 + y^2}\, dy.$$

4°) Deduce the relation (a).

5°) If we suppose moreover that $\mathfrak{F}_0(f)$ belongs to $L^1([0, +\infty[, \frac{2}{\pi} d\lambda)$, prove that

$$f(x) = \frac{2}{\pi} \int_0^\infty \mathfrak{F}_0(f)(\lambda) \cos\lambda x \, d\lambda.$$

Problem 3.III.7 .

Let f be in $L^1([0, +\infty[, dx)$ and $x \in [\, 0, +\infty[$. We suppose that f is continuous and admits a right and left derivatives at the point x. We want to show that

(b) $$f(x) = \lim_{M \to +\infty} \frac{2}{\pi} \int_0^M \mathfrak{F}_0(f)(\lambda) \cos\lambda x \, d\lambda.$$

1°) Show that we can bring the study of the relation (b) to the case $x = 0$.

2°) i) Show that the relation (b) is true for $f(x) = e^{-x}$, $x \geq 0$.

ii) Deduce that we can take f satisfying $f(0) = 0$.

3°) Prove the relation (b).

Problem 3.III.8.

We consider the following space

$$\mathcal{A}([0, +\infty[) = \overline{\mathfrak{F}}_0(L^1([0, +\infty[, \frac{2}{\pi}\, d\lambda)).$$

1°) For f in $L^1([0, +\infty[, dx) \cap \mathcal{A}([0, +\infty[)$ we put

$$|||f||| = ||f||_1 + ||\mathfrak{F}_0(f))||_{L^1} .$$

where

$$||h||_{L^1} = \frac{2}{\pi}\int_0^\infty |h(\lambda)|\, d\lambda, \text{ for } h \in L^1([0, \infty[, \frac{2}{\pi}\, d\lambda).$$

Show that $|||.|||$ is a norm on the space $L^1([0, +\infty[, dx) \cap \mathcal{A}([0, +\infty[)$.

2°) Prove that the space $L^1([0, +\infty[, dx) \cap \mathcal{A}([0, +\infty[)$ provided with the norm $|||.|||$ is a Banach space.

3°) Let f, g be in $L^1([0, +\infty[, dx) \cap \mathcal{A}([0, +\infty[)$. Show that $f.g$ and $f*_0g$ belong to $L^1([0, +\infty[, dx) \cap \mathcal{A}([0, +\infty[)$.

Problem 3.III.9.

1) Calculate $\mathfrak{F}_0((\text{ch}x)^{-\alpha})$, $\alpha \in]0, +\infty[$.

2) Calculate the following integrals

$$\int_0^\infty \mathfrak{F}_0((\text{ch}x)^{-\alpha})(\lambda)d\lambda \quad \text{and} \quad \int_0^\infty [\mathfrak{F}_0((\text{ch}x)^{-\alpha})(\lambda)]^2 d\lambda .$$

4 Generalized Convolution Product Associated With The Bessel Operator

Introduction

J. Delsarte has considered in [21] the operator T_x, with $x \geq 0$, given by

$$\forall\, y \geq 0,\ T_x(f)(y) = \frac{\Gamma(\alpha+1)}{\sqrt{\pi}\,\Gamma(\alpha+\frac{1}{2})}\int_0^{\pi} f(\sqrt{x^2+y^2-2xy\cos\theta}\,)(\sin\theta)^{2\alpha}d\theta.$$

Where f is a continuous function on $[0, +\infty[$.

He has called it, circular average operator. It has been called in this chapter, generalized translation operator associated with the Bessel operator.

This operator is related to the product formula for the normalized Bessel function $j_\alpha(\lambda x)$ studied in Chapter 1. More precisely we have the relation

$$\forall\, x \geq 0,\ y \geq 0,\ T_x(j_\alpha(\lambda.))\,(y) = j_\alpha(\lambda x)\, j_\alpha(\lambda y).$$

In this chapter we study properties of the operators T_x, $x \geq 0$, and we use them to define the generalized convolution product $*_B$ of two regular functions f and g on $[0, +\infty[$, by

$$\forall\, x \geq 0,\ f *_B g(x) = \int_0^{\infty} T_x(f)(y)g(y)d\mu_\alpha(y),$$

where

$$d\mu_\alpha(y) = \frac{y^{2\alpha+1}}{2^\alpha\Gamma(\alpha+1)}dy.$$

We define also the generalized convolution product $*_B$ of complex bounded measures on $[0, +\infty[$ and of even distributions on $\mathbb{R}$.

We establish many properties of this generalized convolution product $*_B$ on different spaces of functions, on the space $M^b([0, +\infty[)$ and on spaces of even distributions. In particular for all $x \geq 0$, $y \geq 0$, we have

$$\delta_x *_B \delta_y(f) = T_x(f)(y),$$

where f is a continuous function on $[0, +\infty[$, with compact support and δ_x the Dirac measure at x. This relation and the properties of the generalized convolution product $*_B$ on $M^b([0, +\infty[)$ imply that the pair $([0, +\infty[, *_B)$ is a hypergroup, called Bessel Kingmann hypergroup. This is a particular case of the Chébli-Trimèche hypergroups (see [8] and [88] Chap.2).

Many authors have generalized the notions studied in this chapter, to the case of a singular differential operator more general than the Bessel operator (see [7] [11] [12] [23] [83]).

4.I. Convolution product of radial functions.

4.I.1. Definition and properties.

We said that a function f defined on $\mathbb{R}^n$, $n \geq 1$, is radial, if its value on each point $x \in \mathbb{R}^n$ depends only from the distance $\|x\|$ of x to the origin. Otherwise say there exists a function φ defined on $[0, +\infty[$ such that

(4.I.1) $$\forall\, x \in \mathbb{R}^n,\ f(x) = \varphi(\|x\|) = \varphi(r),\ \text{with}\ r = \|x\|.$$

- If $n = 1$. This notion is the same that of even functions.

- If $n \geq 2$. A function f is radial if and only if

(4.I.2) $$\forall\, x \in \mathbb{R}^n,\ \forall\ u \in SO(n),\ \ f(u(x)) = f(x),$$

where SO(n) is the orthogonal group.

Examples.

The functions

(4.I.3) $$f(x) = e^{-a\|x\|^2},\quad a \in\ ]0, +\infty[.$$

and

(4.I.4) $$g(x) = \frac{t}{(t^2+\|x\|^2)^{\frac{n+1}{2}}},\ t \in\ ]0, +\infty[.$$

are radial.

We have the following properties for radial functions.

i) Let f be a radial function in $L^1(\mathbb{R}^n)$ (the space of integrable functions on $\mathbb{R}^n$ with respect to the Lebesgue measure). Then the function φ given by the relation (4.I.1) is integrable on $[0, +\infty[$ with respect to the measure $r^{n-1}dr$, and we have

(4.I.5) $$\int_{\mathbb{R}^n} f(x)\,dx = \omega_n \int_0^\infty \varphi(r)\, r^{n-1}\, dr,$$

with

(4.I.6) $$\omega_n = \frac{2\pi^{n/2}}{\Gamma(\frac{n}{2})}.$$

ii) We consider

(4.I.7) $$\Delta = \sum_{j=1}^{n} \frac{\partial^2}{\partial x_j^2},$$

the Laplacian operator on $\mathbb{R}^n$, $n \geq 2$. Using the polar coordinates of $\mathbb{R}^n$, we obtain

(4.I.8) $$\Delta = \ell_{\frac{n-2}{2}} + \frac{1}{r^2}\Delta_0, \quad r \in]0, +\infty[,$$

where

(4.I.9) $$\ell_{\frac{n-2}{2}} = \frac{d^2}{dr^2} + \frac{n-1}{r}\frac{d}{dr},$$

is the Bessel operator, and Δ_0 the Laplacian operator on the unit sphere S^{n-1}.

Thus for a radial and of class C^2 function f on $\mathbb{R}^n$, we have

(4.I.10) $$\forall\, x \in \mathbb{R}^n,\ \Delta f(x) = \ell_{\frac{n-2}{2}}(\varphi)(r),$$

where φ is the function of class C^2 on $[0, +\infty[$, given by the relation (4.I.1).

4.I.2. Convolution product.

Let f and g be two radial continuous functions on $\mathbb{R}^n$, $n \geq 2$, with compact support. From the relation (4.I.1) there exist two functions φ and Ψ continuous on $[0, +\infty[$, with compact support such that

(4.I.11) $$\forall\, x \in \mathbb{R}^n,\ f(x) = \varphi(\|x\|) = \varphi(r), \quad \text{with } r = \|x\|.$$

(4.I.12) $$\forall\, x \in \mathbb{R}^n,\ g(x) = \Psi(\|x\|) = \Psi(r), \quad \text{with } r = \|x\|.$$

The convolution product of f and g is the function $f*g$ given by

(4.I.13) $$\forall\, x \in \mathbb{R}^n,\ f*g(x) = \int_{\mathbb{R}^n} f(x-y)\, g(y) dy.$$

This relation can also be written in the form.

(4.I.14) $$\forall\, x \in \mathbb{R}^n,\ f*g(x) = \int_{\mathbb{R}^n} \varphi(\|x-y\|)\, \Psi(\|y\|)\, dy.$$

For all $u \in SO(n)$ we have

$$\forall\, x \in \mathbb{R}^n,\ f*g(u(x)) = \int_{\mathbb{R}^n} \varphi(\|u(x)-y\|)\, \Psi(\|y\|)\, dy.$$

By making the change of variables $z = u^{-1}(y)$, we obtain

$$\forall\, x \in \mathbb{R}^n,\ f*g(u(x)) = \int_{\mathbb{R}^n} \varphi(\|u(x-z\|)\, \Psi(\|u(z)\|)\, dz.$$

Thus

$$\forall\, x \in \mathbb{R}^n,\ f*g(u(x)) = \int_{\mathbb{R}^n} \varphi(\|x-z\|)\, \Psi(z)\, dz.$$

From the relation (4.I.14) we deduce that

$$\forall\, x \in \mathbb{R},\ f*g(u(x)) = f*g(x).$$

Then the function $f*g$ is radial.

To determine $f*g$ it suffices to know it , at the point $x = (\|x\|, 0,\ldots, 0)$. We have

$$\forall\, x \in \mathbb{R}^n,\ f*g(x) = f*g\,((\|x\|, 0,\ldots, 0)) = F(\|x\|) = F(r), \text{ with } r = \|x\|.$$

Where F is a continuous function on $[0, +\infty[$ with compact support.

Thus from (4.I.14) we have

$$F(r) = \int_{\mathbb{R}^n} \varphi(\|(r-y_1, y_2,\ldots, y_n)\|)\, \Psi(\|(y_1, y_2,\ldots, y_n)\|)\, dy_1\, dy_2 \ldots dy_n.$$

$$F(r) = \int_{\mathbb{R}^n} \varphi(\sqrt{(r-y_1)^2 + \sum_{j=2}^{n} y_j^2})\Psi(\sqrt{y_1^2 + \sum_{j=2}^{n} y_j^2})dy_1\, dy_2 \ldots dy_n\,.$$

Using Fubini's theorem we obtain

$$F(r) = \int_{\mathbb{R}} \Big(\int_{\mathbb{R}^{n-1}} \varphi(\sqrt{(r-y_1)^2 + \sum_{j=2}^{n} y_j^2})\Psi(\sqrt{y_1^2 + \sum_{j=2}^{n} y_j^2})dy_2 \ldots dy_n\Big)\, dy_1\,,$$

but, the function under the second integral sign is radial with respect to the variables $y_2, y_3,\ldots, y_n$. Then using the relation (4.I.5) we deduce that

$$F(r) = \omega_{n-1} \int_{\mathbb{R}} \Big(\int_0^\infty \varphi(\sqrt{(r-y_1)^2 + t^2})\Psi(\sqrt{y_1^2 + t^2})t^{n-2}\, dt\Big)dy_1\,.$$

The Fubini's theorem implies

$$F(r) = \omega_{n-1}\int_{\mathbb{R}} \int_0^{\infty} \varphi((\sqrt{(r-y_1)^2+t^2})\Psi(\sqrt{y_1^2+t^2})t^{n-2}\,dt\,dy_1 .$$

By making the change of variables

$$\begin{cases} y_1 = \rho\cos\theta \\ t = \rho\sin\theta \end{cases}, \quad \rho \in]0, +\infty[,\ \theta \in]0, \pi[,$$

we obtain

$$F(r) = \omega_{n-1}\int_0^{\infty}\int_0^{\pi} \varphi(\sqrt{r^2+\rho^2-2r\rho\cos\theta})\, \Psi(\rho)\, (\sin\theta)^{n-2}\, \rho^{n-1}\, d\rho d\theta.$$

Then from Fubini's theorem we have

(4.I.15) $$F(r) = \omega_{n-1}\int_0^{\infty} (\int_0^{\pi} \varphi(\sqrt{r^2+\rho^2-2r\rho\cos\theta})(\sin\theta)^{n-2}d\theta)\Psi(\rho)\rho^{n-1}d\rho .$$

We put

(4.I.16) $$T_r(\varphi)(\rho) = \frac{\Gamma(\frac{n}{2})}{\sqrt{\pi}\,\Gamma(\frac{n-1}{2})}\int_0^{\pi} \varphi((\sqrt{r^2+\rho^2-2r\rho\cos\theta})(\sin\theta)^{n-2}d\theta.$$

Thus using (4.I.6), the relation (4.I.15) takes the form

(4.I.17) $$F(r) = \omega_n\int_0^{\infty} T_r(\varphi)(\rho)\Psi(\rho)\rho^{n-1}d\rho .$$

The operators T_r, $r \geq 0$, satisfy the properties

(4.I.18) - $\forall\, r \geq 0,\ \rho \geq 0,\ T_r(\varphi)(\rho) = T_\rho(\varphi)(r)$.

(4.I.19) - $\forall\, r \geq 0,\ T_0(\varphi)(r) = \varphi(r)$.

(4.I.20) - For all $r \geq 0$, the function $\rho \dashrightarrow T_r(\varphi)(\rho)$ is continuous on $[0, +\infty[$, with compact support.

- For all f, g, h radial continuous functions on $\mathbb{R}^n$, $n \geq 2$, with compact support, we have

$$f * g = g * f,$$

$$(f*g) * h = f * (g * h).$$

From these relations we deduce that

(4.I.21) $$\forall\, r \geq 0, \int_0^{\infty} T_r(\varphi)(\rho)\, \Psi(\rho)\, \rho^{n-1}\, d\rho = \int_0^{\infty} \varphi(\rho)T_r(\Psi)(\rho)\rho^{n-1}d\rho ,$$

where φ, Ψ are the functions given by the relations (4.I.11) and (4.I.12),

and

(4.I.22) $$\forall\, r \geq 0, s \geq 0,\ T_r \circ T_s = T_s \circ T_r .$$

Using the precedent reasoning with f and g in $\mathcal{D}_*(\mathbb{R})$, we deduce from the fact that

$$\forall\, x \in \mathbb{R}^n,\ \Delta(f * g)(x) = f * (\Delta g)(x) = (\Delta f) * g(x),$$

and the relations (4.I.13), (4.I.14), (4.I.10) and (4.I.17) that

(4.I.23) $$\ell_{\frac{n-2}{2}}(F)(r) = \omega_n \int_0^\infty T_r(\ell_{\frac{n-2}{2}}(\varphi))(\rho)\Psi(\rho)\, \rho^{n-1}\, d\rho.$$

(4.I.24) $$\ell_{\frac{n-2}{2}}(F)(r) = \omega_n \int_0^\infty \ell_{\frac{n-2}{2},r}(T_r(\varphi))(\rho)\ \Psi(\rho)\, \rho^{n-1}\, d\rho.$$

(4.I.25) $$\ell_{\frac{n-2}{2}}(F)(r) = \omega_n \int_0^\infty T_r(\varphi)(\rho)\, \ell_{\frac{n-2}{2}}(\Psi)(\rho)\, \rho^{n-1}\, d\rho.$$

Using the relation

(4.I.26) $$\ell_{\frac{n-2}{2}} = \frac{1}{r^{n-1}} \frac{d}{dr}[r^{n-1} \frac{d}{dr}],$$

and two integrations by parts in (4.I.25), we obtain

(4.I.27) $$\ell_{\frac{n-2}{2}}(F)(r) = \omega_n \int_0^\infty \ell_{\frac{n-2}{2},r}(T_r(\varphi))(\rho)\ \Psi(\rho)\, \rho^{n-1}\, d\rho.$$

From the relations (4.I.23), (4.I.24) and (4.I.27) we deduce

(4.I.28) $$\forall\, r \geq 0, \rho \geq 0,\ \ell_{\frac{n-2}{2}}(T_r(\varphi))(\rho) = T_r(\ell_{\frac{n-2}{2}}(\varphi))(\rho).$$

(4.I.29) $$\forall\, r \geq 0, \rho \geq 0,\ \ell_{\frac{n-2}{2}}(T_r(\varphi))(\rho) = \ell_{\frac{n-2}{2},r}(T_r(\varphi))(\rho).$$

We see that the operators T_r, $r \geq 0$, are related to the Bessel operator $\ell_{\frac{n-2}{2}}$.

4.II. Generalized translation operators associated with the Bessel operator.

In the following sections we shall consider $\alpha \in \mathbb{R}$, $\alpha > -\frac{1}{2}$.

Notation. We denote by $L^p([0, +\infty[, d\mu_\alpha)$, $1 \leq p \leq +\infty$, the space of measurable functions on $[0, +\infty[$, satisfying

(4.II.1) $$\|f\|_{p,\alpha} = (\int_0^\infty |f(x)|^p\, d\mu_\alpha(x))^{1/p},\ 1 \leq p < +\infty,$$

(4.II.2) $$\|f\|_{\infty,\alpha} = \operatorname*{ess\,sup}_{x \geq 0} |f(x)|,$$

where $d\mu_\alpha$ is the measure on $[0, +\infty[$ given by the relation (1.II.24).

Definition 4.II.1. The generalized translation operators T_x, $x \geq 0$, associated with the Bessel operator, are defined for a continuous function on $[0, +\infty[$, by

(4.II.3) $$\forall\ y \geq 0, T_x(f)(y) = \frac{\Gamma(\alpha+1)}{\sqrt{\pi}\,\Gamma(\alpha+\frac{1}{2})}\int_0^{\pi} f(\sqrt{x^2+y^2-2xy\cos\theta})(\sin\theta)^{2\alpha}d\theta.$$

Remarks.

i) For $\alpha = \frac{n-2}{2}$, $n \in \mathbb{N}$, $n \geq 2$, we find the operators given by the reltion (4.I.16).

ii) As for an even function f on $\mathbb{R}$, the function $(x, y) \dashrightarrow \int_0^{\pi} f(\sqrt{x^2+y^2-2xy\cos\theta})(\sin\theta)^{2\alpha}d\theta$ is even on $\mathbb{R}$ with respect to the variables x and y, then we define $T_x(f)(y)$ for $x, y \in \mathbb{R}$, by the same relation (4.II.3).

Proposition 4.II.1. For all f in $C_*(\mathbb{R})$ we have

(4.II.4) $$\forall\ x > 0, y > 0,\ T_x(f)(y) = \int_0^{\infty} f(t)W(x, y, t)\, d\mu_\alpha(t).$$

Where $W(x, y, .)$ is the function defined by the relation (1. II. 25).

Proof.

We deduce the result from the relation (4.II.3) and the change of variables

$$t = \sqrt{x^2+y^2-2xy\cos\theta},\ x, y \in]0, +\infty[,\ \theta \in [0, \pi].$$

Remark.

The properties of the function $W(x, y, t)$ are given in the subsection 1.II.4.

From Definition 4.II.1 and the relations (1.II.22) and (1.II.29) we deduce the following properties of the operators T_x, $x \geq 0$.

Properties.

i) For all f in $C_*(\mathbb{R})$ we have

(4.II.5) $$\forall\ x \geq 0, y \geq 0,\ T_x(f)(y) = T_y(f)(x).$$

(4.II.6) $$\forall\ y \geq 0,\ T_0(f)(y) = f(y).$$

ii) We have

(4.II.7) $$\forall\ x \geq 0, y \geq 0, T_x \circ T_y = T_y \circ T_x.$$

iii) We have the product formula

(4.II.8) $$\forall\ x \geq 0, y \geq 0, \forall\ \lambda \in \mathbb{C}, T_x(j_\alpha(\lambda.))(y) = j_\alpha(\lambda x)j_\alpha(\lambda y).$$

Proposition 4.II.2. For all f in $L^p([0, +\infty[, d\mu_\alpha)$, $1 \le p \le +\infty$, and $x \ge 0$, the function $y \dashrightarrow T_x(f)(y)$ defined almost everywhere on $[0, +\infty[$, belongs to $L^p([0, +\infty[, d\mu_\alpha)$, $1 \le p \le +\infty$, and we have

(4. II.9) $$\|T_x(f)\|_{p,\alpha} \le \|f\|_{p,\alpha}.$$

Proof.

- If $p \in [1, +\infty[$.

Using the relation (4.II.4), Hölder's inequality and the relation (1. II.26) we deduce for $x > 0$ and $y > 0$:

$$\left(\int_0^\infty |f(t)|\, W(x, y, t)\, d\mu_\alpha(t)\right)^p \le \int_0^\infty |f(t)|^p\, W(x, y, t)\, d\mu_\alpha(t).$$

Thus

$$\|T_x(f)\|_{p,\alpha}^p \le \int_0^\infty \left(\int_0^\infty |f(t)|\, W(x, y, t)\, d\mu_\alpha(t)\right)^p d\mu_\alpha(y),$$
$$\le \int_0^\infty \left(\int_0^\infty |f(t)|^p\, W(x, y, t) d\mu_\alpha(t)\right) d\mu_\alpha(y).$$

Applying Fubini-Tonelli's theorem and the relations (1. II.26) and (1. II.28) we obtain the inequality (4.II.9).

- If $p = +\infty$.

From the relations (4.II.5) and (4.II.6) we have

$$\forall\, x \ge 0,\ T_x(f)(0) = f(x).$$

Thus

(4.II.10) $$\forall\, x \ge 0,\ |T_x(f)(0)| \le \|f\|_{\infty,\alpha}.$$

Using the relations (4.II.4) and (1.II.26) we obtain

$$\forall\, x > 0,\ y > 0,\ |T_x(f)(y)| \le \int_0^\infty |f(t)|\, W(x, y, t)\, d\mu_\alpha(t),$$
$$\le \|f\|_{\infty,\alpha} \int_0^\infty W(x, y, t)\, d\mu_\alpha(t).$$

Thus

(4.II.11) $$\forall\, x > 0,\ y > 0,\ |T_x(f)(y)| \le \|f\|_{\infty,\alpha}.$$

The relations (4.II.10) and (4.II.11) imply

$$\forall\, x \ge 0,\ \|T_x(f)\|_{\infty,\alpha} \le \|f\|_{\infty,\alpha}.$$

Proposition. 4.II.3.

i) Let f be in $C_{*,c}(\mathbb{R})$ such that $\operatorname{supp} f \subset [-a, a]$, $a \geq 0$. Then for all $x \geq 0$, the function $y \dashrightarrow T_x(f)(y)$ belongs to $C_{*,c}(\mathbb{R})$ and we have

(4.II.12) $$\operatorname{supp} T_x(f) \subset [-a - x, a + x].$$

ii) For f in $C_*(\mathbb{R})$, the mapping $(x, y) \dashrightarrow T_x(f)(y)$ is continuous on $[0, +\infty[\times [0, +\infty[$.

Proof.

We deduce these results from the relations (4.II.3) and (4.II.4).

Proposition 4.II.4. For all f in $L^p([0, +\infty[, d\mu_\alpha)$, $1 \leq p < +\infty$, the mapping $x \dashrightarrow T_x(f)$ is continuous from $[0, +\infty[$ into $L^p([0, +\infty[, d\mu_\alpha)$, $1 \leq p < +\infty$.

Proof.

a) We prove first the result for f in $C_{*,c}(\mathbb{R})$. From the relation (4.II.3) we see that for all $x_0 \geq 0$, there exists $A > 0$ such that

$$\forall\, x \geq 0 \,/\, |x - x_0| < 1 \Longrightarrow \operatorname{supp} T_x(f) \subset [-A, A].$$

Thus

$$\|T_x(f) - T_{x_0}(f)\|^p_{p,\alpha} \leq \left(\int_0^A d\mu_\alpha(y)\right) \sup_{y \in [0,A]} |T_x(f)(y) - T_{x_0}(f)(y)|.$$

Proposition 4.II. ii) implies that

$$\sup_{y \in [0,A]} |T_x(f)(y) - T_{x_0}(f)(y)| \xrightarrow[x \to x_0]{} 0 .$$

Hence we obtain the desired result.

b) Let f be in $L^p([0, +\infty[, d\mu_\alpha)$, $1 \leq p < +\infty$, and $\varepsilon > 0$.From the density of $C_{*,c}(\mathbb{R})$ in $L^p([0, +\infty[, d\mu_\alpha)$ there exists a function g in $C_{*,c}(\mathbb{R})$ such that

$$\|f - g\|_{p,\alpha} < \varepsilon.$$

For all $x_0 \geq 0$, we have

$$\|T_x(f) - T_{x_0}(f)\|_{p,\alpha} \leq \|T_x(f) - T_x(g)\|_{p,\alpha} + \|T_x(g) - T_{x_0}(g)\|_{p,\alpha} + \|T_{x_0}(g) - T_{x_0}(f)\|_{p,\alpha} .$$

Using Proposition 4.II.2 we obtain

$$\|T_x(f) - T_{x_0}(f)\|_{p,\alpha} \leq \|T_x(g) - T_{x_0}(g)\|_{p,\alpha} + 2\,\|f - g\|_{p,\alpha} .$$

We deduce the result from the a).

Proposition 4.II.5. Let f be in $C_{*,c}(\mathbb{R})$. Then for all $x \geq 0$, the function $y \dashrightarrow T_x(f)(y)$ belongs to $C_{*,c}(\mathbb{R})$, and we have

(4.II.13) $$\|T_x(f)\|_u \leq \|f\|_u,$$

and

(4.II.14) $$\lim_{x \to 0} \|T_x(f) - f\|_u = 0\,.$$

Proof.

We obtain the inequality (4.II.13) from the relation (4.II.4).

To prove (4.II.14) we proceed as for Proposition 4.II.4.

Proposition 4.II.6.

i) Let f be in $\mathcal{E}_*(\mathbb{R})$. Then for all $x \geq 0$, the function $y \dashrightarrow T_x(f)(y)$ belongs to $\mathcal{E}_*(\mathbb{R})$.

ii) For all $x \geq 0$, the operator T_x is continuous from $\mathcal{E}_*(\mathbb{R})$ into itself.

Proof.

i) We consider a function f in $\mathcal{E}_*(\mathbb{R})$. From Lemma 2.II.1 there exists a function g of class C^∞ on $[0, +\infty[$, such that

$$\forall\, x \in \mathbb{R},\ f(x) = g(x^2).$$

Thus from the relation (4.II.3) we have for all $x \geq 0$:

(4.II.15) $$\forall\, y \geq 0,\ T_x(f)(y) = \frac{\Gamma(\alpha+1)}{\sqrt{\pi}\,\Gamma(\alpha+\frac{1}{2})} \int_0^\pi g(x^2 + y^2 - 2xy\cos\theta)\,(\sin\theta)^{2\alpha}\,d\theta.$$

We deduce the result from this relation and the derivation theorem under the integral sign.

ii) We deduce the result from the relations (4.II.15) and (2.II.19).

Corollary 4.II.1. Let f be in $\mathcal{D}_*(\mathbb{R})$. Then for all $x \geq 0$, the function $T_x(f)$ belongs to $\mathcal{D}_*(\mathbb{R})$.

Proof.

We deduce the result from Proposition 4.II.6. i) and Proposition 4.II.3. i).

Proposition 4.II.7. The generalized translation operators T_x, $x \geq 0$, satisfy the following relations.

(4.II.15) $$\ell_\alpha(T_x(f)) = T_x(\ell_\alpha(f)) \quad , \quad f \in \mathcal{E}_*(\mathbb{R}).$$

(4.II.16) $$\ell_{\alpha,x}(T_x(f)) = T_x(\ell_\alpha(f)) \quad , \quad f \in \mathcal{E}_*(\mathbb{R}).$$

Where ℓ_α is the Bessel operator given by

$$\ell_\alpha = \frac{d^2}{dx^2} + \frac{2\alpha+1}{x}\frac{d}{dx}.$$

Proof.

From (4.II.8) and (4.II.6) the relations (4.II.15) and (4.II.16) are true for the functions $x \to j_\alpha(\lambda x)$, $\lambda \geq 0$. We deduce these relations from the density of the functions $j_\alpha(\lambda x)$, $\lambda \geq 0$, in $\mathcal{E}_*(\mathbb{R})$ and Proposition 4.II.6.ii).

Remark.

We give in Chapter 5 other properties of the generalized translation operators T_x, $x \geq 0$.

4.III. Generalized convolution product associated with the Bessel operator.

4.III.1. Generalized convolution product of functions.

Definition 4.III.1. The generalized convolution product associated with the Bessel operator of f and g in $C_{*,c}(\mathbb{R})$ is the function $f *_B g$ defined by

(4.III.1) $$\forall\, x \geq 0, \quad f *_B g\,(x) = \int_0^\infty T_x(f)(y)g(y)d\mu_\alpha(y).$$

Remark.

Let F and G be two radial continuous functions on $\mathbb{R}^n$, $n \geq 2$, with compact support. Then from (4.I.1) there exist two functions f and g continuous on $[0, +\infty[$ with compact support, such that

$$F(x) = f(\|x\|) = f(r), \text{ with } r = \|x\|.$$

$$G(x) = g(\|x\|) = g(r), \text{ with } r = \|x\|.$$

From the relation (4.I.17) we deduce that

(4.III.2) $$\forall\, r \geq 0, \quad F * G((r, 0, ..., 0)) = (2\pi)^{\frac{n}{2}} f *_B g(r).$$

In the following we give the properties of the generalized convolution product $*_B$.

Proposition 4.III.1. For all f, g, h in $C_{*,c}(\mathbb{R})$ we have

(4.III.3) i) $f *_B g = g *_B f$.

(4.III.4) ii) $(f *_B g) *_B h = f *_B (g *_B h)$.

Proof.

i) From the relations (4.III.1) and (4.II.4) we have

$$\forall\ x \geq 0,\ f *_B g(x) = \int_0^\infty T_x(f)(y)g(y)d\mu_\alpha(y),$$
$$= \int_0^\infty g(y) \left(\int_0^\infty f(t)\ W(x, y, t)\ d\mu_\alpha(t)\right) d\mu_\alpha(y).$$

Using Fubini's theorem and the relation (1.II.28) we obtain

$$\forall\ x \geq 0,\ f*_B g(x) = \int_0^\infty \left(\int_0^\infty g(y)\ W(x, t, y)\ d\mu_\alpha(y)\right) f(t)\ d\mu_\alpha(y),$$
$$= \int_0^\infty T_x(g)(t)\ f(t)d\mu_\alpha(t).$$

Then

$$\forall\ x \geq 0,\ f *_B g(x) = g *_B f(x)\ .$$

ii) We deduce the result from the relations (4.III.1) and (4.II.7).

Proposition 4.III.2. Let f, g be in $C_{*,c}(\mathbb{R})$ such that supp f $\subset$ [-a, a], a $\geq$ 0, and suppg $\subset$ [-b, b], b $\geq$ 0. Then the function $f *_B g$ belongs to $C_{*,c}(\mathbb{R})$ and we have

(4.III.5) $$\text{Supp}\ f *_B g \subset [-a - b, a + b].$$

Proof.

- From Proposition 4.II.3 and the dominated convergence theorem we deduce that the function $f *_B g$ is an even continuous function on $\mathbb{R}$.

- If suppf $\subset$ [-a, a], a $\geq$ 0, and Supp g $\subset$ [-b, b] , b $\geq$ 0, then from the relations (4.III.1), (4.II.4), (1.II.30), and Fubini's theorem we have

$$\forall\ x \geq 0,\ f *_B g(x) = \int_0^b \int_{|x-y|}^{x+y} f(t)g(y)\ W(x, y, t)\ d\mu_\alpha(t)d\mu_\alpha(y),$$

but, for $x > a + b$ we have

$$f *_B g(x) = 0.$$

Thus

$$\mathrm{Supp}\, f *_B g \subset [-a-b, a+b].$$

Then we deduce that $f *_B g$ belongs to $C_{*,c}(\mathbb{R})$.

Proposition 4.III.3. We consider f and g be in $L^1([0, +\infty[, d\mu_\alpha)$. Then

- For almost all $x \geq 0$, the mapping $y \dashrightarrow T_x(f)(y)\, g(y)$ belongs to $L^1([0, +\infty[, d\mu_\alpha)$.

- The function $f *_B g$ defined almost everywhere on $[0, +\infty[$ by

$$f *_B g(x) = \int_0^\infty T_x(f)(y) g(y) d\mu_\alpha(y), \tag{4.III.6}$$

belongs to $L^1([0, +\infty[, d\mu_\alpha)$ and we have

$$\|f *_B g\|_{1,\alpha} \leq \|f\|_{1,\alpha} \|g\|_{1,\alpha}. \tag{4.III.7}$$

Proof.

From Fubini-Tonelli's theorem we have

$$\int_0^\infty \int_0^\infty |T_x(f)(y)||\, g(y)|\, d\mu_\alpha(x)\, d\mu_\alpha(y) = \int_0^\infty |g(y)| \left(\int_0^\infty |T_x(f)(y)|\, d\mu_\alpha(x)\right) d\mu_\alpha(y).$$

Then Proposition 4.II.2 implies

$$\int_0^\infty \int_0^\infty |T_x(f)(y)||\, g(y)|\, d\mu_\alpha(x)\, d\mu_\alpha(y) \leq \|f\|_{1,\alpha} \|g\|_{1,\alpha}. \tag{4.III.8}$$

Thus the mapping $(x, y) \dashrightarrow T_x(f)(y)\, g(y)$ is integrable on $[0, +\infty[\times [0, +\infty[$ with respect to the measure $d\mu_\alpha(x)\, d\mu_\alpha(y)$. From Fubini's theorem we deduce that for almost all $x \geq 0$, the function $y \dashrightarrow T_x(f)(y)g(y)$ belongs to $L^1([0, +\infty[, d\mu_\alpha)$.

On the other hand from the relation (4.III.8) we have

$$\|f *_B g\|_{1,\alpha} \leq \int_0^\infty \int_0^\infty |T_x(f)(y)||\, g(y)|\, d\mu_\alpha(x)\, d\mu_\alpha(y).$$

Thus

$$\|f *_B g\|_{1,\alpha} \leq \|f\|_{1,\alpha} \|g\|_{1,\alpha}.$$

Proposition 4.III.4. Let g be in $L^1([0, +\infty[, d\mu_\alpha)$ and f be in $L^p([0, +\infty[, d\mu_\alpha)$, $1 \leq p < +\infty$. Then

- For almost all $x \geq 0$, the mapping $y \to T_x(f)(y)g(y)$ belong to $L^1([0, +\infty[, d\mu_\alpha)$.

- The function $f *_B g$ defined almost every where on $[0, +\infty[$ by

(4.III.9) $$f *_B g(x) = \int_0^\infty T_x(f)(y)g(y)\, d\mu_\alpha(y),$$

belongs to $L^p([0, +\infty[, d\mu_\alpha)$, $1 \le p < +\infty$, and we have

(4.III.10) $$\|f *_B g\|_{p,\alpha} \le \|g\|_{1,\alpha} \|f\|_{p,\alpha}.$$

Proof.

a) If $p = 1$.

The result is proved in Proposition 4.III.3.

b) If $p \in]1, +\infty[$.

Let $q \in]1, +\infty[$ such that $\frac{1}{p} + \frac{1}{q} = 1$, and a measurable subset E of $[0, +\infty[$ satisfying

$$\int_0^\infty \mathbb{1}_E(x)d\mu_\alpha(x) < +\infty,$$

where $\mathbb{1}_E$ is the characteristic function of E.

From Fubini-Tonelli's theorem we have

$$\int_0^\infty \int_0^\infty |T_x(f)(y)|\, |g(y)|\, \mathbb{1}_E(x)d\mu_\alpha(x)d\mu_\alpha(y)$$
$$= \int_0^\infty \int_0^\infty |g(y)|\Big(\int_0^\infty |T_x(f)(y)|\ \mathbb{1}_E(x)d\mu_\alpha(x)\Big)\, d\mu_\alpha(y).$$

Using Hölder's inequality and Proposition 4.II.2, we obtain

$$\int_0^\infty \int_0^\infty |T_x(f)(y)|\, |g(y)|\, \mathbb{1}_E(x)d\mu_\alpha(y)d\mu_\alpha(x)$$
$$\le \|g\|_{1,\alpha}\ \|f\|_{p,\alpha}\Big(\int_0^\infty \mathbb{1}_E(x)d\mu_\alpha(x)\Big)^{1/q} < +\infty.$$

Then the function $(x, y) \to T_x(f)(y)\ g(y)\ \mathbb{1}_E(x)$ is integrable on $[0, +\infty[\times [0, +\infty[$ with respect to the measure $d\mu_\alpha(x)d\mu_\alpha(y)$. From Fubini's theorem we deduce that for almost all $x \in E$, the mapping $y \dashrightarrow T_x(f)(y)g(y)$ belongs to $L^1([0, +\infty[, d\mu_\alpha)$, and the mapping $x \dashrightarrow \mathbb{1}_E(x)\int_0^\infty T_x(f)(y)\ g(y)d\mu_\alpha(y)$ belongs also to $L^1([0, +\infty[, d\mu_\alpha)$. Then the function $x \dashrightarrow \int_0^\infty T_x(f)(y)\ g(y)d\mu_\alpha(y)$ is measurable.

We have for almost all $x \ge 0$:

$$|f *_B g(x)| \le \int_0^\infty |T_x(f)(y)|\, |g(y)|^{1/p}\, |g(y)|^{1/q}\, d\mu_\alpha(y).$$

From Hölder's inequality we have

$$|f *_B g(x)| \leq \left(\int_0^\infty |T_x(f)(y)|^p \, |g(y)| \, d\mu_\alpha(x)\right)^{1/p} \left(\int_0^\infty |g(y)| d\mu_\alpha(y)\right)^{1/q}.$$

Using Fubini's theorem and Proposition 4.II.2, we obtain

$$\|f *_B g\|_{p,\alpha}^p \leq \|g\|_{1,\alpha}^{1+p/q} \|f\|_{p,\alpha}^p .$$

Thus

$$\|f *_B g\|_{p,\alpha} \leq \|g\|_{1,\alpha} \|f\|_{p,\alpha}.$$

Proposition 4.III.5. Let f be in $L^p([0, \infty[, d\mu_\alpha)$, $1 < p < +\infty$, and g be in $L^q([0, +\infty[, d\mu_\alpha)$, $1 < q < +\infty$, such that $\frac{1}{p} + \frac{1}{q} = 1$. Then the function $f *_B g$ is continuous on $[0, +\infty[$, tends to zero at the infinity, and we have

(4.III.11) $$\sup_{x \geq 0} |f *_B g(x)| \leq \|f\|_{p,\alpha} \|g\|_{q,\alpha}.$$

Proof.

From Hölder's inequality and Proposition 4.II.2 we have

(4.III.12) $$\forall\, x \geq 0, \quad |f *_B g(x)| \leq \|f\|_{p,\alpha} \|g\|_{q,\alpha}.$$

a) We suppose first that f and g are in $C_{*,c}(\mathbb{R})$. From Proposition 4.III.2 the function $f *_B g$ belongs to $C_{*,c}(\mathbb{R})$.

b) Let f be in $L^p([0, +\infty[, d\mu_\alpha)$, and g be in $L^q([0, +\infty[, d\mu_\alpha)$, $p, q \in \,]1, +\infty[$, such that $\frac{1}{p} + \frac{1}{q} = 1$. From the density of $C_{*,c}(\mathbb{R})$ in $L^p([0, +\infty[, d\mu_\alpha)$ and $L^q([0, +\infty[, d\mu_\alpha)$ there exist two sequences $\{f_n\}_{n\in\mathbb{N}}$ and $\{g_n\}_{n\in\mathbb{N}}$ in $C_{*,c}(\mathbb{R})$ such that

$$\lim_{n \to +\infty} \|f - f_n\|_{p,\alpha} = 0,$$

and

$$\lim_{n \to +\infty} \|g - g_n\|_{q,\alpha} = 0.$$

We have

$$f_n *_B g_n - f *_B g = (f_n - f) *_B g + (g_n - g) *_B f + (f - f_n) *_B (g - g_n).$$

The relation (4.III.12) implies

$$\sup_{x \geq 0} |f_n *_B g_n(x) - f *_B g(x)| \leq \|f_n - f\|_{p,\alpha} \|g\|_{q,\alpha} + \|f\|_{p,\alpha} \|g_n - g\|_{q,\alpha} + \|f_n - f\|_{p,\alpha} \|g_n - g\|_{q,\alpha} .$$

Then the sequence $\{f_n *_B g_n\}_{n\in\mathbb{N}}$ converges uniformly on $[0, +\infty[$ to $f *_B g$ when $n \to +\infty$.

From this result we deduce that the function $f*_B g$ is continuous on $[0, +\infty[$ and

$$\lim_{x\to+\infty} f*_B g(x) = 0.$$

We obtain (4.III.11) from the relation (4.III.12).

Proposition 4.III.6. Let f be in $L^1([0, +\infty[, d\mu_\alpha)$ and g be in $L^\infty([0, +\infty[, d\mu_\alpha)$. Then the function $f*_B g$ is continuous on $[0, +\infty[$, bounded and we have

$$\sup_{x\geq 0} |f*_B g(x)| \leq \|f\|_{1,\alpha}\|g\|_{\infty,\alpha}. \tag{4.III.13}$$

Proof.

From Proposition 4.II.2, we have

$$\forall\, x \geq 0,\ |f*_B g(x)| \leq \|f\|_{1,\alpha}\ \|g\|_{\infty,\alpha}\,. \tag{4.III.14}$$

On the other hand for all $x, x_0 \geq 0$:

$$|f*_B g(x) - f*_B g(x_0)| \leq \int_0^\infty |T_x(f)(y) - T_{x_0}(f)(y)|\,|g(y)|\, d\mu_\alpha(y)$$
$$\leq \|T_x(f) - T_{x_0}(f)\|_{1,\alpha}\|g\|_{\infty,\alpha}.$$

Using Proposition 4.II.4 we obtain the continuity of the function $f*_B g$ on $[0, +\infty[$.

The inequality (4.III.14) gives (4.III.13).

Proposition 4.III.7. Let f be in $L^1([0, +\infty[, d\mu_\alpha)$ and g be in $C_{*,0}(\mathbb{R})$. Then the function $f*_B g$ belongs to $C_{*,0}(\mathbb{R})$ and we have

$$\|f*_B g\|_u \leq \|f\|_{1,\infty}\ \|g\|_u\,. \tag{4.III.14}$$

Proof.

From Proposition 4.III.1 we have

$$\forall\, x \geq 0,\quad f*_B g\,(x) = \int_0^\infty f(y)\, T_x(g)(y)d\mu_\alpha(y).$$

Then the function $f*_B g$ is an even function on $\mathbb{R}$.

We obtain the other results by a proof analogous to the proof of Proposition 4.III.5.

Proposition 4.III.8. Let f be in $L^p([0, +\infty[, d\mu_\alpha)$, $1 \leq p \leq +\infty$, and g be in $\mathcal{D}_*(\mathbb{R})$. Then the function $f*_B g$ belongs to $\mathcal{E}_*(\mathbb{R})$.

Proof.

From the relation (4.III.14) and Proposition 4.III.5, the function $f*_B g$ is even and continous on $\mathbb{R}$.

On the other hand from Lemma 2.II.1 there exists a function φ of class C^∞ on $[0, +\infty[$, with compact support such that

$$\forall\, x \in \mathbb{R},\ g(x) = \varphi(x^2).$$

Thus using the relation (4.II.3) and Fubini's theorem we write (4.III.1) in the form

$$\forall\, x \in \mathbb{R},\ f *_B g(x) = \int_0^\infty \int_0^\pi \varphi(x^2 + y^2 - 2xy\cos\theta)\, f(y)\, (\sin\theta)^{2\alpha}\, d\theta\, d\mu_\alpha(y).$$

Using the derivation theorem under the integral sign we deduce that the function $f *_B g$ is of class C^∞ on $[0, +\infty[$.

This completes the proof.

Corollary 4.III.1. For f, g in $\mathcal{D}_*(\mathbb{R})$, the function $f *_B g$ belongs to $\mathcal{D}_*(\mathbb{R})$.

Proof.

We deduce the result from Propositions 4.III.2 and 4.III.8.

Theorem 4.III.1. Let v be a positive function in $L^1([0, +\infty[, d\mu_\alpha)$ such that $\|v\|_{1,\alpha} = 1$.

We put for $\varepsilon > 0$:

(4.III.15)
$$v_\varepsilon(x) = \varepsilon^{-2\alpha-2}\, v\left(\frac{x}{\varepsilon}\right).$$

Then

i) For all f in $L^p([0, +\infty[, d\mu_\alpha)$, $1 \le p < +\infty$, we have

(4.III.16)
$$\lim_{\varepsilon \to 0} \|f *_B v_\varepsilon - f\|_{p,\alpha} = 0.$$

ii) For all f in $C_{*,0}(\mathbb{R})$, we have

(4.III.17)
$$\lim_{\varepsilon \to 0} \|f *_B v_\varepsilon - f\|_u = 0.$$

Proof.

i) From Proposition 4.III.4 we have for almost all $x \ge 0$:

(4.III.18)
$$f *_B v_\varepsilon(x) - f(x) = \int_0^\infty v_\varepsilon(y)[T_x(f)(y) - f(x)]\, d\mu_\alpha(y).$$

We shall prove that for this x we have

(4.III.19)
$$|f *_B v_\varepsilon(x) - f(x)|^p \le \int_0^\infty v_\varepsilon(y) |T_x(f)(y) - f(x)|^p\, d\mu_\alpha(y)$$

- If $p = 1$, the result is clear.
- If $p \in\,]1, +\infty[$. Let $q \in\,]1, +\infty[$ such that $\frac{1}{p} + \frac{1}{q} = 1$.

We have

$$|f*_B v_\varepsilon(x) - f(x)| \leq \int_0^\infty v_\varepsilon^{1/p}(y)\, v_\varepsilon^{1/q}(y)\, |T_x(f)(y) - f(x)|\, d\mu_\alpha(y).$$

Using Hölder's inequality we obtain (4.III.19).

From the relation (4.III.19) we obtain

$$\int_0^\infty |f*_B v_\varepsilon(x) - f(x)|^p\, d\mu_\alpha(x) \leq \int_0^\infty \int_0^\infty v_\varepsilon(y)\, |T_x(f)(y) - f(x)|^p d\mu_\alpha(y) d\mu_\alpha(x).$$

By applying Fubini-Tonelli's theorem we deduce that

$$\|f*_B v_\varepsilon - f(x)\|_{p,\alpha}^p \leq \int_0^\infty v_\varepsilon(y)\, \|T_y(f) - \|_{p,\alpha}^p\, d\mu_\alpha(y).$$

The change of variables $t = y/\varepsilon$ gives

$$\|f*_B v_\varepsilon - f\|_{p,\alpha}^p \leq \int_0^\infty v(t)\, \|T_{\varepsilon t}(f) - f\|_{p,\alpha}^p\, d\mu_\alpha(t).$$

From Proposition 4.II.4 and the dominated convergence theorem we deduce that

$$\lim_{\varepsilon \to 0} \|f*_B v_\varepsilon - f\|_{p,\alpha} = 0.$$

ii) From Proposition 4.III.7, we have for all $x \geq 0$:

$$|f*_B v_\varepsilon(x) - f(x)| \leq \int_0^\infty v_\varepsilon(y) |T_y(f)(x) - f(x)| d\mu_\alpha(y).$$

Thus

$$\|f*_B v_\varepsilon - f\|_u \leq \int_0^\infty v_\varepsilon(y) \|T_y(f) - f\|_u\, d\mu_\alpha(y).$$

By the change of variables $t = \dfrac{y}{\varepsilon}$, we have

$$\|f*_B v_\varepsilon - f\|_u \leq \int_0^\infty v(t)\, \|T_{\varepsilon t}(f) - f\|_u\, d\mu_\alpha(t).$$

Thus Proposition 4.II.5 and the dominated convergence theorem give

$$\lim_{\varepsilon \to 0} \|f*_B v_\varepsilon - f\|_u = 0.$$

Corollary 4.III.2.

i) The space $\mathcal{D}_*(\mathbb{R})$ is dense in $L^p([0, +\infty[, d\mu_\alpha)$, $1 \leq p < +\infty$.

ii) The injection of $\mathcal{D}_*(\mathbb{R})$ into $L^p([0, +\infty[, d\mu_\alpha)$, $1 \leq p < +\infty$, is continuous.

Proof.

i) Let f be in $L^p([0, +\infty[, d\mu_\alpha)$, $1 \leq p < +\infty$, and $\varepsilon > 0$. From the density of $C_{*,c}(\mathbb{R})$ in $L^p([0, +\infty[, d\mu_\alpha)$ there exists a function g in $C_{*,c}(\mathbb{R})$ such that

$$\|f-g\|_{p,\alpha} < \varepsilon.$$

We consider a positive function φ in $\mathcal{D}_*(\mathbb{R})$ such that

$$\int_0^\infty \varphi(x)\, d\mu_\alpha(x) = 1.$$

We put for $\eta > 0$:

$$\forall\, x \in \mathbb{R},\ \varphi_\eta(x) = \eta^{-2\alpha-2}\, \varphi(\frac{x}{\eta}).$$

From Corollary 4.III.1 the function $g *_B \varphi_\eta$ belongs to $\mathcal{D}_*(\mathbb{R})$. On the other hand from Theorem 4.III.1 the net $\{g *_B \varphi_\eta\}_{\eta>0}$ converges to g in $L^p([0, +\infty[, d\mu_\alpha)$ when $\eta \to 0$.

But we have

$$\|f - g *_B \varphi_\eta\|_{p,\alpha} \le \|f-g\|_{p,\alpha} + \|g - g *_B \varphi_\eta\|_{p,\alpha}.$$

Then the net $\{g *_B \varphi_\eta\}_{\eta>0}$ converges to g in $L^p([0, +\infty, d\mu_\alpha)$ when $\eta \to 0$.

This completes the proof.

ii) Let $\{f_n\}_{n\in\mathbb{N}}$ be a sequence of $\mathcal{D}_*(\mathbb{R})$ which converges to zero in $\mathcal{D}_*(\mathbb{R})$ when n tends to infinity . i.e.

- There exists $a \in [0, +\infty[$, such that

$$\forall\, n \in \mathbb{N},\ \operatorname{supp} f_n \subset [-a,a].$$

- For all $k \in \mathbb{N}$, the sequence $\{f_n^{(k)}\}_{n\in\mathbb{N}}$ converges to zero uniformly when n tends to infinity.

But we have

$$\|f_n\|_{p,\alpha} = \Big(\int_0^\infty |f_n(x)|^p\, d\mu_\alpha(x)\Big)^{1/p} \le \Big(\int_0^a d\mu_\alpha(x)\Big)^{1/p} \sup_{x\in[-a,a]} |f_n(x)|.$$

Thus the sequence $\{f_n\}_{n\in\mathbb{N}}$ converges to zero in $L^p([0, +\infty[, d\mu_\alpha)$, $1 \le p < +\infty$, when n tends to infinity.

Corollary 4.III.3.

i) The space $\mathcal{S}_*(\mathbb{R})$ is dense in $L^p([0, +\infty[, d\mu_\alpha)$, $1 \le p < +\infty$.

ii) The injection of $\mathcal{S}_*(\mathbb{R})$ into $L^p([0, +\infty[, d\mu_\alpha)$, $1 \le p < +\infty$, is continuous.

Proof.

i) We have

$$\mathcal{D}_*(\mathbb{R}) \subset \mathcal{S}_*(\mathbb{R}) \subset L^p([0, +\infty[, d\mu_\alpha),$$

we deduce the result from Corollary 4.III.2.i).

ii) Let $\{f_n\}_{n\in\mathbb{N}}$ be a sequence of $\mathcal{S}_*(\mathbb{R})$ which converges to zero in $\mathcal{S}_*(\mathbb{R})$ when n tends to infinity.

Thus

$$\sup_{x\in\mathbb{R}} (1+x^2)^{[\alpha]+2}|f_n(x)| = \varepsilon_n \text{ ----------}> 0,$$

where $[\alpha]$ is the entire part of α.

The function $g(x) = \dfrac{1}{(1+x^2)^{[\alpha]+2}}$ belongs to $L^p([0, +\infty[, d\mu_\alpha)$, $1 \le p < +\infty$, and we have

$$\|f_n\|_{p,\alpha} \le \varepsilon_n \|g\|_{p,\alpha}.$$

Thus the sequence $\{f_n\}_{n\in\mathbb{N}}$ converges to zero in $L^p([0, +\infty[, d\mu_\alpha)$, $1 \le p < +\infty$, when n tends to infinity.

4. III.2. Generalized convolution product of measures of $M^b([0, +\infty[)$.

Definition 4.III.2. The generalized convolution product associated with the Bessel operator of μ and ν in $M^b([0, +\infty[)$, is the measure $\mu *_B \nu$ defined by

(4.III.20) $$\mu *_B \nu\, (f) = \int_0^\infty \int_0^\infty T_x(f)(y) d\mu(x)\, d\nu(y), \quad f \in C_c([0, +\infty[).$$

The generalized convolution product $*_B$ possesses the following properties.

Proposition 4.III.9.

i) For μ, ν in $M^b([0, +\infty[)$ the measure $\mu *_B \nu$ belongs to $M^b([0, +\infty[)$ and we have

(4.III.21) $$\|\mu *_B \nu\| \le \|\mu\|\, \|\nu\|.$$

ii) For all μ, ν, ω in $M^b([0, +\infty[)$, we have

(4.III.22) • $\mu *_B \nu = \nu *_B \mu.$

(4.III.23) • $\mu *_B (\nu *_B \omega) = (\mu *_B \nu) *_B \omega.$

Proof.

We deduce these results from Definition 4.III.20 and the relations (4.II.4), (4.II.5) and (4.II.7).

Proposition 4.III.10.

i) For all $x \geq 0$ and $y \geq 0$, we have

(4.III.24) $$\delta_x *_B \delta_y (f) = T_x(f)(y), \quad f \in C_c([0, +\infty[).$$

Where δ_x is the Dirac measure at the point x.

ii) Let ν in $M^b([0, +\infty[)$ and $\omega = f\mu_\alpha$ with f in $L^1([0, +\infty[, d\mu_\alpha)$. We have

(4.III.25) $$\nu *_B \omega = \nu *_B (f\mu_\alpha) = (\nu *_B f)\mu_\alpha.$$

Where $\nu *_B f$ is the function given by

(4.III.26) $$\nu *_B f(x) = \int_0^\infty T_x(f)(y) d\nu(y), \quad x \geq 0.$$

It is called the generalized convolution product of the measure ν and the function f.

iii) If $\nu = g\mu_\alpha$ and $\omega = h\mu_\alpha$, with g, h in $L^1([0, +\infty[, d\mu_\alpha)$, then

(4.III.27) $$\nu *_B \omega = (g\mu_\alpha) *_B (h\, \mu_\alpha) = (g *_B h)\mu_\alpha.$$

Where $g *_B h$ is the generalized convolution product of g and h given by the relation (4.III.1).

Proposition 4.III.11. Let ν be in $M^b([0, +\infty[)$ and f in $L^1([0, +\infty[, d\mu_\alpha)$, $1 \leq p \leq +\infty$. Then the function $\nu *_B f$ belongs to $L^1([0, +\infty[, d\mu_\alpha)$, $1 \leq p \leq +\infty$, and we have

(4.III.28) $$\|\nu *_B f\|_{p,\alpha} \leq \|\nu\| \, \|f\|_{p,\alpha}.$$

Proof.

We deduce these results from the relations (4.III.26) and (4.II.8).

4.III.3. Generalized convolution product of distributions.

Definition 4.III.3. Let S be in $\mathcal{D}'_*(\mathbb{R})$ (resp. $\mathcal{S}'_*(\mathbb{R})$) and Ψ in $\mathcal{D}_*(\mathbb{R})$ (resp. $\mathcal{S}_*(\mathbb{R})$). The generalized convolution product associated with the Bessel operator of S and Ψ is the function $S *_B \Psi$ defined by

(4.III.29) $$\forall\, x \geq 0, \quad S *_B \Psi(x) = \langle S, T_x(\Psi)\rangle.$$

Definition 4.III.4. Let $\mathcal{T}$ be in $\mathcal{D}'_*(\mathbb{R})$ and S in $\mathcal{E}'_*(\mathbb{R})$. The generalized convolution product associated with the Bessel operator of $\mathcal{T}$ and S is the distribution $\mathcal{T} *_B S$ in $\mathcal{D}'_*(\mathbb{R})$ defined by

$$<\mathcal{T} *_B S, \varphi> = <\mathcal{T}_x, <S_y, T_x(\varphi)(y)>>, \ \varphi \in \mathcal{D}_*(\mathbb{R}).$$

Proposition 4.III.12.

i) For S in $\mathcal{D}'_*(\mathbb{R})$ (resp. $\mathcal{S}'_*(\mathbb{R})$) and Ψ in $\mathcal{D}_*(\mathbb{R})$ (resp. $\mathcal{S}_*(\mathbb{R})$), the function $S *_B \Psi$ belongs to $\mathcal{E}_*(\mathbb{R})$.

ii) For $\mathcal{T}$ in $\mathcal{S}'_*(\mathbb{R})$ and S in $\mathcal{E}'_*(\mathbb{R})$, the distribution $\mathcal{T} *_B S$ belongs to $\mathcal{S}'_*(\mathbb{R})$.

4. IV. Problems.

Problem 4.IV.1.

1°) Let f be in $C_*(\mathbb{R})$ and T_x, $x \geq 0$, the generalized translation operators associated with the Bessel operator. Show that

$$\forall\ x > 0, T_x(f)(x) = \frac{2^{2\alpha+1}\Gamma(\alpha+1)}{\sqrt{\pi}\,\Gamma(\alpha+\frac{1}{2})} x^{-4\alpha} \int_0^x f(2t)(x^2 - t^2)^{\alpha-\frac{1}{2}} t^{2\alpha} dt.$$

2°) We consider f in $C_*(\mathbb{R})$. Deduce that if

$$T_x(f)(x) = 0, \ \text{for all} \ x \geq 0,$$

then

$$f(x) = 0, \ \text{for all} \ x \geq 0.$$

Problem 4.IV.2.

For all $k \in \mathbb{N}$, we put

$$\forall\ x \geq 0, \ f_k(x) = x^{2k}.$$

1°) Show that for all $x \geq 0$, $y \geq 0$ with $x \neq y$:

$$T_x(f_k)(y) = (x^2 - y^2)^k R_k^{(\alpha,\alpha)}\left(\frac{x^2 + y^2}{x^2 - y^2}\right).$$

Where $R_k^{(\alpha,\alpha)}$ is the Gengenbauer polynomial of degree k and order α, normalized by $R_k^{(\alpha,\alpha)}(1) = 1$.

2°) Prove that these exists a poisitive constant $k(\alpha)$ such that

$$\forall\ x \geq 0, \ T_x(f_k)(x) = k(\alpha) f_k(x).$$

Give the value of $k(\alpha)$.

Problem 4.IV.3.

We consider a net $\{K_\varepsilon\}_{\varepsilon>0}$ of measurable functions on $[0, +\infty[$, satisfying

i) $\int_0^\infty |K_\varepsilon(x)| d\mu_\alpha(x) \leq C$, (C does not depend of ε).

ii) $\int_0^\infty K_\varepsilon(x)\, d\mu_\alpha(x) = 1$.

iii) For all $a \in]0, +\infty[$, $\int_a^{+\infty} |K_\varepsilon(x)| d\mu_\alpha(x) \xrightarrow[\varepsilon\to 0]{} 0$.

1°) Let K be in $L^1([0, +\infty[, d\mu_\alpha)$ such that $\|K\|_{1,\alpha} = 1$. We put for $\varepsilon > 0$:

$$K_\varepsilon(x) = \varepsilon^{-2\alpha-2} K(\frac{x}{\varepsilon}).$$

Show that K_ε satisfies the properties i), ii), iii).

2°) a) Let f be in $L^\infty([0, +\infty[, d\mu_\alpha)$ and K_ε, $\varepsilon > 0$, a measurable function on $[0, +\infty[$, satisfying the properties i), ii), iii). Prove that at all continuity point x of f we have

$$\lim_{\varepsilon\to 0} f *_B K_\varepsilon(x) = f(x).$$

b) Let f be in $L^\infty([0, +\infty[, d\mu_\alpha)$ and g be in $L^1([0, +\infty[, d\mu_\alpha)$. Deduce that

$$\lim_{\varepsilon\to 0} \int_0^\infty (f *_B K_\varepsilon)(x) g(x)\, d\mu_\alpha(x) = \int_0^\infty f(x)g(x) d\mu_\alpha(x) .$$

3°) Let f be in $L^p([0, +\infty[, d\mu_\alpha)$, $1 \leq p < +\infty$, and K_ε, $\varepsilon > 0$, a measurable function on $[0, +\infty[$, satisfying the properties i), ii), iii). Show that

$$\lim_{\varepsilon\to 0} \|f *_B K_\varepsilon - f\|_{p,\alpha} = 0.$$

Problem 4.IV.4.

Let $t \in]0, +\infty[$. We consider the functions k_t and p_t defined by

$$\forall\ x \geq 0,\ k_t(x) = \frac{e^{-\frac{x^2}{4t}}}{(2t)^{\alpha+1}} .$$

$$\forall\ x \geq 0,\ p_t(x) = \frac{2^{\alpha+1}\,\Gamma(\alpha + \frac{3}{2})}{\sqrt{\pi}} \frac{t}{(t^2 + x^2)^{\alpha+\frac{3}{2}}}.$$

1°) Prove the relation

$$\forall\ x \geq 0,\ p_t(x) = \frac{1}{\sqrt{\pi}} \int_0^\infty \frac{e^{-u}}{\sqrt{u}}\, k_{\frac{t^2}{4u}}(x) du.$$

2°) Establish the relations

$$k_t *_B k_s = k_{t+s} \quad , \quad t, s \in]0, +\infty[.$$

$$p_t *_B p_s = p_{t+s} \quad , \quad t, s \in]0, +\infty[.$$

3°) Show that the functions k_t and p_t satisfy the properties i), ii), iii) of Problem 4.IV.3.

Problem 4.IV.5.

1°) We consider f in $\mathcal{D}_*(\mathbb{R})$. We put

$$v(x, t) = k_t *_B f(x) \quad , \quad x \geq 0, \quad t > 0.$$

$$u(v, t) = p_t *_B f(x) \quad , \quad x \geq 0, \quad t > 0.$$

where k_t and p_t the functions given in Problem 4.IV.4

Prove that the functions $v(x, t)$ and $u(x, t)$ are related by the relation

$$\forall \; x \geq 0, \; p_t *_B f(x) = \frac{1}{\sqrt{\pi}} \int_0^\infty \frac{e^{-u}}{\sqrt{u}} k_{\frac{t^2}{4u}} *_B f(x) du.$$

2°) Show that there exists a positive constant C such that

i) $|u(x, t)| \leq \dfrac{C}{(x^2 + t^2)^{\alpha+1}}$.

ii) $\left|\dfrac{\partial u}{\partial t}(x,t)\right| + \left|\dfrac{\partial u}{\partial x}(x,t)\right| \leq \dfrac{C}{(x^2 + t^2)^{\alpha+\frac{3}{2}}}$.

iii) $\left|\dfrac{\partial^2 u}{\partial t^2}(x,t)\right| + \left|\dfrac{\partial^2 u}{\partial x^2}(x,t)\right| \leq \dfrac{C}{(x^2 + t^2)^{\alpha+2}}$.

Problem 4.IV.6.

1°) Let f be in $L^p([0, +\infty[, d\mu_\alpha)$, $1 \leq p < +\infty$. Show that the formula

$$A(f)(a, .) = \frac{1}{a} \int_0^a T_y(f)(.) dy$$

where T_y, $y \geq 0$, are the generalized translation operators associated with the Bessel operator, define for all $a \in]0, +\infty[$, an element of $L^p([0, +\infty[, d\mu_\alpha)$, $1 \leq p < +\infty$.

2°) Let f be in $L^p([0, +\infty[, d\mu_\alpha)$, $1 \leq p < +\infty$.

Prove that there exists a negligible set $E(f)$ such that

i) $\forall\, a \in]0, +\infty[,\ \forall\, x \notin E(f)$:

$$A(f)(a, x) = \frac{1}{a}\int_0^a T_y(f)(x)dy.$$

ii) For all $x \notin E(f)$, the mapping $a \dashrightarrow A(f)(a, x)$ is continuous on $]0, +\infty[$

iii) The mapping $(a, x) \dashrightarrow A(f)(a, x)$ is measurable.

(See [1] and [25].

3°) i) Let f be in $L^p([0, +\infty[, d\mu_\alpha)$, $1 \le p < +\infty$. Show that the mapping $M(f) : [0, +\infty[\dashrightarrow [0, +\infty]$ defined almost everywhere on $[0, +\infty[$ by

$$M(f)(x) = \operatorname*{Sup}_{a>0} \left| \frac{1}{a}\int_0^a T_y(f)(x)dy \right| ,$$

is measurable on $[0, +\infty[$.

ii) Prove that for all f in $L^p([0, +\infty[, d\mu_\alpha)$, $1 < p < +\infty$, the mapping $M(f)$ belongs to $L^p([0, +\infty[, d\mu_\alpha)$, $1 < p < +\infty$, and we have

$$\|M(f)\|_{p,\alpha} \le 2\left(\frac{p}{p-1}\right)^{1/p} \|f\|_{p,\alpha} .$$

(See [1]).

Problem 4.IV.7.

Let f be in $\mathcal{D}_*(\mathbb{R})$. We consider the function

$$F(x) = \sup_{t>0} |p_t *_B f(x)| ,$$

where p_t is the function given in Problem 4.IV.4.

1°) Show that F is measurable on $[0, +\infty[$.

2°) Prove that there exists a positive constant C which does not depend of f and p such that

$$\forall\ x \ge 0, \quad F(x) \le CM(f)(x) ,$$

where $M(f)$ is the function given in Problem 4.IV.6.

3°) Deduce that F belongs to $L^p([0, +\infty[, d\mu_\alpha)$, $1 < p < +\infty$, and there exists a positive constant B_p which does not depend of f such that

$$\|F\|_{p,\alpha} \leq B_p \|f\|_{p,\alpha}.$$

(See [1]).

Problem 4.IV.8.

Let f be in $\mathcal{D}_*(\mathbb{R})$. We consider the function

$$G(x) = \operatorname*{Sup}_{t>0}\left(t\left|\frac{\partial}{\partial t} p_t\right| *_B |f|(x)\right),$$

where p_t is the function given in Problem 4.IV.4.

1°) Show that G is a measurable function on $[0, +\infty[$.

2°) Prove the following inequality

$$\forall\, x \geq 0, \quad G(x) \leq kM(|f|)(x),$$

where k is a positive constant which does not depend of f and p, and $M(|f|)$ the function given in Problem 4.IV.6.

3°) Deduce that G belongs to $L^p([0, +\infty[, d\mu_\alpha)$, $1 < p < +\infty$, and we have

$$\|G\|_{p,\alpha} \leq C_p \|f\|_{p,\alpha},$$

where C_p is a positive constant which does not depend on f.

(See [1]).

Problem 4.IV.9.

Let f be a locally integrable function on $[0, +\infty[$.

We call $x \in [0, +\infty[$ a Lebesgue point of f if

$$\lim_{\substack{t\to 0\\ t>0}} \frac{1}{t^{2\alpha+2}} \int_0^t |T_x(f)(y) - f(x)|\, d\mu_\alpha(y) = 0.$$

1°) Show that if f is continuous at x_0, then x_0 is a Lebesgue point of f.

2°) If f is locally integrable, prove that for almost all $x \in [0, +\infty[$, x is a Lebesgue point of f.

3°) Let g be a measurable function on $[0, +\infty[$ such that $\operatorname*{ess\,sup}_{t\geq x}|g(t)| \in L^p([0, +\infty[, d\mu_\alpha)$. Let $f \in L^p([0, +\infty[, d\mu_\alpha)$, $1 \leq p \leq +\infty$. Show that for each Lebesgue point x of f we have

$$\lim_{\substack{a\to 0\\ a>0}} f *_B g_a(x) = f(x) \int_0^\infty g(t)d\mu_\alpha(t),$$

where g_a is the function given by

$$g_a(x) = \frac{1}{a^{2\alpha+2}} g(\frac{x}{a}), \quad x \geq 0.$$

(See [65]).

5. Fourier-Bessel Transform

Introduction.

We define the Fourier-Bessel transform of a regular function f on $[0, +\infty[$, by

$$\forall\, \lambda \geq 0,\ \mathfrak{F}_B(f)(\lambda) = \int_0^{\infty} f(x) j_\alpha(\lambda x) d\mu_\alpha(x),$$

where $j_\alpha(\lambda x)$ is the normalized Bessel function studied in Chapter 1 and

$$d\mu_\alpha(x) = \frac{x^{2\alpha+1}}{2^\alpha \Gamma(\alpha+1)} dx.$$

This transform has been called in [15] [69] [78] Hankel-Stieltjes transform and in [44] [46] [50] Hankel transform.

The transform $\mathfrak{F}_B$ is related to the Fourier-cosine transform $\mathfrak{F}_0$ studied in Chapter 3 by the relation

$$\mathfrak{F}_B(f) = 2^{-\alpha}(\Gamma(\alpha+1))^{-1}\, \mathfrak{F}_0 \circ W_\alpha(f).$$

Where W_α is the Weyl integral transform given in Chapter 2.

Using the expressions of the inverse transforms W_α^{-1} and $\mathfrak{F}_0^{-1}$, we prove in this chapter that the inverse transform $\mathfrak{F}_B^{-1}$ is equal to $\mathfrak{F}_B$.

We study also the Fourier-Bessel transform on $L^2([0, +\infty[, d\mu_\alpha)$(the space of square integrable functions on $[0, +\infty[$, with respect to the measure $d\mu_\alpha$), and we prove for this transform an inversion formula, a Plancherel formula and a Plancherel theorem.

We define also the Fourier-Bessel transform on the space of complex bounded measures on $[0, +\infty[$ and on spaces of even distributions on $\mathbb{R}$, and we study its properties. A.H.Zemanian has considered in [92] a Hankel transform on generalized functions, which is of the same type as the Fourier-Bessel transform.

The results obtained in this chapter and in the precedents constitute the Harmonic Analysis associated with the Bessel operator. Many applications of the classical Harmonic Analysis on $\mathbb{R}$ have been extended to this Harmonic Analysis, such as

- Maximal ideal spaces of some algebras of functions and measures (see Problems 5.VIII.6, 5.VIII.7 and [78]).

- Positive definite functions associated with the Bessel operator (see Problem 5.VIII.12, 5.VIII.13 and [15]).

- Pseudo-differential operators associated with the Bessel operator (see Problems 5.VIII.17, 5.VIII.18 and [49] [71] [72] [73]).

- Almost-periodic functions associated with the Bessel operator (see Problems 5.VIII. 19, 5.VIII.20, 5.VIII.21, and [21]).

- Generalized mean-periodic functions (see Problems 5.VIII.22, 5.VIII.23 and 5.VIII.24).

- Multipliers associated with the Bessel operator (see [1] [81]).

- Fourier-Bessel series (see [91] chap. XVIII).

- The generalized heat equation associated with the Bessel operator (see [14]).

- Renewal theorem associated with the Bessel operator (see [38]).

In [2] [7] [11] [12] [13] [18] [33] [38] [54] [83] [84] [85] [86] [92] we find other Harmonic Analysis associated with a singular differential operator more general than the Bessel operator, and many applications.

5.I. Fourier transform of radial functions.

Notation. We denote by $L^1(\mathbb{R}^n)$ the space of integrable functions on $\mathbb{R}^n$ with respect to the Lebesgue measure.

Proposition 5.I.1. Let f be in $L^1(\mathbb{R}^n)$. Then

i) for all automorphism u of $\mathbb{R}^n$ we have

$$\forall \lambda \in \mathbb{R}^n, \ \mathscr{F}(f \circ u)(\lambda) = \frac{1}{|\det u|}[\mathscr{F}(f) \circ (u^{-1})^*](\lambda), \tag{5.I.1}$$

where $(u^{-1})^*$ is the adjoint operator of u^{-1} and $\mathfrak{F}$ is the classical Fourier transform on $\mathbb{R}^n$.

ii) for all $u \in SO(n)$ we have

$$\mathfrak{F}(f \circ u) = \mathfrak{F}(f) \circ u. \tag{5.I.2}$$

Proof.

i) We have

$$\forall \lambda \in \mathbb{R}^n, \ \mathfrak{F}(f \circ u)(\lambda) = \int_{\mathbb{R}^n} f \circ u(x)\, e^{-i(\lambda,x)}\, dx.$$

where $(.,.)$ is the scalar product on $\mathbb{R}^n$.

By making the change of variables $y = u(x)$ we obtain

$$\forall \lambda \in \mathbb{R}^n, \ \mathfrak{F}(f \circ u)(\lambda) = \frac{1}{|\det u|}\int_{\mathbb{R}^n} f(y)\, e^{-i((u^{-1})^*(\lambda),y)} dy,$$
$$= \frac{1}{|\det u|}\mathfrak{F}(f)((u^{-1})^*(\lambda)).$$

ii) We deduce the result from the i).

Theorem 5.III.1. Let f be a radial function in $L^1(\mathbb{R}^n)$. Then

i) The function $\mathfrak{F}(f)$ is radial.

ii) The functions φ and Ψ defined on $[0, +\infty[$ by

$$f(x) = \varphi(\|x\|) = \varphi(r), \text{ with } r = \|x\|.$$

$$\mathfrak{F}(f)(\lambda) = \Psi(\|\lambda\|) = \Psi(\rho), \text{ with } \rho = \|\lambda\|.$$

are related by the relation

$$\forall \rho \geq 0, \Psi(\rho) = (2\pi)^{n/2}\int_0^\infty \varphi(r)\, j_{\frac{n-2}{2}}(r\rho)\, d\mu_{\frac{n-2}{2}}(r). \tag{5.I.3}$$

Proof.

i) We deduce the result from the definition of a radial function and Proposition 5.I.1.

ii) From the i), to determine $\mathfrak{F}(f)$ it suffices to know it at the point $\lambda = (\|\lambda\|, 0, ..., 0)$.

We have

$$\forall \lambda \in \mathbb{R}^n, \mathfrak{F}(f)(\lambda) = \mathfrak{F}(f)((\|\lambda\|, 0, ..., 0)) = \Psi(\|\lambda\|) = \Psi(\rho),$$

with $\rho = \|\lambda\|$.

Thus

$$\forall \rho \geq 0, \Psi(\rho) = \int_{\mathbb{R}^n} \varphi(\|x\|)\, e^{-ix_1\rho} dx_1\, dx_2 \ldots dx_n,$$

$$= \int_{\mathbb{R}^n} \varphi\Big(\sqrt{x_1^2 + \sum_{j=2}^{n} x_j^2}\Big)\, e^{-ix_1\rho} dx_1\, dx_2 \ldots dx_n .$$

From Fubini's theorem we have

$$\forall \rho \geq 0, \Psi(\rho) = \int_{\mathbb{R}} \Big(\int_{\mathbb{R}^{n-1}} \varphi\Big(\sqrt{x_1^2 + \sum_{j=2}^{n} x_j^2}\Big)\, dx_2 \ldots dx_n\Big)\; e^{-ix_1\rho} dx_1 .$$

But the function $\varphi\Big(\sqrt{x_1^2 + \sum_{j=2}^{n} x_j^2}\Big)$ is radial with respect to the variables $x_2, \ldots, x_n$. Then from the relation (4.I.5) we have

$$\forall \rho \geq 0, \Psi(\rho) = \int_{\mathbb{R}} \Big(\omega_{n-1} \int_0^{\infty} \varphi\big(\sqrt{x_1^2 + t^2}\big) t^{n-2} dt\Big) e^{-ix_1\rho} dx_1 .$$

Using Fubini's theorem we obtain

$$\forall \rho \geq 0, \Psi(\rho) = \omega_{n-1} \int_{\mathbb{R}} \int_0^{\infty} \varphi\big(\sqrt{x_1^2 + t^2}\big) e^{-ix_1\rho}\, t^{n-2}\, dt dx_1 .$$

By making the change of variables

$$\begin{cases} x_1 = r\cos\theta \\ t = r\sin\theta \end{cases}, \quad r \in\,]0, +\infty[, \quad \theta \in\,]0, \pi[,$$

we have

$$\forall \rho \geq 0, \Psi(\rho) = \omega_{n-1} \int_0^{\infty} \int_0^{\pi} \varphi(r)\, e^{-ir\rho\cos\theta}\, (\sin\theta)^{n-2}\, r^{n-1}\, dr d\theta.$$

But from the relation (1.II.13) we have

$$\int_0^{\pi} e^{-ir\rho\cos\theta}\, (\sin\theta)^{n-2}\, d\theta = \frac{\sqrt{\pi}\, \Gamma(\frac{n-1}{2})}{\Gamma(\frac{n}{2})} j_{\frac{n-2}{2}}(r\rho),$$

where $j_{\frac{n-2}{2}}(r\rho)$ is the normalized Bessel function of first kind and order $\dfrac{n-2}{2}$ defined by the relation (1.II.1).

Using Fubini's theorem and the fact that

$$\omega_{n-1} = \frac{2\pi^{\frac{n-1}{2}}}{\Gamma(\frac{n-1}{2})},$$

we deduce that

$$\forall\ \rho \geq 0,\ \Psi(\rho) = (2\pi)^{n/2} \int_0^\infty \varphi(r)\, j_{\frac{n-2}{2}}(r\rho)\, d\mu_{\frac{n-2}{2}}(r).$$

5.II. The Fourier-Bessel transform on $L^1([0, +\infty[, d\mu_\alpha)$.

In the following sections we consider $\alpha \in \mathbb{R}$, $\alpha > -\frac{1}{2}$.

Definition 5.II.1. Let f be in $L^1([0,+\infty[, d\mu_\alpha)$. The Fourier-Bessel transform of f is the function $\mathfrak{F}_B(f)$ defined by

(5.II.1) $$\forall\ \lambda \geq 0,\ \mathfrak{F}_B(f)(\lambda) = \int_0^\infty f(x)\, j_\alpha(\lambda x)\, d\mu_\alpha(x),$$

where $j_\alpha(\lambda x)$ is the normalized Bessel function of first kind and order α studied in the section 1.II.

Remark.

For f in $\mathcal{D}_*(\mathbb{R})$, the function $\mathfrak{F}_B(f)(\lambda)$ is well defined for all $\lambda \in \mathbb{C}$.

Example.

Let $t \in]0, +\infty[$. The function $\dfrac{e^{-\frac{x^2}{4t}}}{(2t)^{\alpha+1}}$ belongs to $L^1([0,+\infty[, d\mu_\alpha)$ and we have

(5.II.2) $$\forall\ \lambda \geq 0,\ \mathfrak{F}_B(f)(\lambda) = e^{-t\lambda^2}.$$

(See Problem 1.III.10).

Proposition 5.II.1. Let f be in $L^1([0,+\infty[, d\mu_\alpha)$. Then

i) the Fourier-Bessel transform of f belongs to $C_{*,0}(\mathbb{R})$.

ii) We have

(5.II.3) $$\|\mathfrak{F}_B(f)\|_u \leq \|f\|_{1,\alpha}.$$

To prove this proposition we need the following Lemma.

Lemma 5.II.1. For all $x \in]0, +\infty[$, we have

(5.II.4) $$\lim_{\lambda \to +\infty} j_\alpha(\lambda x) = 0.$$

Proof.

From the relation (1.II.14) we have

(5.II.5) $$\forall\, x > 0,\ \forall\, \lambda \geq 0,\ j_\alpha(\lambda x) = \int_0^\infty K(x, y) \cos\lambda y\, dy,$$

with

(5.II.6) $$K(x, y) = \frac{2\Gamma(\alpha+1)}{\sqrt{\pi}\,\Gamma(\alpha+\frac{1}{2})}\, x^{-2\alpha}(x^2 - y^2)^{\alpha - \frac{1}{2}} \mathbb{1}_{]0,x[}(y),$$

where $\mathbb{1}_{]0,x[}$ is the characteristic function of the interval $]0, x[$.

The function $y \dashrightarrow K(x, y)$ belongs to $L^1([0,+\infty[, dx)$.Then from Riemann-Lebesgue's Lemma for the Fourier-cosine transform we deduce that

$$\lim_{\lambda\to+\infty} j_\alpha(\lambda x) = \lim_{\lambda\to+\infty} \int_0^\infty K(x, y) \cos\lambda y\, dy = 0.$$

Proof of Proposition 5.II.1.

i) The parity of the function $\lambda \dashrightarrow j_\alpha(\lambda x)$ implies that the function $\mathfrak{F}_B(f)$ given by the relation (5.II.1), is even on $\mathbb{R}$.

On the other hand from the relation (1.II.18) we have

(5.II.7) $$\forall\, x \geq 0,\ \forall\, \lambda \geq 0,\ |j_\alpha(\lambda x)| \leq 1.$$

Using this inequality, Lemma 5.II.1 and dominated convergence theorem, we deduce from the relation (5.II.1) that $\mathfrak{F}_B(f)$ belongs to $C_{*,0}(\mathbb{R})$.

ii) From the relation (5.II.1) we have

$$\forall\, \lambda \geq 0,\ |\mathfrak{F}_B(f)(\lambda)| \leq \int_0^\infty |f(x)|\, |j_\alpha(\lambda x)| d\mu_\alpha(x).$$

The relation (5.II.7) implies

$$\forall\, \lambda \geq 0,\ |\mathfrak{F}_B(f)(\lambda)| \leq \int_0^\infty |f(x)| d\mu_\alpha(x).$$

Thus

$$\|\mathfrak{F}_B(f)\|_u \leq \|f\|_{1,\alpha}.$$

Proposition 5.II.2.

i) Let f be in $L^1([0, +\infty[, d\mu_\alpha)$. Then for all $x \geq 0$, we have

(5.II.8) $$\forall\, \lambda \geq 0,\ \mathfrak{F}_B(T_x(f))(\lambda) = j_\alpha(\lambda x)\ \mathfrak{F}_B(f)(\lambda).$$

ii) Let f and g be in $L^1([0, +\infty[, d\mu_\alpha)$. Then

(5.II.9) $$\forall\, \lambda \geq 0,\ \mathfrak{F}_B(f *_B g)(\lambda) = \mathfrak{F}_B(f)(\lambda) \,.\, \mathfrak{F}_B(g)(\lambda).$$

Proof.

i) From Proposition 4.II.2, for all $x \geq 0$, the function $y \dashrightarrow T_x(f)(y)$ belongs to $L^1([0, +\infty[, d\mu_\alpha)$. Then from the relations (5.II.1), (4.II.4) and Fubini theorem we have

$$\begin{aligned}\forall \lambda \geq 0,\ \mathfrak{F}_B(T_x(f))(\lambda) &= \int_0^\infty T_x(f)\,(y)\, j_\alpha(\lambda y) d\mu_\alpha(y),\\ &= \int_0^\infty \left(\int_0^\infty f(t) W(x, y, t) d\mu_\alpha(t)\right) j_\alpha(\lambda y) d\mu_\alpha(y),\\ &= \int_0^\infty f(t)\left(\int_0^\infty j_\alpha(\lambda y) W(x, y, t)\, d\mu_\alpha(y)\right) d\mu_\alpha(t).\end{aligned}$$

Applying the relations (1.II.23) and (1.II.28) we obtain

$$\forall \lambda \geq 0,\ \mathfrak{F}_B(T_x(f))(\lambda) = \int_0^\infty f\,(t) j_\alpha(\lambda x)\, j_\alpha(\lambda t) d\mu_\alpha(t).$$

Thus

$$\forall \lambda \geq 0,\ \mathfrak{F}_B(T_x(f))(\lambda) = j_\alpha(\lambda x)\mathfrak{F}_B(f)(\lambda).$$

ii) From Proposition 4.III.3 the function $f *_B g$ belongs to $L^1[0, +\infty[, d\mu_\alpha)$. Then from the relations (5.II.1), (4.III.1) and Fubini theorem we have

$$\begin{aligned}\forall \lambda \geq 0,\ \mathfrak{F}_B(f *_B g)(\lambda) &= \int_0^\infty f *_B g\,(x)\, j_\alpha(\lambda x)\ d\mu_\alpha(x),\\ &= \int_0^\infty \left(\int_0^\infty T_x(f)(y) g(y) d\mu_\alpha(y)\right) j_\alpha(\lambda x)\ d\mu_\alpha(x),\\ &= \int_0^\infty g(y)\left(\int_0^\infty T_x(f)(y)\, j_\alpha(\lambda x)\ d\mu_\alpha(x)\right) d\mu_\alpha(y).\end{aligned}$$

Using the relation (5.II.8) we obtain

$$\forall \lambda \geq 0,\ \mathfrak{F}_B(f *_B g)(\lambda) = \int_0^\infty g(y)\, j_\alpha(\lambda x)\mathfrak{F}_B(f)(\lambda)\ d\mu_\alpha(x).$$

Thus

$$\forall \lambda \geq 0,\ \mathfrak{F}_B(f *_B g)(\lambda) = \mathfrak{F}_B(f)(\lambda).\mathfrak{F}_B(g)(\lambda).$$

Proposition 5.II.3.

i) Let f be an even C^2-function on $\mathbb{R}$, such that $f'(0) = 0$, and with compact support. Then we have

(5.II.10) $$\forall \lambda \geq 0,\ \mathfrak{F}_B(\ell_\alpha(f))(\lambda) = -\lambda^2\mathfrak{F}_B(f)(\lambda),$$

where ℓ_α is the Bessel operator given by

(5.II.11) $$\ell_\alpha = \frac{d^2}{dx^2} + \frac{2\alpha + 1}{x}\frac{d}{dx}.$$

ii) Let f be in $\mathcal{D}_*(\mathbb{R})$. For all $p \in \mathbb{N} \setminus \{0\}$, we have

((5.II.12) $$\forall \lambda \geq 0,\ \mathcal{F}_B(\ell_\alpha^p(f))(\lambda) = (-1)^p \lambda^{2p} \mathcal{F}_B(f)(\lambda).$$

(5.II.13) $$\forall \lambda \geq 0,\ \lambda^{2p} |\mathcal{F}_B(f)(\lambda)| \leq \|\ell_\alpha^p(f)\|_{1,\alpha}.$$

Proof.

i) From the relation (5.II.1) we have

$$\forall \lambda \geq 0,\ \mathcal{F}_B(\ell_\alpha(f))(\lambda) = \int_0^\infty \ell_\alpha(f)(x)\, j_\alpha(\lambda x)\, d\mu_\alpha(x).$$

As

$$\ell_\alpha(f)(x) = \frac{1}{x^{2\alpha+1}} \frac{d}{dx}[x^{2\alpha+1} \frac{d}{dx} f(x)],$$

then by making two integrations by parts, we obtain

$$\forall \lambda \geq 0,\ \mathcal{F}_B(\ell_\alpha(f))(\lambda) = \int_0^\infty f(x) \ell_\alpha(j_\alpha(\lambda .))(x)\, d\mu_\alpha(x).$$

But from the relation (1.II.8) we have

$$\ell_\alpha(j_\alpha(\lambda .))(x) = -\lambda^2 j_\alpha(\lambda x).$$

Thus

$$\forall \lambda \geq 0,\ \mathcal{F}_B(\ell_\alpha(f))(\lambda) = -\lambda^2 \mathcal{F}_B(f)(\lambda).$$

ii) Using the i) we obtain for all $p \in \mathbb{N} \setminus \{0\}$:

$$\forall \lambda \geq 0,\ \mathcal{F}_B(\ell_\alpha^p(f))(\lambda) = (-1)^p \lambda^{2p} \mathcal{F}_B(f)(\lambda).$$

We deduce (5.II.13) from the inequality (5.II.3).

Proposition 5.II.4. For all f in $\mathcal{D}_*(\mathbb{R})$, the function $\mathcal{F}_B(f)$ is an even entire function on $\mathbb{C}$, and we have

(5.II.14) $$\forall \lambda \in \mathbb{C},\ \mathcal{F}_B(f)(\lambda) = 2^{-\alpha}(\Gamma(\alpha+1))^{-1} \mathcal{F}_0 \circ W_\alpha(f)(\lambda),$$

where $\mathcal{F}_0$ is the Fourier-cosine transform defined by the relation (3.II.1) and W_α the Weyl integral transform studied on $\mathcal{D}_*(\mathbb{R})$ in the section 2.II.

Proof.

Let f be in $\mathcal{D}_*(\mathbb{R})$. From Definition 5.II.1 we have

$$\forall \lambda \in \mathbb{C},\ \mathcal{F}_B(f)(\lambda) = \int_0^\infty f(x)\, j_\alpha(\lambda x)\, d\mu_\alpha(x).$$

The parity of the function $\lambda \dashrightarrow j_\alpha(\lambda x)$ shows that $\mathcal{F}_B(f)$ is even on $\mathbb{C}$. On the other hand using the relation (1.II.17), and making a derivation under the integral sign, we see that the function $\mathcal{F}_B(f)$ is entire on $\mathbb{C}$.

We prove now the relation (5.II.14). Theorem 2.II.2 shows that for f in $\mathcal{D}_*(\mathbb{R})$ the function $W_\alpha(f)$ belongs to $\mathcal{D}_*(\mathbb{R})$.

From the relation (2.II.2) we have

$$\forall\, \lambda \in \mathbb{C},\ \int_0^\infty W_\alpha(f)(y)\cos\lambda y\,dy = 2^\alpha\,\Gamma(\alpha+1)\int_0^\infty \mathcal{R}_\alpha(\cos(\lambda .))(x)f(x)\,d\mu_\alpha(x).$$

Using the relation (2.I.15) :

$$\forall\, x \geq 0,\ \forall\, \lambda \in \mathbb{C},\ \mathcal{R}_\alpha(\cos(\lambda .))(x) = j_\alpha(\lambda x),$$

we obtain

$$\forall\, \lambda \in \mathbb{C},\ \mathcal{F}_0(W_\alpha(f))(\lambda) = 2^\alpha\,\Gamma(\alpha+1)\,\mathcal{F}_B(f)(\lambda).$$

Theorem 5.II.1. Let f be in $\mathcal{D}_*(\mathbb{R})$. Then we have the inversion formula

(5.II.15) $$\forall\, x \geq 0,\ f(x) = \int_0^\infty \mathcal{F}_B(f)(\lambda)j_\alpha(\lambda x)d\mu_\alpha(\lambda).$$

To prove this theorem we need the following Lemma.

Lemma 5.II.1. We have the following relation

(5.II.16) $$\forall\, k \in \mathbb{N},\ \forall\, y \geq 0,\ \left(\frac{d}{dy^2}\right)^{k+1}(\cos\lambda y) = \frac{(-1)^{k+1}\sqrt{\pi}}{2^{2k+2}\,\Gamma(k+\frac{3}{2})}\lambda^{2k+2}\, j_{k+\frac{1}{2}}(\lambda y).$$

Proof.

We have

$$\left(\frac{d}{dy^2}\right)^{k+1}(\cos\lambda y) = -\lambda^2\left(\frac{d}{dy^2}\right)^k\left(\frac{\sin\lambda y}{\lambda y}\right),$$

$$= -\frac{\lambda^2}{2^{k+1}}\left(\left(\frac{1}{y}\frac{d}{dy}\right)^k j_{\frac{1}{2}}(\lambda y)\right),$$

but, if we put $u = \lambda y$, we obtain

(5.II.17) $$\frac{1}{y}\frac{d}{dy} = \lambda^2\,\frac{1}{u}\frac{d}{du}.$$

By applying the derivation formula (1.II.4) :

$$\frac{1}{u}\frac{d}{du}\, j_\alpha(u)\,/_{u=\lambda y} = -\frac{1}{2(\alpha+1)}\, j_{\alpha+1}(\lambda y),$$

we obtain

$$(\frac{1}{y}\frac{d}{dy})^k (j_{\frac{1}{2}}(\lambda y)) = \frac{(-1)^k \lambda^{2k} j_{k+\frac{1}{2}}(\lambda y)}{2^k(\frac{1}{2}+1)(\frac{3}{2}+1)\ldots((k-\frac{1}{2})+1)}.$$

But

$$\Gamma(k+\tfrac{3}{2}) = \sqrt{\pi}\,.\,\frac{3 \times 5 \times \ldots \times (2k+1)}{2^{k+1}}.$$

Thus

$$(\frac{d}{dy^2})^{k+1}(\cos\lambda y) = \frac{(-1)^{k+1}\sqrt{\pi}}{2^{2k+2}\Gamma(k+\frac{3}{2})}\lambda^{2k+2}\, j_{k+\frac{1}{2}}(\lambda y).$$

Proof of Theorem 5.II.1.

Let f be in $\mathcal{D}_*(\mathbb{R})$.. From Theorem 2.II.2 and Proposition 5.II.4 the function $W_\alpha(f)$ belongs to $\mathcal{D}_*(\mathbb{R})$ and we have

$$\forall \lambda \geq 0,\ \mathcal{F}_B(f)(\lambda) = 2^{-\alpha}(\Gamma(\alpha+1))^{-1}\,\mathcal{F}_0 \circ W_\alpha(f)(\lambda).$$

Using the inversion formula (3.II.7) for the Fourier-cosine transform $\mathcal{F}_0$ we obtain

(5.II.18) $$\forall x \geq 0,\ W_\alpha(f)(x) = \frac{2^{\alpha+1}\Gamma(\alpha+1)}{\pi}\int_0^\infty \mathcal{F}_B(f)(\lambda)\cos\lambda x\, d\lambda.$$

i) If $\alpha = k + \frac{1}{2}$, $k \in \mathbb{N}$.

By applying Theorem 2.II.2 to the relation (5.II.18), we obtain

$$\forall x \geq 0,\ f(x) = W_\alpha^{-1}\left[\frac{2^{\alpha+1}\Gamma(\alpha+1)}{\pi}\int_0^\infty \mathcal{F}_B(f)(\lambda)\cos(\lambda .)\, d\lambda\right](x),$$

$$= \frac{(-1)^{k+1}2^{\alpha+1}\Gamma(\alpha+1)C_2}{\pi}(\frac{d}{dx^2})^{k+1}\int_0^\infty \mathcal{F}_B(f)(\lambda)\cos\lambda x\, d\lambda.$$

Using the relation (5.II.17) and (5.II.13) we see that we can derive under the integral sign, and we have

$$\forall x \geq 0,\ f(x) = \frac{(-1)^{k+1}2^{\alpha+1}\Gamma(\alpha+1)C_2}{\pi}\int_0^\infty \mathcal{F}_B(f)(\lambda)(\frac{d}{dx^2})^{k+1}(\cos\lambda x)\, d\lambda.$$

From Lemma 5.II.1 and the fact that

$$C_2 = \frac{\sqrt{\pi}}{\Gamma(k+\frac{3}{2})}$$

we deduce that

(5.II.19) $$\forall\, x \geq 0,\ f(x) = \int_0^\infty \mathfrak{F}_B(f)(\lambda)\ j_{k+\frac{1}{2}}(\lambda x)\ d\mu_{k+\frac{1}{2}}(\lambda).$$

ii) If $\alpha = k + r,\ k \in \mathbb{N},\ r \in]-\frac{1}{2}, \frac{1}{2}[$.

By applying Theorem 2.II.2 to the relation (5.II.18) we obtain

$$\forall\, x \geq 0,\ f(x) = W_\alpha^{-1}\left[\frac{2^{\alpha+1}\Gamma(\alpha+1)}{\pi}\int_0^\infty \mathfrak{F}_B(f)(\lambda)\cos(\lambda .)\,d\lambda\right](x),$$

$$= \frac{(-1)^{k+1}2^{\alpha+1}\,\Gamma(\alpha+1)C_1}{\pi}\int_x^\infty (y^2 - x^2)^{-r-\frac{1}{2}}\ y\left(\frac{d}{dy^2}\right)^{k+1}$$

$$\times \left[\int_0^\infty \mathfrak{F}_B(f)(\lambda)\ (x)\ \cos\lambda y\ d\lambda\right] dy.$$

By derivation under the integral sign and by applying Lemma 5.II.1, we obtain

$$\forall\, x \geq 0,\ f(x) = \frac{2^{\alpha-2k-1}\,\Gamma(\alpha+1)C_1}{\sqrt{\pi}\,\Gamma(k+\frac{3}{2})}\int_x^\infty (y^2 - x^2)^{-r-\frac{1}{2}}\ y\Big(\int_0^\infty \mathfrak{F}_B(f)\ (\lambda)$$

$$\times\ j_{k+\frac{1}{2}}(\lambda\, y)\lambda^{2k+2}\ d\lambda\Big)dy.$$

The change of variables $y = tx$ implies

(5.II.20) $$\forall\, x > 0,\ f(x) = \frac{2^{\alpha-2k-1}\,\Gamma(\alpha+1)C_1\,x^{1-2r}}{\sqrt{\pi}\,\Gamma(k+\frac{3}{2})}\int_1^\infty (t^2-1)^{-r-\frac{1}{2}}\Big(\int_0^\infty \mathfrak{F}_B(f)(\lambda)$$

$$\times\ j_{k+\frac{1}{2}}(\lambda t x)\ \lambda^{2k+2}\ d\lambda\Big)dt.$$

But, from the relations (1.II.3) and (1.I. 19) we deduce that there exists a positive constant M such that

$$\forall\, u \geq 1,\ |j_\alpha(u)| \leq M\ u^{-\alpha-\frac{1}{2}}.$$

This result implies that for all $t \geq 1,\ x > 0,\ \lambda \geq \frac{1}{x}$, we have

$$\int_1^\infty\int_{\frac{1}{x}}^\infty t^{-k-1}(t^2-1)^{-r-\frac{1}{2}}\ |\mathfrak{F}_B(f)(\lambda)|\ \lambda^{k+1}\ d\lambda\ dt < +\infty.$$

Thus, by applying Fubini's theorem to the second member of (5.II.20), we obtain

$$\forall\, x > 0,\ f(x) = \frac{2^{\alpha-2k-1}\,\Gamma(\alpha+1)C_1\,x^{1-2r}}{\sqrt{\pi}\,\Gamma(k+\frac{3}{2})}\int_0^\infty \mathfrak{F}_B(f)\ (\lambda)\ \Big(\int_1^\infty (t^2-1)^{-r-\frac{1}{2}}$$

$$\times\ j_{k+\frac{1}{2}}(\lambda t x)\ dt\Big)\ \lambda^{2k+2}\ d\lambda.$$

But, by the change of variables $y = tx$, and the relation (1.II.35), we have

$$x^{1-2r}\int_1^{\infty}(t^2-1)^{-r-\frac{1}{2}}\, j_{k+\frac{1}{2}}(\lambda t x)\,dt = \int_y^{\infty}(y^2-x^2)^{-r-\frac{1}{2}} j_{k+\frac{1}{2}}(\lambda y)\, y\, dy.$$

$$= \frac{2^{-2r}\,\Gamma(\frac{1}{2}-r)\Gamma(k+\frac{3}{2})}{\Gamma(\alpha+1)}\lambda^{2r-1}\, j_\alpha(\lambda x).$$

Thus

$$\forall\, x>0,\ f(x) = \frac{2^{-\alpha-1}\, C_1\Gamma(\frac{1}{2}-r)}{\sqrt{\pi}}\int_0^{\infty}\mathcal{F}_B(f)\,(\lambda) j_\alpha(\lambda x)\lambda^{2\alpha+1}\,d\lambda.$$

Using the fact that

$$C_1 = \frac{2\sqrt{\pi}}{\Gamma(\alpha+1)\Gamma(\frac{1}{2}-r)},$$

we deduce that

$$\forall\, x>0,\ f(x) = \int_0^{\infty}\mathcal{F}_B(f)\,(\lambda) j_\alpha(\lambda x)\, d\mu_\alpha(\lambda).$$

As the two members are continuous on $[0,+\infty[$, then

(5.II.21) $$\forall\, x\geq 0,\ f(x) = \int_0^{\infty}\mathcal{F}_B(f)\,(\lambda)\, j_\alpha(\lambda x)\, d\mu_\alpha(\lambda).$$

We deduce (5.II.15) from the relations (5.II.19) and (5.II.21).

Theorem 5.II.2. Let f be in $L^1([0,+\infty[,\ d\mu_\alpha)$ such that $\mathcal{F}_B(f)$ belongs to $L^1([0,+\infty[,\ d\mu_\alpha)$. Then we have the following inversion formula

(5.II.22) $$f(x) = \int_0^{\infty}\mathcal{F}_B(f)\,(\lambda)\, j_\alpha(\lambda x)\, d\mu_\alpha(\lambda),\ \text{a.e.}$$

Proof.

Let V be a positive function in $\mathcal{D}_*(\mathbb{R})$ with support in the interval $[-1, 1]$ and satisfying $\|V\|_{1,\alpha} = 1$. For $\varepsilon\in\,]0,+\infty[$, we put

$$\forall\, x\geq 0,\ V_\varepsilon(x) = \varepsilon^{-2\alpha-2}\, V(\frac{x}{\varepsilon}).$$

The function V_ε belongs to $\mathcal{D}_*(\mathbb{R})$ and from Theorem 5.II.1 we have

$$\forall\, x\geq 0,\ V_\varepsilon(x) = \int_0^{\infty}\mathcal{F}_B(V_\varepsilon)(\lambda)\, j_\alpha(\lambda x)\, d\mu_\alpha(\lambda).$$

Then

$$\forall\, x\geq 0,\ f *_B V_\varepsilon(x) = \int_0^{\infty} T_x(f)(y)\, V_\varepsilon(y) d\mu_\alpha(y).$$

$$= \int_0^{\infty} T_x(f)(y)\Big(\int_0^{\infty}\mathcal{F}_B(V_\varepsilon)(\lambda)\, j_\alpha(\lambda x)\, d\mu_\alpha(\lambda)\Big)\, d\mu_\alpha(y).$$

Using Fubini's theorem and the relation (5.II.8), we obtain

$$\forall\, x \geq 0,\ f *_B V_\varepsilon(x) = \int_0^\infty \mathcal{F}_B(f)(\lambda)\mathcal{F}_B(V_\varepsilon)(\lambda)\, j_\alpha(\lambda x)\, d\mu_\alpha(\lambda).$$

Then from dominated convergence theorem, we deduce that

$$(5.II.23) \qquad \forall\, x \geq 0,\ \lim_{\varepsilon\to 0} f *_B V_\varepsilon(x) = \int_0^\infty \mathcal{F}_B(f)(\lambda) j_\alpha(\lambda x)\, d\mu_\alpha(\lambda).$$

But from Theorem 4.III.1. i) we have

$$\lim_{\varepsilon\to 0} \| f *_B V_\varepsilon - f \|_{1,\alpha} = 0,$$

then we can extract from $\{f *_B V_\varepsilon\}_{\varepsilon>0}$ a subsequence which converges to f almost everywhere.

Thus using (5.II..23) we deduce that

$$f(x) = \int_0^\infty \mathcal{F}_B(f)(\lambda)\, j_\alpha(\lambda x)\, d\mu_\alpha(\lambda), \text{ a.e.}$$

5.III. The Fourier-Bessel transform on $\mathcal{S}_*(\mathbb{R})$ and $\mathcal{D}_*(\mathbb{R})$.

Proposition 5.III.1. For all f in $\mathcal{S}_*(\mathbb{R})$, the function $\mathcal{F}_B(f)$ belongs to $\mathcal{E}_*(\mathbb{R})$ and we have

$$(5.III.1) \qquad \forall\, \lambda \geq 0,\ \mathcal{F}_B(f)(\lambda) = 2^{-\alpha}(\Gamma(\alpha+1))^{-1}\, \mathcal{F}_0 \circ W_\alpha(f)(\lambda).$$

Proof.

Le f be in $\mathcal{S}_*(\mathbb{R})$. From Definition 5.II.1 we have

$$\forall\, \lambda \geq 0,\ \mathcal{F}_B(f)(\lambda) = \int_0^\infty f(x)\, j_\alpha(\lambda x)\, d\mu_\alpha(x).$$

The parity of the function $\lambda \to j_\alpha(\lambda x)$ shows that the function $\mathcal{F}_B(f)$ is even on $\mathbb{R}$. On the other hand the relation (1.II.17) and derivations under the integral sign imply that $\mathcal{F}_B(f)$ belongs to $\mathcal{E}_*(\mathbb{R})$.

We obtain the relation (5.III.1) by the same method used in the proof of Proposition 5.II.4.

Theorem 5.III.1. The Fourier-Bessel transform $\mathcal{F}_B$ is a topological isomorphism from

- $\mathcal{D}_*(\mathbb{R})$ onto the space $\mathbb{H}_*(\mathbb{C})$ defined in 3.II.2 .
- $\mathcal{S}_*(\mathbb{R})$ onto itself.

The inverse transform $\mathfrak{F}_B^{-1}$ is given by

(5.III.2) $$\forall\, x \geq 0,\ \mathfrak{F}_B^{-1}(f)(x) = \int_0^\infty f(\lambda) j_\alpha(\lambda x)\, d\mu_\alpha(\lambda).$$

Proof.

From Proposition 5.II.4 and 5.III.1 we have for all f in $\mathcal{D}_*(\mathbb{R})$ (resp. $\mathcal{S}_*(\mathbb{R})$)

$$\mathfrak{F}_B(f) = 2^{-\alpha}(\Gamma(\alpha+1))^{-1}\, \mathfrak{F}_0 \circ W_\alpha(f).$$

We deduce the results from Theorems 2.II.2, 2.III.1, 3.II.2, 3.II.3 and 5.II.1.

Remarks.

i) From the relation (5.III.2) we deduce that

(5.III.3) $$\mathfrak{F}_B^{-1} = \mathfrak{F}_B.$$

ii) From Proposition 2.II.3. i) and Theorem 2.II.2, we deduce that a function f in $\mathcal{D}_*(\mathbb{R})$ has its support in the interval $[-a, a]$, $a \geq 0$, if and only if the function $W_\alpha(f)$ has its support in $[-a,a]$.

Using this result, the relation (5.II.14) and Theorem 3.II.2 we deduce that the support of f is in $[-a, a]$ if and only if the function $\mathfrak{F}_B(f)$ is of type a. i.e. there exists a positive constant M such that

(5.III.4) $$\forall\, \lambda \in \mathbb{C},\ |\mathfrak{F}_B(f)(\lambda)| \leq M e^{a|\mathrm{Im}\lambda|}.$$

Corollary 5.III.1. The space $\mathcal{S}_*(\mathbb{R})$ is an algebra for the generalized convolution product $*_B$, an algebra for the multiplication, and the Fourier-Bessel transform $\mathfrak{F}_B$ is a topological isomorphism from the first onto the second.

Proof.

We deduce these results from Theorem 5.III.1 and Proposition 5.II.2. ii).

Proposition 5.III.2.

i) Let f and g be in $\mathcal{S}_*(\mathbb{R})$. Then we have

(5.III.5) $$\int_0^\infty f(x)\, \overline{g(x)}\, d\mu_\alpha(x) = \int_0^\infty \mathfrak{F}_B(f)(\lambda)\, \overline{\mathfrak{F}_B(g)(\lambda)}\, d\mu_\alpha(\lambda).$$

ii) Let f be in $\mathcal{S}_*(\mathbb{R})$. Then we have

(5.III.6) $$\int_0^\infty |f(x)|^2\, d\mu_\alpha(x) = \int_0^\infty |\mathfrak{F}_B(f)(\lambda)|^2\, d\mu_\alpha(\lambda).$$

Proof.

i) We put

$$h = f *_B \bar{g}.$$

From Corollary 5.III.1 the function h belongs to $\mathcal{S}_*(\mathbb{R})$ and we have

$$\forall \lambda \geq 0, \mathfrak{F}_B(h)(\lambda) = \mathfrak{F}_B(f)(\lambda).\overline{\mathfrak{F}_B(g)(\lambda)}.$$

Using Theorem 5.III.1, we have

(5.III.7) $$h(0) = \int_0^{\infty} \mathfrak{F}_B(f)(\lambda).\overline{\mathfrak{F}_B(g)(\lambda)}\,d\mu_\alpha(\lambda).$$

On the other hand from the definition of the function h we have

(5.III.8) $$h(0) = \int_0^{\infty} f(x)\ \overline{g(x)}\ d\mu_\alpha(x).$$

The relations (5.III.7) and (5.III.8) give (5.III.5).

ii) We obtain (5.III.6) by taking g = f in the relation (5.III.5).

Remark.

The relation (5.III.6) can also be written in the form

(5.III.9) $$\|\mathfrak{F}_B(f)\|_{2,\alpha} = \|f\|_{2,\alpha}.$$

Proposition 5.III.3. For all f and g in $\mathcal{S}_*(\mathbb{R})$ we have

(5.III.10) $$\forall \lambda \geq 0, \mathfrak{F}_B(f.g)(\lambda) = \mathfrak{F}_B(f) *_B \mathfrak{F}_B(g)(\lambda).$$

Proof.

From the relations (5.II.9) and (5.III.3) we have

$$\mathfrak{F}_B[\mathfrak{F}_B(f) *_B \mathfrak{F}_B(g)] = \mathfrak{F}_B \circ \mathfrak{F}_B(f).\ \mathfrak{F}_B \circ \mathfrak{F}_B(g) = f.g.$$

We deduce the result from Theorem 5.III.1 and the relation (5.III.3).

Proposition 5.III.4. Let f be in $\mathcal{S}_*(\mathbb{R})$. Then for all $x \geq 0$, the function $T_x(f)$ belongs to $\mathcal{S}_*(\mathbb{R})$.

Proof.

From Proposition 4.II.2. for all $x \geq 0$, the function $T_x(f)$ belongs to $L^1([0, +\infty[, d\mu_\alpha)$. Then using the relation (5.II.8) we have

(5.III.11) $$\forall \lambda \geq 0, \mathfrak{F}_B(T_x(f))\ (\lambda) = j_\alpha(\lambda x)\ \mathfrak{F}_B(f)(\lambda).$$

But from the relation (1.II.17) and the fact that the function $\mathcal{F}_B(f)$ is in $\mathcal{S}_*(\mathbb{R})$, we deduce that the function $j_\alpha(\lambda x)\,\mathcal{F}_B(f)(\lambda)$ belongs to $\mathcal{S}_*(\mathbb{R})$. We obtain the desired result from (5.III.11) and Theorem 5.III.1.

Proposition 5.III.5. Let f be in $\mathcal{D}_*(\mathbb{R})$ (resp. $\mathcal{S}_*(\mathbb{R})$). Then the function $u(x, y) = T_x(f)(y)$ is the unique solution of the equation

$$(5.III.12)\qquad \begin{cases} \ell_{\alpha,x}\, u(x,y) = \ell_{\alpha,y}\, u(x, y), \\ u(x, 0) = f(x)\,, \ \dfrac{\partial}{\partial y} u(x, 0) = 0, \quad \text{for all } x \geq 0, \end{cases}$$

where ℓ_α is the Bessel operator given by the relation (5.II.11).

Proof.

From Proposition 4.II.7 we deduce that the function $u(x, y) = T_x(f)(y)$ is a solution of the equation (5.III.12). It remainders to prove that this equation admits a unique solution in $\mathcal{D}_*(\mathbb{R})$ (resp. $\mathcal{S}_*(\mathbb{R})$).

We consider the equation

$$\begin{cases} \ell_{\alpha,x}\, v(x, y) = \ell_{\alpha,y}\, v(x,y), \\ v(x, 0) = 0, \ \dfrac{\partial}{\partial y} v(x, 0) = 0, \quad \text{for all } x \geq 0. \end{cases}$$

From Corollary 4.II.1 and Proposition 5.III.4 for all $y \geq 0$, the function $x \dashrightarrow v(x,y)$ belongs to $\mathcal{D}_*(\mathbb{R})$ (resp. $\mathcal{S}_*(\mathbb{R})$).

We take the Fourier-Bessel transform with respect to the variable x, of this equation, and using (5.II.12) we obtain the following equation

$$\begin{cases} \ell_{\alpha,y}(\mathcal{F}_B(v(.,y))(\lambda)) = -\lambda^2\, \mathcal{F}_B(v(.,y))(\lambda)\,, \\ \mathcal{F}_B(v(.,0))(\lambda) = 0, \ \dfrac{\partial}{\partial y}\, \mathcal{F}_B(v(.,0))(\lambda) = 0. \end{cases}$$

But from (1.II.8) the function $\mathcal{F}_B(v(.,y))(\lambda) = 0$ is the unique solution of this equation. Then from Theorem 5.III.1 we deduce that

$$u \equiv 0.$$

This completes the proof of the proposition.

Proposition 5.III.6. Let f be in $\mathscr{D}_*(\mathbb{R})$ (resp. $\mathscr{S}_*(\mathbb{R})$). Then the generalized translation operators T_x, $x \geq 0$, are related to the translation operators σ_x, $x \geq 0$, of Chapitre 3, by the relations

(5.III.13) $$\forall\, y \geq 0,\quad T_x(f)(y) = \mathcal{R}_{\alpha,x}\, \mathcal{R}_{\alpha,y}\, [\sigma_x\, \mathcal{R}_\alpha^{-1}(f)(y)].$$

(5.III.14) $$\forall\, y \geq 0,\quad T_x(f)(y) = \mathcal{R}_{\alpha,x} W_{\alpha,y}^{-1}[\sigma_x\, W_\alpha(f)(y)].$$

Where $\mathcal{R}_\alpha$ and W_α are respectively the Riemann-Liouville and Weyl integral transforms.

Proof.

We deduce these relations from Proposition 5.III.5, and the properties of the operators $\mathcal{R}_\alpha$, W_α and σ_x given respectively in Chapters 2 and 3.

Proposition 5.III.7. Let f and g be in $\mathscr{D}_*(\mathbb{R})$ (resp. $\mathscr{S}_*(\mathbb{R})$). Then we have the following relation

(5.III.15) $$f *_B g = 2^{-\alpha}(\Gamma(\alpha+1))^{-1}\ W_\alpha^{-1}[W_\alpha(f) *_0 W_\alpha(g)].$$

Proof.

From Corollaries 4.III.1 and 5.III.1 the function $f *_B g$ belongs to $\mathscr{D}_*(\mathbb{R})$ (resp. $\mathscr{S}_*(\mathbb{R})$).

Using the relation (5.II.9), (5.II.14), (5.III.1) and (3.II.4) we obtain

$$\forall\, \lambda \geq 0,\ \mathscr{F}_B(f *_B g)(\lambda) = \mathscr{F}_B(f)(\lambda)\, \mathscr{F}_B(g)(\lambda).$$

$$\forall\, \lambda \geq 0,\ \mathscr{F}_0\, (W_\alpha(f *_B g))(\lambda) = 2^{-\alpha}\, (\Gamma(\alpha+1))^{-1}\, \mathscr{F}_0(W_\alpha(f))(\lambda).\ \mathscr{F}_0(W_\alpha(g))(\lambda).$$

Then

$$\forall\, \lambda \geq 0,\ \mathscr{F}_0\, (W_\alpha(f *_B g))(\lambda) = 2^{-\alpha}\, (\Gamma(\alpha+1))^{-1}\, \mathscr{F}_0[W_\alpha(f) *_0 W_\alpha(g)](\lambda).$$

From Theorems 2.II.2, 2.III.1 and Corollary 3.I.1 we deduce that

$$\forall\, x \geq 0,\ W_\alpha(f *_B g)(x) = 2^{-\alpha}\, (\Gamma(\alpha+1))^{-1}\, W_\alpha(f) *_0 W_\alpha(g)(x).$$

Thus

$$\forall\, x \geq 0,\ f *_B g(x) = 2^{-\alpha}\, (\Gamma(\alpha+1))^{-1}\ W_\alpha^{-1}[W_\alpha(f) *_0 W_\alpha(g)](x).$$

5.IV. The Fourier-Bessel transform on $L^2([0, +\infty[, d\mu_\alpha)$.

Theorem 5.IV.1. We consider.

- H a hilbert space .
- V a dense subspace of H.
- u a linear mapping of V into H, continuous when we provide V with the norm induced by her of H.

Then u extends to a linear continuous mapping of H into itself. In particular if u is an isometry, it extends by continuity to an isometry of H into itself.

Theorem 5.IV.2 (of Plancherel). There exists an isometric isomorphism Φ_B of $L^2([0, +\infty[, d\mu_\alpha)$ onto itself which coincides with the Fourier-Bessel transform $\mathcal{F}_B$ on $\mathcal{S}_*(\mathbb{R})$. The inverse transform Φ_B^{-1} satisfies

(5.IV.1) $$\Phi_B^{-1} = \Phi_B.$$

Proof.

We take
$$H = L^2([0, +\infty[, d\mu_\alpha), \; V = \mathcal{S}_*(\mathbb{R}), \; u = \mathcal{F}_B .$$
From the relation (5.III.9) the mapping u is an isometry of $\mathcal{S}_*(\mathbb{R})$ into H. Using Corollary 4.III.3 i) and Theorem 5.IV.1 we deduce that u extends to an isometry of H into itself.

We shall show that this isometry is onto. It is clear that u(H) is a closed subspace of H. On the other hand u(H) coutains $u(\mathcal{S}_*(\mathbb{R}))$. But from Theorem 5.III.1 and Corollary 4.III.3.i),
$$u\,(\mathcal{S}_*(\mathbb{R})) = \mathcal{S}_*(\mathbb{R}).$$
and $\mathcal{S}_*(\mathbb{R})$ is dense in H. Then u(H) is dense in H. Thus
$$u(H) = H.$$
This completes the proof of the theorem.

Remark.

The mapping Φ_B is called Fourier-Bessel-Plancherel transform.

By using Proposition 5.III.2 and Theorem 5.IV.2 we obtain the following results.

Theorem 5.IV.3. (Plancherel formulas). For all f and g in $L^2([0, +\infty[, d\mu_\alpha)$ we have

(5.IV.2)
$$\int_0^\infty f(x)\,\overline{g(x)}\,d\mu_\alpha(x) = \int_0^\infty \Phi_B(f)(\lambda)\,\overline{\Phi_B(g)(\lambda)}\,d\mu_\alpha(\lambda)\,.$$

(5.IV.3)
$$\|\Phi_B(f)\|_{2,\alpha} = \|f\|_{2,\alpha}\,.$$

Proposition 5.IV.1. Let f be in $(L^1 \cap L^2)([0, +\infty[, d\mu_\alpha)$. Then we have

(5.IV.4)
$$\mathfrak{F}_B(f) = \Phi_B(f).$$

(More precisely $\mathfrak{F}_B(f)$ is a representative of the class $\Phi_B(f)$).

Proof.

Let φ be a positive function in $\mathcal{D}_*(\mathbb{R})$ such that $\|\varphi\|_{1,\alpha} = 1$. For all $n \in \mathbb{N}$, we put

$$\forall\, x \geq 0,\ \ \varphi_n(x) = n^{2\alpha+2}\,\varphi(nx).$$

We consider the sequence $\{f_n\}_{n\in\mathbb{N}}$ defined by

$$\forall\, x \geq 0,\ \ f_n(x) = (f\,\mathbb{1}_{[0,n]}) *_B \varphi_n(x),$$

where $\mathbb{1}_{[0,n]}$ is the characteristic function of the interval $[0, n]$.

From Propositions 4.III.8, 4.III.2 and Theorem 4.III.1 the function f_n belongs to $\mathcal{D}_*(\mathbb{R})$ and we have

$$\lim_{n\to+\infty} \|f_n - f\|_{p,\alpha} = 0, \quad p = 1, 2,$$

but, the relation (5.II.3) implies

$$\operatorname{Sup}_{\lambda\geq 0} |\mathfrak{F}_B(f_n)(\lambda) - \mathfrak{F}_B(f)(\lambda)| \leq \|f_n - f\|_{1,\alpha}\,.$$

Thus the sequence $\{\mathfrak{F}_B(f_n)\}_{n\in\mathbb{N}}$ converges to $\mathfrak{F}_B(f)$ uniformly on $[0, +\infty[$. On the other hand from Theorem 5.IV.1, the sequence $\{\Phi_B(f_n)\}_{n\in\mathbb{N}}$ converges to $\Phi_B(f)$ in $L^2([0, +\infty[, d\mu_\alpha)$. Then we can extract a subsequence $\{\Phi_B(f_{n_k})\}_{k\in\mathbb{N}}$ which converges to $\Phi_B(f)$ almost everywhere on $[0, +\infty[$.

But from Theorem 5.IV.1 we have

$$\mathfrak{F}_B(f_n) = \Phi_B(f_n)\,.$$

Thus

$$\mathfrak{F}_B(f)(\lambda) = \Phi_B(f)(\lambda), \quad \text{a.e. on } [0, +\infty[.$$

Proposition 5.IV.2. Let f be in $L^2([0, +\infty[, d\mu_\alpha)$. We put for $N \in \mathbb{N}$:

(5.IV.5)
$$g_N(\lambda) = \int_0^N f(x) j_\alpha(\lambda x)\, d\mu_\alpha(x).$$

Then the sequence $\{g_N\}_{N\in\mathbb{N}}$ converges to $\Phi_B(f)$ in $L^2([0, +\infty[, d\mu_\alpha)$.

Proof.

We consider for $N \in \mathbb{N}$ the function

$$f_N = f.\, \mathbb{1}_{[0, N]},$$

where $\mathbb{1}_{[0, N]}$ is the characteristic function of the interval $[0, N]$.

The function f_N belongs to $(L^1 \cap L^2)([0, +\infty[, d\mu_\alpha)$. As the sequece $\{f_N\}_{N\in\mathbb{N}}$ converges to f in $L^2([0, +\infty[, d\mu_\alpha)$, then from Theorem 5.IV.1 the sequence $\{\Phi_B(f_N)\}_{N\in\mathbb{N}}$ converges to $\Phi_B(f)$ in $L^2([0, +\infty[, d\mu_\alpha)$. But from Proposition 5.IV.1 we have

$$g_N = \mathfrak{F}_B(f_N) = \Phi_B(f_N).$$

Thus

$$\lim_{N\to+\infty} g_N = \Phi_B(f),$$

the limit is in $L^2([0, +\infty[, d\mu_\alpha)$.

Proposition 5.IV.3. Let f be in $L^1([0, +\infty[, d\mu_\alpha)$ and g in $L^2([0, +\infty[, d\mu_\alpha)$.

Then we have

(5.IV.6)
$$\Phi_B(f *_B g) = \mathfrak{F}_B(f).\Phi_B(g).$$

Proof.

We consider a sequence $\{g_n\}_{n\in\mathbb{N}}$ in $\mathcal{S}_*(\mathbb{R})$ such that

$$\lim_{n\to+\infty} \|g_n - g\|_{2,\alpha} = 0.$$

From Propositions 4.III.4, 5.IV.1 and the relation (5.II.9) we have

$$\Phi_B(f *_B g_n) = \mathfrak{F}_B(f *_B g_n) = \mathfrak{F}_B(f)\, \mathfrak{F}_B(g_n),$$
$$= \mathfrak{F}_B(f)\, \Phi_B(g_n).$$

On the other hand from Proposition 4.III.4 the sequence $\{f *_B g_n\}_{n\in\mathbb{N}}$ converges to $f *_B g$ in $L^2([0, +\infty[, d\mu_\alpha)$, and from Hölder's inequality the sequence $\{\mathcal{F}_B(f) \,.\, \Phi_B(g_n)\}_{n\in\mathbb{N}}$ converges to $\mathcal{F}_B(f) \,.\Phi_B(g)$ in $L^2([0, +\infty[, d\mu_\alpha)$.

Thus

$$\Phi_B(f *_B g) = \mathcal{F}_B(f)\Phi_B(g).$$

Proposition 5.IV.4. Let f and g be in $L^2([0, +\infty[, d\mu_\alpha)$. Then we have

$$f *_B g = \mathcal{F}_B(\Phi_B(f).\Phi_B(g)). \tag{5.IV.7}$$

Proof.

We consider $\{f_n\}_{n\in\mathbb{N}}$ and $\{g_n\}_{n\in\mathbb{N}}$ two sequences in $\mathcal{S}_*(\mathbb{R})$ such that

$$\lim_{n\to+\infty} \|f_n - f\|_{2,\alpha} = 0 ,$$

and

$$\lim_{n\to+\infty} \|g_n - g\|_{2,\alpha} = 0.$$

From Corollary 5.III.1 and the relation (5.III.3) we have

$$f_n *_B g_n = \mathcal{F}_B(\mathcal{F}_B(f_n).\mathcal{F}_B(g_n)).$$

Using Proposition 5.IV.1 we obtain

$$f_n *_B g_n = \mathcal{F}_B(\Phi_n(f_n).\Phi_n(g_n)).$$

But from Proposition 4.III.5 the sequence $\{f_n *_B g_n\}_{n\in\mathbb{N}}$ converges to $f *_B g$ uniformly on $[0, +\infty[$. On the other hand from Hölder's inequality the sequence $\{\Phi_B(f_n).\Phi_B(g_n)\}_{n\in\mathbb{N}}$ converges to $\Phi_B(f).\Phi_B(g)$ in $L^1([0, +\infty[, d\mu_\alpha)$. Or from the relation (5.II.3) we have

$$\sup_{\lambda\geq 0} |\mathcal{F}_B(\Phi_B(f_n).\Phi_B(g_n) - \Phi_B(f)\Phi_B(g))(\lambda)| \leq \|\Phi_B(f_n).\Phi_B(g_n) - \Phi_B(f)\Phi_B(g)\|_{1,\alpha} .$$

Thus the sequence $\{\mathcal{F}_B(\Phi_B(f_n) \,.\, \Phi_B(g_n))\}_{n\in\mathbb{N}}$ converges to $\mathcal{F}_B(\Phi_B(f) \,.\, \Phi_B(g))$ uniformly on $[0, +\infty[$.

Hence we obtain the relation (5.IV.7).

Corollary 5.IV.1. We have

$$(L^2 *_B L^2)\,([0, +\infty[, d\mu_\alpha) = \mathcal{F}_B(L^1([0, +\infty[, d\mu_\alpha)). \tag{5.IV.8}$$

Proof.

From Proposition 5.IV.4 we deduce that

$$(L^2 * L^2)\ ([0, +\infty[, d\mu_\alpha) \subset \mathfrak{F}_B(L^1([0, +\infty[, d\mu_\alpha)).$$

We prove now the other inclusion. Let h be in $\mathfrak{F}_B(L^1([0, +\infty[, d\mu_\alpha))$. Then there exists f in $L^1([0, +\infty[, d\mu_\alpha)$ such that

$$h = \mathfrak{F}_B(f).$$

We consider the functions f_1 and f_2 defined by

$$f_1(x) = \sqrt{|f(x)|}.$$

$$f_2(x) = \begin{cases} \dfrac{f(x)}{\sqrt{|f(x)|}} & \text{if} \quad f(x) \neq 0, \\ 0 & \text{if} \quad f(x) = 0, \end{cases}$$

These functions belongs to $L^2([0, +\infty[, d\mu_\alpha)$ and we have

$$f = f_1 f_2 .$$

We apply now the relation (5.IV.7) to the functions $\Phi_B(f_1)$ and $\Phi_B(f_2)$ and we use the relation (5.IV.1) to obtain

$$\Phi_B(f_1) *_B \Phi_B(f_2) = \mathfrak{F}_B(f_1.f_2).$$

Thus

$$\Phi_B(f_1) *_B \Phi_B(f_2) = \mathfrak{F}_B(f) = h.$$

Hence

$$\mathfrak{F}_B(L^1([0, +\infty[, d\mu_\alpha)) \subset (L^2 *_B L^2)\ ([0, \infty[, d\mu_\alpha).$$

This completes the proof of the proposition.

Lemma 5.IV.1. For all f and g in $L^2([0, +\infty[, d\mu_\alpha)$ and Ψ in $\mathcal{S}_*(\mathbb{R})$ we have the relation

(5.IV.9) $$\int_0^\infty f *_B g(x)\, \Psi(x)\, d\mu_\alpha(x) = \int_0^\infty \Phi_B(f)(\lambda)\, \Phi_B(g)(\lambda)\, \mathfrak{F}_B(\Psi)(\lambda)\, d\mu_\alpha(\lambda).$$

Proof.

Let g be in $L^2([0, +\infty[, d\mu_\alpha)$ and Ψ in $\mathcal{S}_*(\mathbb{R})$. We consider the operators ℓ_1 and ℓ_2 defined on $L^2([0, +\infty[, d\mu_\alpha)$ by

$$\ell_1(f) = \int_0^\infty f *_B g(x)\, \Psi(x)\, d\mu_\alpha(x)\ .$$

$$\ell_2(f) = \int_0^\infty \Phi_B(f)(\lambda)\Phi_B(g)(\lambda)\, \mathcal{F}_B(\Psi)(\lambda)\, d\mu_\alpha(\lambda).$$

From Propositions 4.III.4, 5.IV.3 and the relation (5.IV.2), we deduce that for all f in $(L^1 \cap L^2)([0, +\infty[, d\mu_\alpha)$ we have

$$\ell_1(f) = \ell_2(f)\ .$$

Using Hölder's inequality Proposition 4.III.5 and the relation (5.IV.3) we obtain

$$|\ell_1(f)| \le \|f *_B g\|_{\infty,\alpha}\ \|\Psi\|_{1,\alpha},$$
$$\le \|f\|_{2,\alpha}\ \|g\|_{2,\alpha}\ \|\Psi\|_{1,\alpha},$$

and

$$|\ell_2(f)| \le \|\Phi_B(f)\ \Phi_B(g)\|_{1,\alpha}\ \|\mathcal{F}_B(\Psi)\|_{\infty,\alpha},$$
$$\le \|\Phi_B(f)\|_{2,\alpha}\ \|\Phi_B(g)\|_{2,\alpha} \|\mathcal{F}_B(\Psi)\|_{\infty,\alpha},$$
$$\le \|f\|_{2,\alpha} \|(g)\|_{2,\alpha}\ \|\mathcal{F}_B(\Psi)\|_{\infty,\alpha}.$$

This completes the proof of the Lemma.

Theorem 5.IV.4. Let f and g be in $L^2([0, +\infty[, d\mu_\alpha)$. Then the function $f *_B g$ belongs to $L^2([0, +\infty[, d\mu_\alpha)$ if and only if the function $\Phi_B(f).\Phi_B(g)$ belongs to $L^2([0, +\infty[, d\mu_\alpha)$ and we have

(5.IV.10) $$\Phi_B(f *_B g) = \Phi_B(f).\ \Phi_B(g).$$

Proof.

i) We suppose that $f *_B g$ belongs to $L^2([0, +\infty[, d\mu_\alpha)$. As f and g are in $L^2([0, +\infty[, d\mu_\alpha)$, then from Theorem 5.IV.2 the function $\Phi_B(f).\ \Phi_B(g)$ belongs to $L^1([0, +\infty[, d\mu_\alpha)$, thus for all Ψ in $\mathcal{S}_*(\mathbb{R})$ the function $\Phi_B(f).\ \Phi_B(g).\Psi$ is in $L^1([0, +\infty[, d\mu_\alpha)$. From the relation (5.IV.9), Hölder's inequality and the relations (5.IV.3), (5.III.3), we deduce

$$\left|\int_0^\infty \Phi_B(f)(\lambda)\ \Phi_B(g)(\lambda)\ \Psi(\lambda)\ d\mu_\alpha(\lambda)\right| \le \|f *_B g\|_{2,\alpha}\ \|\mathcal{F}_B(\Psi)\|_{2,\alpha}$$
$$\le \|f *_B g\|_{2,\alpha}\ \|\Psi\|_{2,\alpha}\ .$$

Thus the quantity

$$\operatorname{Sup}\{|\int_0^\infty \Phi_B(f)(\lambda)\,\Phi_B(g)(\lambda)\,\Psi(\lambda)\,d\mu_\alpha(\lambda)|,\ \Psi\in\mathcal{D}_*(\mathbb{R}),\ \|\Psi\|_{2,\alpha}=1\},$$

is finite. From the converse of Hölder's inequality, (see [35] p. 181), we deduce that $\Phi_B(f).\Phi_B(g)$ belongs to $L^2([0, +\infty[, d\mu_\alpha)$. We obtain the relation (5.IV.10) from the relations (5.IV.9), (5.IV.2) and Corollary 4.III.3.

ii) We suppose that $\Phi_B(f).\Phi_B(g)$ belongs to $L^2([0, +\infty[, d\mu_\alpha)$. As f and g are in $L^2([0, +\infty[, d\mu_\alpha)$, then from Proposition 4.III.5 the function $f*_Bg$ belongs to $L^\infty([0, +\infty[, d\mu_\alpha)$, thus for all Ψ in $\mathcal{D}_*(\mathbb{R})$ the functions $f *_B g.\Psi$ is in $L^1([0, +\infty[, d\mu_\alpha)$. From the relation (5.IV.9), Hölder's inequality and the relation (5.IV.3) we obtain

$$|\int_0^\infty f*_Bg(x)\,\Psi(x)\,d\mu_\alpha(x)| \le \|\Phi_B(f)\,\Phi_B(g)\|_{2,\alpha}\,\|\mathcal{F}_B(\Psi)\|_{2,\alpha},$$

$$\le \|\Phi_B(f)\,\Phi_B(g)\|_{2,\alpha}\,\|\Psi\|_{2,\alpha}.$$

We deduce the results by the same proof as for the i).

Corollary 5.IV.2. For all f and g in $L^2([0, +\infty[, d\mu_\alpha)$, we have

$$\text{(5.IV.11)} \qquad \int_0^\infty |f*_Bg(x)|^2\,d\mu_\alpha(x) = \int_0^\infty |\Phi_B(f)(\lambda)|^2\,|\Phi_B(g)(\lambda)|^2 d\mu_\alpha(\lambda),$$

both members are finite or infinite.

Proof.

For $f*_Bg$ in $L^2([0, +\infty[, d\mu_\alpha)$ the relation (5.IV.11) can be deduced from Theorem 5.IV.3 and the relation (5.IV.3). For the other case the two members of the relation (5.IV.11) are infinite.

Corollary 5.IV.3. Let f_i, g_i, $i = 1, 2$, be in $L^2([0, +\infty[, d\mu_\alpha)$ such that $f_i *_B g_i$, $i = 1, 2$, belong to $L^2([0, +\infty[, d\mu_\alpha)$. Then we have

$$\text{(5.IV.12)} \qquad \int_0^\infty (f_1 *_B g_1)(x)\,\overline{(f_2 *_B g_2)(x)}\,d\mu_\alpha(x) =$$

$$\int_0^\infty \Phi_B(f_1)(\lambda)\,\Phi_B(g_1)(\lambda)\,\overline{\Phi_B(f_2)(\lambda)}\,\overline{\Phi_B(g_2)(\lambda)}\,d\mu_\alpha(\lambda).$$

Remark.

Some times the transform Φ_B is also denoted $\mathcal{F}_B$.

5.V. The Fourier-Bessel transform on $L^p([0, +\infty[, d\mu_\alpha)$, $1 \leq p \leq 2$.

- For $p = 1$ and $p = 2$, the Fourier-Bessel transform has been studied respectively in the sections 5.III. and 5.IV. In particular we have proved the following relations

(5.V.1) $$\|\mathcal{F}_B(f)\|_{\infty,\alpha} \leq \|f\|_{1,\alpha} \quad , \quad \text{for } f \in L^1([0, +\infty[, d\mu_\alpha).$$

(5.V.2) $$\|\mathcal{F}_B(f)\|_{2,\alpha} = \|f\|_{2,\alpha} \quad , \quad \text{for } f \in L^2([0, +\infty[, d\mu_\alpha).$$

- For $p \in]1, 2[$, the Fourier-Bessel transform defined on $\mathcal{S}_*(\mathbb{R})$ extends to a bounded linear operator on $L^p([0, +\infty[, d\mu_\alpha)$, $1 < p < 2$, and

(5.V.3) $$\|\mathcal{F}_B(f)\|_{p',\alpha} \leq \|f\|_{p,\alpha} .$$

Here p' denotes the dual exponent of p given by $\frac{1}{p} + \frac{1}{p'} = 1$.

This result can be obtained by using the Riesz convexity theorem to interpolate between the endpoint estimates for $L^1([0, +\infty[, d\mu_\alpha)$ and $L^2([0, +\infty[, d\mu_\alpha)$.

In [31] A. Fitouhi has obtained a sharp $L^p([0, +\infty[, d\mu_\alpha)$ inequality for the Fourier-Bessel transform called Babenko's inequality. More precisely he has proved the following result.

Theorem 5.V.1. Let $p \in]1, 2]$ and $p' = \frac{p}{p-1}$ be the dual exponent of p. For all f in $L^p([0, +\infty[, d\mu_\alpha)$, the function $\mathcal{F}_B(f)$ belongs to $L^{p'}([0, +\infty[, d\mu_\alpha)$, and we have

(5.V.4) $$\|\mathcal{F}_B(f)\|_{p',\alpha} \leq A_p \|f\|_{p,\alpha},$$

with

(5.V.5) $$A_p = [p^{1/p} / (p')^{1/p'}]^{\alpha+1}.$$

Remark.

The equality is attained in (5.V.4) for Gaussian functions

$$f(x) = e^{-tx^2} \quad , \quad t \in]0, +\infty[.$$

Using Theorem 5.V.1 we deduce the following result for the generalized convolution product $*_B$.

Corollary 5.V.1. Let $p, q \in]1, 2]$ and $r \in]2, +\infty]$ such that $\frac{1}{r} = \frac{1}{p} + \frac{1}{q} - 1$. For all f in $L^p([0, +\infty[, d\mu_\alpha)$ and g in $L^q([0, +\infty[, d\mu_\alpha)$ we have the relation

$$f *_B g = \mathfrak{F}_B(\mathfrak{F}_B(f)\,\mathfrak{F}_B(g)). \tag{5.V.6}$$

Proof.

i) For f and g in $\mathcal{S}_*(\mathbb{R})$ we deduce from Corollary 5.III.1 and the relation (5.III.3), that the relation (5.V.6) is true.

ii) Let f be in $L^p([0, +\infty[, d\mu_\alpha)$ and g in $L^q([0, +\infty[, d\mu_\alpha)$ with $p, q \in]1, 2]$. We consider two sequences $\{f_n\}_{n\in\mathbb{N}}$ and $\{g_n\}_{n\in\mathbb{N}}$ in $\mathcal{S}_*(\mathbb{R})$ which converge to f and g respectively in $L^p([0, +\infty[, d\mu_\alpha)$ and $L^q([0, +\infty[, d\mu_\alpha)$. From the i) we have

$$f_n *_B g_n = \mathfrak{F}_B(\mathfrak{F}_B(f_n).\mathfrak{F}_B(g_n)).$$

Let $r' = \frac{r}{r-1}$ the dual exponent of r which belongs to $]1, 2]$. We have

$$\frac{1}{r'} = \frac{1}{p'} + \frac{1}{q'} \tag{5.V.7}$$

with $p' = \frac{p}{p-1}$ and $q' = \frac{q}{q-1}$ the dual exponents of p and q respectively, which belong to $[2, +\infty[$. From Theorem 5.V.1 and Hölder's inequality we have

$$\begin{aligned}\|\mathfrak{F}_B(f_n).\mathfrak{F}_B(g_n) - \mathfrak{F}_B(f)\,\mathfrak{F}(g)\|_{r',\alpha} \leq {}& \|\mathfrak{F}_B(f)\|_{p',\alpha}\,\|\mathfrak{F}_B(g_n) - \mathfrak{F}_B(g)\|_{q',\alpha} + \\ & \|\mathfrak{F}_B(g)\|_{q',\alpha}\,\|\mathfrak{F}_B(f_n) - \mathfrak{F}_B(f)\|_{p',\alpha} + \\ & \|\mathfrak{F}_B(f_n) - \mathfrak{F}_B(f)\|_{p',\alpha}\,\|\mathfrak{F}_B(g_n) - \mathfrak{F}_B(g)\|_{q',\alpha}.\end{aligned}$$

We deduce that the sequence $\{\mathfrak{F}_B(f_n)\,\mathfrak{F}_B(g_n)\}_{n\in\mathbb{N}}$ converges to $\mathfrak{F}_B(f).\mathfrak{F}_B(g)$ in $L^{r'}([0, +\infty[, d\mu_\alpha)$.Theorem 5.V.1 implies that the sequence$\{\mathfrak{F}_B(\mathfrak{F}_B(f_n)\,.\mathfrak{F}_B(g_n))\}_{n\in\mathbb{N}}$ converges to $\mathfrak{F}_B(\mathfrak{F}_B(f)\,.\mathfrak{F}_B(g))$ in $L^r([0, +\infty[, d\mu_\alpha)$.

On the other hand from Propositions 4.III.3, 4.III.4 and 4.III.5 we have

$$\|f_n *_B g_n - f *_B g\|_{r,\alpha} \leq \|f\|_{p,\alpha}\,\|g_n - g\|_{q,\alpha} + \|g\|_{q,\alpha}\,\|f_n - f\|_{p,\alpha} + \|f_n - f\|_{p,\alpha}\,\|g_n - g\|_{q,\alpha}.$$

Thus the sequence $\{f_n *_B g_n\}_{n\in\mathbb{N}}$ converges to $f *_B g$ in $L^r([0, +\infty[, d\mu_\alpha)$.

Thus

$$f *_B g = \mathfrak{F}_B(\mathfrak{F}_B(f).\mathfrak{F}_B(g)).$$

The following result given in [31] is a sharp Young's inequality for the generalized convolution product $*_B$.

Corollary 5.V.2. Let $p, q \in]1, 2[$ and $r \in]2, +\infty[$ such that $\frac{1}{r} = \frac{1}{p} + \frac{1}{q} - 1$. For all f in $L^p([0, +\infty[, d\mu_\alpha)$, and g in $L^q([0, +\infty[, d\mu_\alpha)$ we have

(5.V.8) $$\|f *_B g\|_{r,\alpha} \leq A_p A_q A_{r'} \|f\|_{p,\alpha} \|g\|_{q,\alpha}.$$

Where r' is the dual exponent of r and

(5.V.9) $$A_m = [m^{1/m} / (m')^{1/m'}]^{\alpha+1}.$$

Proof.

From Corollary 5.V.1, Theorem 5.V.1 and Hölder's inequality we have

$$\begin{aligned}\|f *_B g\|_{r,\alpha} &= \|\mathfrak{F}_B(\mathfrak{F}_B(f)\,(\mathfrak{F}_B(g))\|_{r,\alpha},\\ &\leq A_{r'} \|\mathfrak{F}_B(f)\,\mathfrak{F}_B(g)\|_{r',\alpha},\\ &\leq A_{r'} \|\mathfrak{F}_B(f)\|_{p',\alpha} \|\mathfrak{F}_B(g)\|_{q',\alpha},\end{aligned}$$

where p' and q' are given by the relation (5.V.7).

Thus

$$\|f *_B g\|_{r,\alpha} \leq A_p A_q A_{r'} \|f\|_{p,\alpha} \|g\|_{q,\alpha}.$$

5.VI. The Fourier-Bessel transform on $M^b([0, +\infty[)$.

Definition 5.VI.1. The Fourier-Bessel transform of ν in $M^b([0, +\infty[)$ is the function $\mathfrak{F}_B(\nu)$ defined by

(5.VI.1) $$\forall \lambda \geq 0,\ \mathfrak{F}_B(\nu)(\lambda) = \int_0^\infty j_\alpha(\lambda x) d\nu(x).$$

The properties of this transform are given in the following propositions.

Proposition 5.VI.1. Let ν be in $M^b([0, +\infty[)$. Then

i) The Fourier-Bessel transform of ν belongs to $C_*(\mathbb{R})$.

ii) We have

(5.VI.2) $$\lim_{\lambda \to +\infty} \mathfrak{F}_B(\nu)(\lambda) = \nu(\{0\}).$$

iii) We have

(5.VI.3) $$\forall \lambda \geq 0,\ |\mathfrak{F}_B(\nu)(\lambda)| \leq \|\nu\|.$$

iv) If $\nu = f\mu_\alpha$, with f in $L^1([0, +\infty[, d\mu_\alpha)$, then $\mathfrak{F}_B(\nu)$ is the Fourier-Bessel transform of f given by the relation (5.II.1).

Proof.

i) The parity of the function $\lambda \to j_\alpha(\lambda x)$ implies that the function $\mathfrak{F}_B(\nu)$ given by the relation (5.VI.1) is even on $\mathbb{R}$.

On the other hand using the relation (5.II.7) and dominated convergence theorem we deduce that $\mathfrak{F}_B(\nu)$ is continuous on $[0, +\infty[$. Then the function $\mathfrak{F}_B(\nu)$ belongs to $C_*(\mathbb{R})$.

ii) We have the unique decomposition of ν :

(5.VI.4) $$\nu = a\delta_0 + \nu'$$

where δ_0 is the Dirac measure at the point 0 and ν' a measure in $M^b([0, +\infty[)$ concentrated in $]0, +\infty[$.

From the relation (5.II.7), Lemma 5.II.1 and dominated convergence theorem we deduce that

(5.VI.5) $$\lim_{\lambda \to +\infty} \mathfrak{F}_B(\nu')(\lambda) = 0.$$

As

(5.VI.6) $$\mathfrak{F}_B(\delta_0)(\lambda) = \langle \delta_0, j_\alpha(\lambda.)\rangle = 1.$$

Then from the relations (5.VI.1), (5.VI.5) and (5.VI.6) we have

$$\lim_{\lambda \to +\infty} \mathfrak{F}_B(\nu)(\lambda) = a = \nu(\{0\}).$$

iii) From the relation (5.VI.1) we have

$$\forall \lambda \geq 0, |\mathfrak{F}_B(\nu)(\lambda)| \leq \int_0^\infty |j_\alpha(\lambda x))| \, d|\nu|(x).$$

The relation (5.II.7) implies

$$\forall \lambda \geq 0, |\mathfrak{F}_B(\nu)(\lambda)| \leq \int_0^\infty d|\nu|(x).$$

Thus

$$\forall \lambda \geq 0, |\mathfrak{F}_B(\nu)(\lambda)| \leq \|\nu\|.$$

iv) We have

$$\forall \lambda \geq 0, \mathcal{F}_B(\nu)(\lambda) = \int_0^\infty f(x)\, j_\alpha(\lambda x)\, d\mu_\alpha(x),$$
$$= \mathcal{F}_B(f)(\lambda)\,.$$

Proposition 5.VI.2.

i) For all ν, $\mathcal{T}$ in $M^b([0, +\infty[)$ we have

(5.VI.7) $$\forall \lambda \geq 0, \mathcal{F}_B(\nu *_B \mathcal{T})(\lambda) = \mathcal{F}_B(\nu)(\lambda).\mathcal{F}_B(\mathcal{T})(\lambda).$$

ii) For ν in $M^b([0, +\infty[)$ and f in $L^p([0, +\infty[, d\mu_\alpha)$, $p = 1, 2$, we have

(5.VI.8) $$\mathcal{F}_B(\nu *_B f) = \mathcal{F}_B(\nu)\,.\mathcal{F}_B(f)\,.$$

Proof.

i) We have

$$\forall \lambda \geq 0, \mathcal{F}_B(\nu *_B \mathcal{T})(\lambda) = \int_0^\infty j_\alpha(\lambda x) d(\nu *_B \mathcal{T})(x).$$

Using Definition 4.III.2, the relation (4.II.6) and Fubini theorem we obtain

$$\forall \lambda \geq 0, \mathcal{F}_B(\nu *_B \mathcal{T})(\lambda) = \int_0^\infty \int_0^\infty T_y\, (j_\alpha(\lambda.))(x)\, d\nu(x) d\mathcal{T}(y)\,,$$
$$= \int_0^\infty \int_0^\infty j_\alpha(\lambda x)\, j_\alpha(\lambda y)\, d\nu(x) d\mathcal{T}(y)\,,$$
$$= \left(\int_0^\infty j_\alpha(\lambda x)\, d\nu(x)\right)\left(\int_0^\infty j_\alpha(\lambda y) d\mathcal{T}(y)\right).$$

Thus

$$\forall \lambda \geq 0, \mathcal{F}_B(\nu *_B \mathcal{T})(\lambda) = \mathcal{F}_B(\nu)(\lambda).\, \mathcal{F}_B(\mathcal{T})(\lambda).$$

ii) - If $f \in L^1([0, +\infty[, d\mu_\alpha)$.

We deduce the result from the i), the relation (4.III.27) and Proposition 5.VI.1.iv).

- If $f \in L^2([0, +\infty[, d\mu_\alpha)$.

The precedent result, the density of $\mathcal{S}_*(\mathbb{R})$ in $L^2([0, +\infty[, d\mu_\alpha)$ and the relations (5.VI.3), (4.III.28) give the result.

Proposition 5.VI.3. The transform $\mathcal{F}_B$ is injective on $M^b([0, +\infty[)$.

Proof.

Let ν be a measure in $M^b([0, +\infty[)$ such that

$$\forall\, \lambda \geq 0,\ \mathfrak{F}_B(\nu)(\lambda) = 0.$$

We take f in $\mathcal{D}_*(\mathbb{R})$. From Theorem 5.III.1 we have

$$\forall\, x \geq 0,\ f(x) = \int_0^\infty \mathfrak{F}_B(f)(\lambda) j_\alpha(\lambda x)\, d\mu_\alpha(\lambda).$$

We consider

$$\int_0^\infty f(x) d\nu(x) = \int_0^\infty \left(\int_0^\infty \mathfrak{F}_B(f)(\lambda) j_\alpha(\lambda x) d\mu_\alpha(\lambda)\right) d\nu(x).$$

As f belongs to $\mathcal{D}_*(\mathbb{R})$, then from Theorem 5.III.1 there exists a positive constant M such that

(5.VI.9) $$\forall\, \lambda \geq 0, |\mathfrak{F}_B(f)(\lambda)| \leq \frac{M}{(1+|\lambda|^2)^{[\alpha]+2}},$$

where $[\alpha]$ is the entire part of α.

Using (5.VI.9), (5.II.7) and Fubini's theorem we obtain

$$\int_0^\infty f(x)\, d\nu(x) = \int_0^\infty \mathfrak{F}_B(f)(\lambda)\left(\int_0^\infty j_\alpha(\lambda x)\, d\nu(x)\right) d\mu_\alpha(\lambda),$$
$$= \int_0^\infty \mathfrak{F}_B(f)(\lambda) \mathfrak{F}_B(\nu)(\lambda)\, d\mu_\alpha(\lambda) = 0.$$

Thus

$$\int_0^\infty f(x)\, d\nu(x) = 0, \quad \text{for all } f \in \mathcal{D}_*(\mathbb{R}).$$

Hence

$$\nu \equiv 0.$$

5.VII. The Fourier-Bessel transform on $\mathcal{E}_*'(\mathbb{R})$ and $\mathcal{S}'_*(\mathbb{R})$.

Definition 5.VII.7.

i) The Fourier-Bessel transform of S in $\mathcal{S}'_*(\mathbb{R})$ is the distribution $\mathfrak{F}_B(S)$ defined by

(5.VII.1) $$<\mathfrak{F}_B(S), \varphi> = <S, \mathfrak{F}_B(\varphi)>, \quad \varphi \in \mathcal{S}_*(\mathbb{R}).$$

ii) The Fourier-Bessel transform of U in $\mathcal{E}'_*(\mathbb{R})$ is the function $\mathfrak{F}_B(U)$ defined by

(5.VII.2) $$\forall\, \lambda \geq 0,\ \mathfrak{F}_B(U)(\lambda) = <U, j_\alpha(\lambda .)>.$$

Remark.

For U in $\mathcal{E}'_*(\mathbb{R})$, the function $\mathcal{F}_B(U)(\lambda)$ is well defined for $\lambda \in \mathbb{C}$.

Examples.

i) The distribution $\mathcal{T}_1$ given by the constant function equal to 1 belongs to $\mathcal{S}'_*(\mathbb{R})$, and we have

$$<\mathcal{F}_B(\mathcal{T}_1), \varphi> = <\mathcal{T}_1, \mathcal{F}_B(\varphi)>, \quad \varphi \in \mathcal{S}_*(\mathbb{R}).$$

Thus

$$<\mathcal{F}_B(\mathcal{T}_1), \varphi> = \int_0^\infty \mathcal{F}_B(\varphi)(\lambda)\, d\lambda .$$

From Theorem 5.III.1 we deduce that

$$\begin{aligned} <\mathcal{F}_B(\mathcal{T}_1), \varphi> &= 2^\alpha \Gamma(\alpha+1)\varphi(0), \\ &= < 2^\alpha \Gamma(\alpha+1)\, \delta_0, \varphi>. \end{aligned}$$

Then

$$\mathcal{F}_B(\mathcal{T}_1) = 2^\alpha \Gamma(\alpha+1)\delta_0 .$$

ii) Let $a \in]0, +\infty[$. The distribution $U = \frac{1}{2}(\delta_a + \delta_{-a})$ belongs to $\mathcal{E}'_*(\mathbb{R})$, and we have

$$\forall\, \lambda \geq 0, \mathcal{F}_B(U)(\lambda) = < \frac{1}{2}(\delta_a + \delta_{-a}) , j_\alpha(\lambda .)>$$

Then

$$\forall\, \lambda \geq 0, \mathcal{F}_B(U)(\lambda) = j_\alpha(\lambda a).$$

The transforms given by the relations (5.VII.1) and (5.VII.2) possess properties given by the following propositions.

Proposition 5.VII.1

i) For all S in $\mathcal{S}'_*(\mathbb{R})$ the distribution $\mathcal{F}_B(S)$ belongs to $\mathcal{S}'_*(\mathbb{R})$.

ii) For all U in $\mathcal{E}'_*(\mathbb{R})$ the function $\mathcal{F}_B(U)$ is even and entire on $\mathbb{C}$, and we have

$$\forall\, \lambda \in \mathbb{C}, \mathcal{F}_B(U)(\lambda) = 2^{-\alpha}(\Gamma(\alpha+1))^{-1} \mathcal{F}_0 \circ W_\alpha(U)(\lambda). \quad (5.VII.3)$$

Where W_α is the Weyl transform defined by the relation (2.III.13) and $\mathcal{F}_0$ the Fourier-cosine transform given by Definition 3.II.3.ii).

Proposition 5.VII.2. Let S be in $\mathcal{S}'_*(\mathbb{R})$ and U be in $\mathcal{E}'_*(\mathbb{R})$. Then we have

(5.VII.4) $$\mathcal{F}_B(S *_B U) = \mathcal{F}_B(U).\mathcal{F}_B(S).$$

Theorem 5.VII.1. The Fourier-Bessel transform $\mathcal{F}_B$ is a topological isomorphism from

- $\mathcal{E}'_*(\mathbb{R})$ onto the space $\mathcal{H}_*(\mathbb{C})$ given in 3.II.5.
- $\mathcal{S}'_*(\mathbb{R})$ onto itself.

Proof.

We deduce there results from the relations (5.VII.1), (5.VII.3) and Theorems 2.IV.1, 5.III.1.

Proposition 5.VII.3. Let f and g be two measurable functions on $[0, +\infty[$ satisfying : There exist p, q $\in \mathbb{N}$ such that the functions $(1+x^2)^{-p} f$ and $(1+\lambda^2)^{-q} g$ belong to $L^1([0, +\infty[, d\mu_\alpha)$. We suppose that for all Ψ in $\mathcal{S}_*(\mathbb{R})$ we have

(5.VII.5) $$\int_0^\infty f(x)\, \mathcal{F}_B(\Psi)(x)\, d\mu_\alpha(x) = \int_0^\infty g(\lambda)\Psi(\lambda)d\mu_\alpha(\lambda).$$

Then the function f belongs to $L^2([0, +\infty[, d\mu_\alpha)$ if and only if the function g belongs to $L^2([0, +\infty[, d\mu_\alpha)$ and we have

(5.VII.6) $$\mathcal{F}_B(f) = g.$$

Proof.

a) We suppose that f belongs to $L^2([0, +\infty[, d\mu_\alpha)$. For all Ψ in $\mathcal{S}_*(\mathbb{R})$ we deduce from the relation (5.VII.5) and Theorem 5.IV.3 that

$$\int_0^\infty \mathcal{F}_B(f)(\lambda)\Psi(\lambda)d\mu_\alpha(\lambda) = \int_0^\infty g(\lambda)\Psi(\lambda)d\mu_\alpha(\lambda).$$

Thus the distributions of $\mathcal{S}'_*(\mathbb{R})$ given by the functions $\mathcal{F}_B(f)(\lambda)\dfrac{\lambda^{2\alpha+1}}{2^\alpha\Gamma(\alpha+1)}$ and $g(\lambda)\dfrac{\lambda^{2\alpha+1}}{2^\alpha\Gamma(\alpha+1)}$ are equal. Then

$$\mathcal{F}(f) = g.$$

b) Conversely, we suppose that g belongs to $L^2([0, +\infty[, d\mu_\alpha)$. For all Ψ in $\mathcal{S}_*(\mathbb{R})$ we obtain from the relation (5.VII.5) and Theorem 5.IV.3 :

$$\int_0^\infty f(x)\, \mathcal{F}_B(\Psi)(x)d\mu_\alpha(x) = \int_0^\infty \mathcal{F}_B(g)(x)\, \mathcal{F}_B(\Psi)(x)d\mu_\alpha(x).$$

This relation and Theorem 5.III.1 implies that the distributions of $\mathcal{S}'_*(\mathbb{R})$ given by the functions $f(x) \dfrac{x^{2\alpha+1}}{2^\alpha \Gamma(\alpha+1)}$ and $\mathfrak{F}_B(f)(x) \dfrac{x^{2\alpha+1}}{2^\alpha \Gamma(\alpha+1)}$ are equal. Thus

$$f = \mathfrak{F}_B(g).$$

Remark.

It is easy to deduce Theorem 5.IV.4 by using Proposition 5.VII.3 and Lemma 5.IV.1.

Theorem 5.VII.2. Let U be in $\mathcal{E}'_*(\mathbb{R})$ and $a \in [0, +\infty[$. Then the singular support of U given by Definition 3.II.4 satisfies

(5.VII.7) $$\text{singsupp } U \subset [-a,a],$$

if and only if there exist a positive integer N and a sequence of positive constants $\{C_m\}_{m\in\mathbb{N}^*}$ such that for $m = 1, 2,...$ we have

(5.VII.8) $$|\mathfrak{F}_B(U)(\lambda)| \leq C_m(1 + |\lambda|^2)^N e^{a|\text{Im}\lambda|}, \text{ if } |\text{Im}\lambda| \leq m \log(1+|\lambda|^2),\ \lambda \in \mathbb{C}.$$

Proof.

a) We prove that the condition (5.VII.8) is necessary.

For all $m \in \mathbb{N}^*$, we write

$$U = S_m + T_{x^{2\alpha+1} f_m},$$

where $S_m \in \mathcal{E}'_*(\mathbb{R})$, $\text{Supp}S_m \subset [-a - \frac{1}{m}, a + \frac{1}{m}]$, $f_m \in \mathcal{D}_*(\mathbb{R})$ and $T_{x^{2\alpha+1} f_m}$ is the distribution given by the function $x^{2\alpha+1} f_m$.

Thus

$$\forall \lambda \in \mathbb{C}, \mathfrak{F}_B(U)(\lambda) = \mathfrak{F}_B(S_m)(\lambda) + \mathfrak{F}_B(f_m)(\lambda).$$

From Theorem 5.VII.1 there exist a positive integer N and a positive constant b_m such that

$$\forall \lambda \in \mathbb{C}, |\mathfrak{F}_B(S_m)(\lambda)| \leq b_m (1 + |\lambda|^2)^{N-1} e^{(a+\frac{1}{m})|\text{Im}\lambda|}.$$

Then if

$$|\text{Im}\lambda| \leq m \text{ Log}(1 + |\lambda|^2),$$

we have

$$\forall \lambda \in \mathbb{C}, |\mathcal{F}_B(S_m)(\lambda)| \leq b_m (1 + |\lambda|^2)^N e^{a|\mathrm{Im}\lambda|}.$$

On the other hand from Theorem 5.III.1 there exist positive constants r and C_j^m, $j \in \mathbb{N}$, such that

$$\forall j \in \mathbb{N}, \forall \lambda \in \mathbb{C}, |\mathcal{F}_B(f_m)(\lambda)| \leq C_j^m (1 + |\lambda|^2)^{-j} e^{r|\mathrm{Im}\lambda|},$$

if

$$|\mathrm{Im}\lambda| \leq m \,\mathrm{Log}(1 + |\lambda|^2),$$

then

$$\forall j \in \mathbb{N}, \forall \lambda \in \mathbb{C}, |\mathcal{F}_B(f_m)(\lambda)| \leq C_j^m (1+|\lambda|^2)^{rm-j}.$$

This completes the proof of the a).

b) We prove that the condition (5.VII.8) is sufficent.

We put

$$\forall \lambda \in \mathbb{C}, \Psi(\lambda) = \mathcal{F}_B(U)(\lambda).$$

From Theorem 3.II.5 there exists a distribution $\mathfrak{T}$ in $\mathcal{E}'_*(\mathbb{R})$ such that

$$\forall \lambda \in \mathbb{C}, \Psi(\lambda) = \mathcal{F}_0(\mathfrak{T})(\lambda).$$

The function $\mathcal{F}_0(\mathfrak{T})$ satisfies the condition (5.VII.8). Then from Theorem 3.II.6 we have

$$\mathrm{singsupp}\ \mathfrak{T} \subset [-a, a].$$

On the other hand for all $p \in \mathbb{N}^*$ we can write Ψ in the form

$$\Psi(\lambda) = s_p(\lambda) + f_p(\lambda),$$

with s_p and f_p satisfying

- there exist a positive integer N and a positive constant d_p such that

$$\forall \lambda \in \mathbb{C}, |s_p(\lambda)| \leq d_p (1 + |\lambda|^2)^N e^{(a+\frac{1}{p})|\mathrm{Im}\lambda|}.$$

- there exist positives constants r and b_j^p, $j \in \mathbb{N}$, such that

$$\forall j \in \mathbb{N}, \forall \lambda \in \mathbb{C}, |f_p(\lambda)| \leq b_j^p (1 + |\lambda|^2)^{-j} e^{r|\mathrm{Im}\lambda|}.$$

Then from Theorems 5.III.1 and 5.VII.1, the functions s_p and f_p are respectively the Fourier-Bessel transform of a distribution S_p in $\mathcal{E}'_*(\mathbb{R})$ which has its support in $[-a - \frac{1}{p}, a + \frac{1}{p}]$, and of a function F_p in $\mathcal{D}_*(\mathbb{R})$. This completes the proof of the b).

Proposition 5.VII.4. Let U be a distribution in $\mathcal{E}'_*(\mathbb{R})$. The distribution $\ell_\alpha(U)$ in $\mathcal{E}'_*(\mathbb{R})$ defined by

(5.VII.9) $$<\ell_\alpha(U), \varphi> = <U, \ell_\alpha(\varphi)> \ , \ \varphi \in \mathcal{E}_*(\mathbb{R}),$$

where ℓ_α is the Bessel operator given by the relation (5.II.11), satisfies

(5.VII.10) $$\forall \lambda \in \mathbb{C}, \ \mathcal{F}_B(\ell_\alpha(U))(\lambda) = -\lambda^2 \mathcal{F}_B(U)(\lambda).$$

Proof.

From Definition 5.VII.1 ii) and the relation (1.II.8) we have

$$\forall \lambda \in \mathbb{C}, \ \mathcal{F}_B(\ell_\alpha(U))(\lambda) = <\ell_\alpha(U), j_\alpha(\lambda .)>,$$
$$= <U, \ell_\alpha(j_\alpha(\lambda .))>,$$
$$= <U, -\lambda^2 j_\alpha(\lambda .)>.$$

Thus

$$\forall \lambda \in \mathbb{C}, \ \mathcal{F}_B(\ell_\alpha(U))(\lambda) = -\lambda^2 \mathcal{F}_B(U)(\lambda) .$$

Theorem 5.VII.3. Let U be in $\mathcal{E}'_*(\mathbb{R})$ and P(X) a polynomial with constant coefficients such that

(5.VII.11) $$\text{singsupp } P(\ell_\alpha) U \subset [-a, a] \text{ (resp. void)}.$$

Then

$$\text{singsupp } U \subset [-a, a] \text{ (resp. void)}.$$

Proof.

We put

$$S = P(\ell_\alpha)U.$$

The distribution S belongs to $\mathcal{E}'_*(\mathbb{R})$ and from Proposition 5.VII.4 we have

$$\forall \lambda \in \mathbb{C}, \ \mathcal{F}_B(S)(\lambda) = P(-\lambda^2)\mathcal{F}_B(U)(\lambda).$$

Thus the distributions S and U have the same type. Then from Theorem 5.VII.2 the singular supports of S and U are contained in the interval [-a, a].

In the case where S is given by a function in $\mathcal{D}_*(\mathbb{R})$, then from Theorem 5.III.1 there exist positive constants r and d_j, $j \in \mathbb{N}$, such that

$$\forall j \in \mathbb{N}, \forall \lambda \in \mathbb{C}, \ |\mathcal{F}_B(S)(\lambda)| \leq d_j(1 + |\lambda|^2)^{-j} e^{r|\mathrm{Im}\lambda|}.$$

Hence there exists a positive constant k_j, $j \in \mathbb{N}$, such that

$$\forall j \in \mathbb{N}, \forall \lambda \in \mathbb{C}, \ |\mathcal{F}_B(U)(\lambda)| \leq k_j(1 + |\lambda|^2)^{-j} e^{r|\mathrm{Im}\lambda|}.$$

Using now Theorem 5.III.1 we deduce that the distribution U is given by a function in $\mathcal{D}_*(\mathbb{R})$.

5.VIII. Problems.

Problem 5.VIII.1.

Let $t \in]0, +\infty[$. We consider the function $k_t : [0, +\infty[\to \mathbb{R}$ defined by

$$k_t(x) = \frac{e^{-\frac{x^2}{4t}}}{(2t)^{\alpha+1}}.$$

1) Show

i) $\frac{\partial}{\partial t} k_t(x) = \ell_\alpha(k_t)(x), \quad t > 0, x > 0,$

where ℓ_α is the Bessel operator defined by

$$\ell_\alpha = \frac{d^2}{dx^2} + \frac{2\alpha+1}{x}\frac{d}{dx}.$$

ii) $\|k_t\|_{1,\alpha} = 1.$

iii) $\forall \lambda \geq 0, \mathcal{F}_B(k_t)(\lambda) = e^{-t\lambda^2}.$

iv) $k_t *_B k_s = k_{t+s}$, $t > 0$, $s > 0$.

2) Let f be in $\mathcal{D}_*(\mathbb{R})$. We consider the function $u(x,t)$ defined by

$$\forall x \geq 0, \ \forall t > 0, u(x, t) = k_t *_B f(x).$$

i) Show that

$$\forall \lambda \geq 0, \mathcal{F}_B(u(., t))(\lambda) = e^{-t\lambda^2}.\mathcal{F}_B(f)(\lambda).$$

ii) Prove that u is a solution of the equation

$$\begin{cases} \frac{\partial}{\partial t} u(x,t) = \ell_\alpha(u(.,t))(x), \\ u(x, 0) = f(x), \text{ for all } x \geq 0. \end{cases}$$

Problem 5.VIII.2.

Let $t \in]0, +\infty[$. We consider the function $p_t : [0, +\infty[\to \mathbb{R}$ defined by

$$p_t(x) = \frac{2^{\alpha+1}\,\Gamma(\alpha+\frac{3}{2})}{\sqrt{\pi}} \cdot \frac{t}{(t^2+x^2)^{\alpha+3/2}}.$$

1) Show

i) $\|p_t\|_{1,\alpha} = 1.$

ii) $\forall\, \lambda \geq 0,\ \mathfrak{F}_B(p_t)(\lambda) = e^{-t\lambda}.$

iii) $p_t *_B p_s = p_{t+s},\ t > 0,\ s > 0.$

2) Let f be in $\mathcal{D}_*(\mathbb{R})$. We consider the function $v(x,t)$ defined by

$$\forall\, x \geq 0,\ \forall\, t > 0,\ v(x,t) = p_t *_B f(x).$$

i) Show that

$$\forall\, \lambda \geq 0,\ \mathfrak{F}_B(v(.,t))(\lambda) = e^{-t\lambda}\,.\,\mathfrak{F}_B(f)(\lambda).$$

ii) Prove that v is a solution of the equation

$$\begin{cases} (\dfrac{\partial^2}{\partial t^2} + \ell_\alpha)v(x,t) = 0, \\ v(x,0) = f(x), \text{ for all } x \geq 0. \end{cases}$$

where ℓ_α is the Bessel operator given in Problem 5.VIII.1.

Problem 5.VIII.3.

i) Let f and g be in $L^1([0,+\infty[, d\mu_\alpha)$. Prove the relation

$$\int_0^\infty f(x)\mathfrak{F}_B(g)(x)\, d\mu_\alpha(x) = \int_0^\infty \mathfrak{F}_B(f)(\lambda)\, g(\lambda)\, d\mu_\alpha(\lambda).$$

ii) We consider f and g in $L^2([0,+\infty[, d\mu_\alpha)$. Establish the relation

$$\int_0^\infty f(x)\Phi_B(g)(x) d\mu_\alpha(x) = \int_0^\infty \Phi_B(f)(\lambda)\, g(\lambda)\, d\mu_\alpha(\lambda).$$

Where Φ_B is the Fourier-Bessel-Plancherel transform.

Problem 5.VIII.4.

Let f be in $L^1([0,+\infty[, d\mu_\alpha)$ continuous at the point $x \in [0,+\infty[$. We want to show that

(a) $$f(x) = \lim_{\substack{\varepsilon \to 0 \\ \varepsilon > 0}} \int_0^\infty \mathfrak{F}_B(f)(\lambda)\, e^{-\varepsilon\lambda^2} j_\alpha(\lambda x)\, d\mu_\alpha(\lambda).$$

1) Study the existence of the integral

$$\int_0^\infty \mathfrak{F}_B(f)(\lambda)\, e^{-\varepsilon\lambda^2} j_\alpha(\lambda x)\, d\mu_\alpha(\lambda).$$

2) Show that we can bring the study of the relation (a) to the case $x = 0$.

3) Prove the relation

$$\int_0^\infty \mathfrak{F}_B(f)(\lambda)\, e^{-\varepsilon\lambda^2} d\mu_\alpha(\lambda) = \int_0^\infty f(x) \frac{e^{-\frac{x^2}{4t}}}{(2t)^{\alpha+1}}\, d\mu_\alpha(x).$$

4) Deduce the relation (a).

5) If we suppose moreover that $\mathfrak{F}_B(f)$ belongs to $L^1([0, +\infty[, d\mu_\alpha)$, prove that

$$f(x) = \int_0^\infty \mathfrak{F}_B(f)(\lambda)\, j_\alpha(\lambda x)\, d\mu_\alpha(\lambda).$$

Problem 5.VIII.5.

We consider the space

$$\mathcal{A}_B([0, +\infty[) = \mathfrak{F}_B(L^1([0, +\infty[, d\mu_\alpha)).$$

1°) For f in $L^1([0, +\infty[, d\mu_\alpha) \cap \mathcal{A}_B([0, +\infty[)$ we put

$$|||f||| = \|f\|_{1,\alpha} + \| \mathfrak{F}_B(f)\|_{1,\alpha}.$$

Show that $|||.|||$ is a norm on the space $L^1([0, +\infty[, d\mu_\alpha) \cap \mathcal{A}_B([0, +\infty[)$.

2) Prove that the space $L^1([0, +\infty[, d\mu_\alpha) \cap \mathcal{A}_B([0, +\infty[)$ provided with the norm $|||.|||$ is a Banach space.

3) Let f, g be in $L^1([0, +\infty[, d\mu_\alpha) \cap \mathcal{A}_B([0, +\infty[)$. Show that $f.g$ and $f *_B g$ belong to $L^1([0, +\infty[, d\mu_\alpha) \cap \mathcal{A}_B([0, +\infty[)$.

Problem 5.VIII.6.

I. Show that the Fourier-Bessel transform $\mathfrak{F}_B : L^1([0, +\infty[, d\mu_\alpha) \to \mathbb{R}$ satisfies the following properties

i) $\mathfrak{F}_B$ is linear and continuous.

ii) $|||\mathfrak{F}_B||| \leq 1$.

iii) $\forall$ f, g $\in L^1([0, +\infty[, d\mu_\alpha)$, $\mathfrak{F}_B(f *_B g) = \mathfrak{F}_B(f) . \mathfrak{F}_B(g)$.

II. We consider a homomorphism $\chi : L^1([0, +\infty[, d\mu_\alpha) \to \mathbb{R}$ verifying the conditions

i) $\chi \neq 0$.

ii) χ is continuous.

iii) $|||\chi||| \leq 1$.

iv) $\forall$ f, g $\in L^1([0, +\infty[, d\mu_\alpha)$, $\chi(f *_B g) = \chi(f). \chi(g)$.

1) Show that there exists Ψ in $L^\infty([0, +\infty[, d\mu_\alpha)$ such that

- $\forall f \in L^1([0, +\infty[, d\mu_\alpha), \chi(f) = \int_0^\infty f(x)\, \Psi(x)\, d\mu_\alpha(x)$.
- $||\Psi||_{\infty,\alpha} \leq 1$.

We extend Ψ on $\mathbb{R}$, by parity.

2) i) Prove that for all f and g in $L^1([0, +\infty[, d\mu_\alpha)$ we have

$$\chi(f *_B g) = \int_0^\infty (f *_B \Psi)(x) g(x)\, d\mu_\alpha(x).$$

ii) Deduce that

$$\forall f \in L^1([0, +\infty[, d\mu_\alpha), \quad f *_B \Psi(x) = \chi(f)\, \Psi(x), \text{ a.e.}$$

3) i) Let f be in $\mathcal{D}_*(\mathbb{R})$. Show that $f *_B \Psi$ is a C^∞-function on $[0, +\infty[$.

ii) Deduce that Ψ is almost everywhere equal to a C^∞-function.

In the following we suppose that Ψ is of class C^∞ on $[0, +\infty[$.

4) Let v be a positive function in $\mathcal{D}_*(\mathbb{R})$ such that supp$v \subset [-1, 1]$ and $||v||_{1,\alpha} = 1$. For $\varepsilon > 0$, we put

$$\forall x \geq 0, \; v_\varepsilon(x) = \varepsilon^{-2\alpha-2}\, v\left(\frac{x}{\varepsilon}\right) .$$

i) Show that the family $\{f *_B v_\varepsilon\}_{\varepsilon>0}$ converges to f in $L^1([0, +\infty[, d\mu_\alpha)$.

ii) Prove that

$$\lim_{\varepsilon \to 0} \chi(v_\varepsilon) = \Psi(0).$$

iii) Deduce that

$$\Psi(0) = 1.$$

5) Let f be in $\mathcal{D}_*(\mathbb{R})$ such that $\chi(f) \neq 0$.

i) Using the question 2) ii) show that

$$\chi(f)\, \ell_\alpha(\Psi) = \chi(\ell_\alpha(f))\, \Psi,$$

where ℓ_α is the Bessel operator given by

$$\ell_\alpha = \frac{d^2}{dx^2} + \frac{2\alpha+1}{x}\frac{d}{dx}.$$

ii) Prove that Ψ is the unic solution of the equation

$$\begin{cases} \ell_\alpha u(x) = -\lambda^2 u(x) \ , \ \lambda \in \mathbb{R}, \\ u(0) = 1, u'(0) = 0. \end{cases}$$

iii) Deduce that there exists $\lambda \in \mathbb{R}$ such that

$$\forall\, x \geq 0, \ \Psi(x) = j_\alpha(\lambda x).$$

6) Conclude.

Problem 5.VIII.7.

We denote by $M_0^b([0, +\infty[)$ the subspace of $M^b([0, +\infty[)$ consisting of measures ν concentrated in $]0, +\infty[$.

1) i) Show that for ν in $M_0^b([0, +\infty[)$ we have

$$\lim_{\lambda \to +\infty} \mathcal{F}_B(\nu)(\lambda) = 0.$$

ii) Deduce that for ν in $M^b([0, +\infty[)$ we have

$$\lim_{\lambda \to +\infty} \mathcal{F}_B(\nu)(\lambda) = \nu(\{0\}).$$

We put

$$\mathcal{F}_B(\nu)(+\infty) = \lim_{\lambda \to +\infty} \mathcal{F}_B(\nu)(\lambda).$$

iii) Prove that a measure ν belongs to $M_0^b([0, +\infty[)$ if and only if

$$\mathcal{F}_B(\nu)(+\infty) = 0.$$

2) Let ν, ω be in $M_0^b([0, +\infty[)$. Prove that the measure $\nu *_0 \omega$ is absolutely continuous with respect to the Lebesgue measure on $[0, +\infty[$.

3) We denote by S^* the set of all homomorphisms of $M^b([0, +\infty[)$. Show that the mapping $\nu \to \mathfrak{F}_B(\nu)(\lambda)$ defined on $M^b([0, +\infty[)$, belongs to S^*.

4) Let χ be in S^*. We suppose that there exists ν in $M_0^b([0, +\infty[)$ such that $\chi(\nu) \neq 0$.

i) Show that for all ω in $M_0^b([0, +\infty[)$ there exists a unic $\lambda \in [0, +\infty[$ such that

$$\chi(\omega *_B \nu *_B \nu) = \mathfrak{F}_B(\omega *_B \nu *_B \nu)(\lambda),$$

and

$$\chi(\nu *_B \nu) = \mathfrak{F}_B(\nu *_B \nu)(\lambda).$$

(See Problem 5.VIII.6).

ii) Prove that there exist a unic $\lambda \in [0, +\infty[$ such that

$$\forall\, \omega \in M_0^b([0, +\infty[),\ \chi(\omega) = \mathfrak{F}_B(\omega)(\lambda).$$

iii) Deduce that there exists a unic $\lambda \in [0, +\infty[$ such that

$$\forall\, \omega \in M^b([0, +\infty[),\ \chi(\omega) = \mathfrak{F}_B(\omega)(\lambda).$$

5) Let χ be in S^*. We suppose that for all ν in $M_0^b([0, +\infty[)$ we have $\chi(\nu) = 0$. Show that

$$\forall\, \omega \in M^b([0, +\infty[),\ \chi(\nu) = \mathfrak{F}_B(\nu)(+\infty).$$

6) Deduce that to each homomorphism $\chi : M^b([0, +\infty[) \to \mathbb{R}$, corresponds a unic element $\lambda \in [0, +\infty]$ such that

$$\forall\, \nu \in M^b([0, +\infty[),\ \chi(\nu) = \mathfrak{F}_B(\nu)(\lambda).$$

(See [78]).

Problem 5.VIII.8.

We consider the space

$$B_\alpha([0, +\infty[) = \mathfrak{F}_B(M^b([0, +\infty[)).$$

We provide this space with the norm $\|.\|_{B_\alpha}$ defined by

$$\|\mathfrak{F}_B(\nu)\|_{B_\alpha} = \|\nu\|.$$

Show that the following two properties are equivalent

1) $\phi \in B_\alpha([0, +\infty[$ and $\|\phi\|_{B_\alpha} \leq k$.

2) ϕ is continuous at the point 0 and

$$|\sum_{j=1}^{n} c_j \phi(t_j)| \leq k \, \|f\|_{\infty,\alpha}$$

for all f of the form $f(\lambda) = \sum_{j=1}^{n} c_j j_\alpha(\lambda t_j)$, with $c_j \in \mathbb{C}$, $t_j > 0$, $j = 1, 2, ..., n$.

(To prove that 2) implies 1) consider the space

$$E = \{f \,/\, f(\lambda) = \sum_{j=1}^{n} c_j j_\alpha(\lambda t_j), \text{ with } n \in \mathbb{N}^*, c_j \in \mathbb{C}, t_j > 0\}$$

and the linear operator L defined on E by

$$L(f) = \sum_{j=1}^{n} c_j \phi(t_j) ,$$

and apply Hahn-Banach theorem).

(See [44]).

Problem 5.VIII.9.

Let $\{\phi_n\}_{n\in\mathbb{N}}$ be a sequence of functions in the space $B_\alpha([0, +\infty[)$ defined in Problem 5.VIII.8. We suppose that there exists a positive constant k such that

$$\forall n \in \mathbb{N}, \|\phi_n\|_{B_\alpha} \leq k ,$$

and

$$\forall \lambda \geq 0, \lim_{n\to+\infty} \phi_n(\lambda) = \phi(\lambda),$$

with ϕ continuous on $[0, +\infty[$.

Show that ϕ belongs to $B_\alpha([0, +\infty[)$ and we have

$$\|\phi\|_{B_\alpha} \leq k .$$

(See [44]).

Problem 5.VIII.10.

We consider the space

$$\mathcal{A}_B([0, +\infty[) = \mathfrak{F}_B(L^1([0, +\infty[, d\mu_\alpha)).$$

We provide it with the norm

$$\|\mathfrak{F}_B(f)\|_{\mathcal{A}_B} = \|f\|_{1,\alpha}.$$

1) Let ϕ be a function defined on $[0, +\infty[$ such that for all f in $\mathcal{A}_B([0, +\infty[)$ the function ϕf belongs to the space $B_\alpha([0, +\infty[)$ defined in Problem 5.VIII.8.

We consider the linear operator $L : f \to \phi . f$ from $\mathcal{A}_B([0, +\infty[)$ into $B_\alpha([0, +\infty[)$. Using the closed graph theorem show that L is continuous.

2) Let $f_r(\lambda) = \exp(- \frac{\lambda}{r})$, $\lambda \geq 0$, $r > 0$. Show that

i) f_r belongs to $\mathcal{A}_B([0, +\infty[)$ and calculate $\|f_r\|_{\mathcal{A}_B}$.

ii) the family $\{\phi . f_r\}_{r>0}$ converges pointwise to ϕ as $r \to +\infty$.

3) Deduce that ϕ belongs to $B_\alpha([0, +\infty[)$.

(Use Problem 5.VIII.9).

(See [44]).

Problem 5.VIII.11.

We consider a continuous linear operator $\mathfrak{L}$ from $L^1([0, +\infty[, d\mu_\alpha)$ into itself, satisfying the condition

$$\forall\, x \geq 0,\ \forall\, f \in L^1([0, +\infty[, d\mu_\alpha),\ \mathfrak{L} T_x(f) = T_x(\mathfrak{L}(f)).$$

where T_x, $x \geq 0$, are the generalized translation operators defined by the relation (4.II.3).

1) Show that for all f and g in $L^1([0, +\infty[, d\mu_\alpha)$, we have

$$g *_B \mathfrak{L}(f) = f *_B \mathfrak{L}(g).$$

2) Deduce that there exists a continuous function ϕ on $[0, +\infty[$, such that for all f in $L^1([0, +\infty[, d\mu_\alpha)$:

$$\mathfrak{F}_B(\mathfrak{L}(f)) = \phi\, \mathfrak{F}_B(f).$$

3) Using Problem 5.VIII.8 show that ϕ belongs to $B_\alpha([0, +\infty[)$.

4) Deduce that there exists a measure ν in $M^b([0, +\infty[)$ such that

$$\forall f \in L^1([0, +\infty[, d\mu_\alpha), \ \mathfrak{L}(f) = \nu *_B f,$$

with

$$\|\nu\| = |||\mathfrak{L}|||.$$

(See [44]).

Problem 5.VIII.12.

We denote by P_α the set of functions $f : [0, +\infty[\to \mathbb{R}$ which belong to $L^\infty([0, +\infty[, d\mu_\alpha)$ and satisfying

(a)
$$\sum_{i=1}^{n} \sum_{j=1}^{n} \gamma_i \bar{\gamma}_j \, T_{\lambda_i}(f)(\lambda_j) \geq 0 ,$$

for all $\lambda_1, \lambda_2, \ldots, \lambda_n$ in $]0, +\infty[$ and complex numbers $\gamma_1, \gamma_2, \ldots, \gamma_n$. Where T_{λ_i} are the generalized translation operators defined by the relation (4.II.3).

(We said that P_α is the set of positive definite functions associated with the Bessel operator).

1) i) Show that a function f in $L^\infty([0, +\infty[, d\mu_\alpha)$ satisfies (a) if and only if for all g in $L^1([0, +\infty[, d\mu_\alpha)$ we have

(b)
$$\int_0^\infty \int_0^\infty g(\lambda_1) \overline{g(\lambda_2)} \, T_{\lambda_1}(f)(\lambda_2) d\mu_\alpha(\lambda_1) \, d\mu_\alpha(\lambda_2) \geq 0.$$

ii) Deduce that (b) is equivalent to the following relation

$$\forall g \in L^1([0, +\infty[, d\mu_\alpha), \ \int_0^\infty f(\lambda) \, (g *_B \bar{g})(\lambda) \, d\mu_\alpha(\lambda) \geq 0.$$

2) Prove the following results

i) $f \in P_\alpha \cap C([0, +\infty[) \Rightarrow |f(\lambda)| \leq f(0)$, for all $\lambda \geq 0$.

ii) $f, g \in P_\alpha, C > 0 \Rightarrow Cf, \ f + g, \ f.g \in P_\alpha$.

iii) If the sequence $\{f_n\}_{n \in \mathbb{N}} \subset P_\alpha$ converges to a function f almost every where and if $\sup_{n \in \mathbb{N}} f_n(0) < +\infty$, then $f \in P_\alpha$.

3) Let ν be a positive measure in $M^b([0, +\infty[)$. Show that the function $f = \mathfrak{F}_B(\nu)$ belongs to P_α.

4) Let f be in P_α. We consider the linear operator L on $L^1([0, +\infty[, d\mu_\alpha)$ defined by

$$L(g) = \int_0^\infty f(\lambda)g(\lambda)\ d\mu_\alpha(\lambda).$$

i) Show that L is continuous on $L^1([0, +\infty[, d\mu_\alpha)$.

ii) Prove the relation

$$\forall\ g \in L^1([0, +\infty[, d\mu_\alpha),\ L(g *_B \bar{g}) \geq 0.$$

iii) Deduce that there exists a positive measure in $M^b([0, +\infty[$, such that for all g in $L^1([0, +\infty[, d\mu_\alpha)$ we have

$$L(g) = \int_0^\infty \mathfrak{F}_B(g)(x)d\nu(x).$$

(Use the Helglotz-Bochner-Weil-Raikov theorem, (see [60] p. 97)).

iv) Prove that

$$f(\lambda) = \mathfrak{F}_B(\nu)(\lambda), \text{ a.e. on } [0, +\infty[.$$

v) We suppose that there exists another positive measure ω in $M^b([0, +\infty[)$ satisfying

$$f(\lambda) = \mathfrak{F}_B(\omega)(\lambda), \text{ a.e. on } [0, +\infty[.$$

Show that

$$\omega = \nu.$$

(See [15]).

Problem 5.VIII.13.

1) We consider a positive measure ν in $M^b([0, +\infty[)$ and $f = \mathfrak{F}_B(\nu)$.

i) Prove that f belongs to $L^\infty([0, +\infty[, d\mu_\alpha) \cap C([0, +\infty[)$.

ii) Show that for all g in $L^1([0, +\infty[, d\mu_\alpha)$ such that $\mathfrak{F}_B(g) \geq 0$, we have

$$\int_0^\infty f(\lambda)\ g(\lambda)\ d\mu_\alpha(\lambda) \geq 0.$$

2) Let f be in $L^{\infty}([0,+\infty[, d\mu_\alpha) \cap C([0,+\infty[)$ satisfying the condition of the question 1) ii).

i) Show that f belongs to the set P_α given in Problem 5.VIII.12.

ii) Deduce from Problem 5.VIII.12 that there exists a positive measure ν in $M^b([0,+\infty[)$ such that

$$\forall\, \lambda \geq 0,\ f(\lambda) = \mathcal{F}_B(\nu)(\lambda).$$

(See [15]).

Problem 5.VIII.14.

We consider $M^1([0,+\infty[)$ the set of probability measures on $[0,+\infty[$ i.e. the set of positive measures ν in $M^b([0,+\infty[)$ such that $\|\nu\| = 1$.

1) Show that for every ν in $M^1([0,+\infty[)$ we have

i) $\mathcal{F}_B(\nu)(0) = 1$.

ii) $|\mathcal{F}_B(\nu)(\lambda)| \leq 1$, for all $\lambda \geq 0$.

2) Let ν in $M^b([0,+\infty[)$ and suppose that there exists $x_0 > 0$ such that $[0, x_0] \subset \operatorname{supp} \nu$.

i) Prove that $\lambda = 0$ is the unique solution of the equation

$$|\mathcal{F}_B(\nu)(\lambda)| = 1.$$

ii) Show that for all $\eta > 0$, there exists $a = a(\eta) < 1$, such that

$$|\mathcal{F}_B(\nu)(\lambda)| \leq a, \text{ for all } \lambda \geq \eta.$$

3) Prove that if ν belongs to $M^1([0,+\infty[)$ and if there is $x_0 \in \operatorname{Supp}\nu$ such that $x_0 \neq 0$, then

$$[0, 2x_0] \subset \operatorname{supp} \nu *_B \nu .$$

4) Let ν in $M^1([0,+\infty[)$, $\nu \neq \delta_0$, satisfying

$$\int_0^\infty x^2 d\nu(x) < +\infty.$$

Prove that for all f in $\mathcal{D}_*(\mathbb{R})$ we have

$$\lim_{n\to+\infty} n^{\alpha+1} \int_0^\infty f(x)\, d\nu^{*_B n}(x) = C_f,$$

where

$$\nu^{*_B n} = \nu *_B \nu *_B \cdots *_B \nu,$$

and C_f a positive constant. Give the value of C_f .

(See [78]).

Problem 5.VIII.15.

In this problem we suppose $\alpha \in]0, \frac{1}{2}[$. Let ν be in $M^1([0, +\infty[)$, $\nu \neq \delta_0$, satisfying

$$\int_0^\infty x^2 \, d\nu(x) < +\infty,$$

and f in $\mathcal{D}_*(\mathbb{R})$.

1) Show that for all $x \geq 0$, the series

$$\Gamma(f)(x) = \sum_{n=0}^{\infty} \mu^{*_B n} *_B f(x),$$

converge

2) Prove that the function $\mathfrak{F}_B(\nu)$ is of class C^2 on $[0, +\infty[$ and we have

$$\mathfrak{F}_B(\nu)(\lambda) = 1 - d_2(\nu) \frac{\lambda^2}{2} + 0(\lambda^2), \quad \text{as } \lambda \to 0.$$

where

$$d_2(\nu) = \frac{d^2}{d\lambda^2} \mathfrak{F}_B(\nu)(0).$$

3) Let $n \in \mathbb{N}$. Show that there exists a positive constant k such that

$$\forall \lambda > 0, \left| \frac{1 - (\mathfrak{F}_B(\nu)(\lambda))^{n+1}}{1 - \mathfrak{F}_B(\nu)(\lambda)} \right| \leq k \begin{cases} \lambda^{2\alpha - 1}, & \text{if } \lambda \in]0, 1], \\ \lambda^{2\alpha+1}, & \text{if } \lambda \in [1, +\infty[. \end{cases}$$

4) Prove that

$$\forall x \geq 0, \Gamma(f)(x) = \int_0^\infty \frac{\mathfrak{F}_B(f)(\lambda)}{1 - \mathfrak{F}_B(\nu)(\lambda)} j_\alpha(\lambda x) \, d\mu_\alpha(\lambda).$$

5) We put for all $\lambda > 0$:

$$V(\lambda) = \frac{1 - \mathfrak{F}_B(\nu)(\lambda)}{\lambda^2}, \quad g(\lambda) = \frac{\mathfrak{F}_B(f)(\lambda)}{V(\lambda)}.$$

i) Show that

$$\forall\, \lambda \geq 0,\ \ V(\lambda) \neq 0.$$

ii) Deduce that g is continuous on $[0, +\infty[$, bounded and satisfying

$$g(0) = \frac{2\, \mathfrak{F}_B(f)(0)}{d_2(\nu)}.$$

6) Prove that

$$\forall\, x > 0,\ x^{2\alpha}\Gamma(f)(x) = \int_0^\infty g(\frac{\lambda}{x}) j_\alpha(\lambda)\, \lambda^{-2} d\mu_\alpha(\lambda),$$

and deduce the value of the limit

$$\lim_{x \to +\infty} x^{2\alpha}\Gamma(f)(x).$$

(See [38]).

Problem 5.VIII.16.

1) Let K be a compact set in $[0, +\infty]$ and λ_0 in $[0, +\infty] \setminus K$. Show that there exists a measure ν in $M^b([0, +\infty[)$ such that

$$\mathfrak{F}_B(\nu)(K) = 1 \quad \text{and} \quad \mathfrak{F}_B(\nu)(\lambda_0) = 0.$$

2) Let K be a compact set in $[0, +\infty]$. Prove that the set

$$I_K = \{\nu \in M^b([0, +\infty[) \;/\; \mathfrak{F}_B(\nu)(K) = \{0\}\},$$

is a closed ideal in $M^b([0, +\infty[)$.

3) Let K be a compact set in $[0, +\infty]$. Show that the set $A_K = M^b([0, +\infty[) / I_K$ is a commutative and unitary Banach algebra with neutral element $\delta_0 + I_K$.

4) If K is a compact set in $[0, +\infty]$ and ν a measure in $M^b([0, +\infty[)$ satisfying $\mathfrak{F}_B(\nu)(\lambda) \neq 0$ for all $\lambda \in K$. Prove that there exists a measure ω in $M^b([0, +\infty[)$ such that

$$\forall\, \lambda \in K,\ \mathfrak{F}_B(\nu)(\lambda).\mathfrak{F}_B(\omega)(\lambda) = 1\,.$$

(Use Problem 5.VIII.7).

(See [77] and [54]).

Problem 5.VIII.17.

Let m be a real.

- The function $p : \mathbb{R} \times \mathbb{C} \to \mathbb{C}$ is called a symbol in the class S_0^m if it satisfies.

a) For fixed $x \geq 0$, the function $\lambda \to p(x, \lambda)$ is even and C^∞ on $[0, +\infty[$.

b) For fixed $\lambda \geq 0$, the function $x \to p(x, \lambda)$ is an even C^∞-function on $\mathbb{R}$.

c) For all $k, n \in \mathbb{N}$, there exists $C_{k,m,n} > 0$, such that for every $x \geq 0$ and $\lambda \geq 0$, we have

$$\left|\frac{\partial^k}{\partial x^k} \frac{\partial^n}{\partial \lambda^n} p(x,\lambda)\right| \leq C_{k,m,n} (1+\lambda)^{m-n}.$$

- The function $p : \mathbb{R} \times \mathbb{C} \to \mathbb{C}$ is called a symbol in the class S^m if it satisfies a), b) precedend and

d) For all $k, \ell, n \in \mathbb{N}$, there exists $C_{k,m,n} > 0$, such that for every $x \geq 0$ and $\lambda \geq 0$, we have

$$(1+x)^\ell \left|\frac{\partial^k}{\partial x^k} \frac{\partial^n}{\partial \lambda^n} p(x,\lambda)\right| \leq C_{k,\ell,m,n} (1+\lambda)^{m-n}.$$

1) i) Show that $S^m \subset S_0^m$.

ii) Prove that the symbol p defined by $p(x, \lambda) = \lambda^{2m}$, $m \geq 0$, is an element of S_0^{2m}, but it does not belong to S^{2m}.

2) Let $p \in S^m$. Show that

- For fixed $\lambda \geq 0$, the function $x \to p(x, \lambda)$ belongs to $\mathcal{S}_*(\mathbb{R})$.

- For fixed $x \geq 0$, the function $\lambda \to p(x, \lambda)$ is in $\mathcal{S}_*(\mathbb{R})$.

3) Let $p \in S_0^m$. The pseudo-differential operator associated with the symbol p is defined on $\mathcal{S}_*(\mathbb{R})$, by

$$\forall\ u \in \mathcal{S}_*(\mathbb{R}),\ \forall\ x \geq 0,\ h_p(u)(x) = \int_0^\infty p(x, \lambda)\, \mathcal{F}_B(u)(\lambda)\, j_\alpha(\lambda x)\, d\mu_\alpha(\lambda).$$

i) If $p(x, \lambda) = -\lambda^2$, show that

$$h_p = \ell_\alpha,$$

where ℓ_α is the Bessel operator

$$\ell_\alpha = \frac{d^2}{dx^2} + \frac{2\alpha+1}{x}\frac{d}{dx}.$$

ii) If

$$p(x,\lambda) = \sum_{k=0}^{N} (-1)^k p_k(x)\lambda^{2k},$$

deduce that

$$h_p(u)(x) = \sum_{k=0}^{N} p_k(x)\ell_\alpha^k(u)(x).$$

4) Let $p \in S_0^m$. Prove that the pseudo-differential operator h_p associated with p is a linear continuous mapping from $\mathcal{S}_*(\mathbb{R})$ into itself.

(See [71] [72] and [18]).

Problem 5.VIII.18.

We consider in this problem the spaces S_0^m and S^m, and the pseudo-differential operator h_p associated with the symbol $p \in S_0^m$, defined in Problem 5.VIII.17.

1) Let $p \in S^m$. We denote by

- $U_\lambda(x) = j_\alpha(\lambda x)\, p(x,\lambda)$,
- $P(\lambda_1, \lambda_2) = \mathcal{F}_B(U_{\lambda_1}(.))(\lambda_2), \quad \lambda_1 \geq 0,\ \lambda_2 \geq 0.$

Show that for all $\lambda_1 \geq 0$, $\lambda_2 \geq 0$, we have

$$P(\lambda_1, \lambda_2) = \int_0^\infty \mathcal{F}_B(p(.,\lambda_1))W(\lambda_1, \lambda_2, \lambda)d\mu_\alpha(\lambda).$$

Where $W(\lambda_1, \lambda_2, .)$ is the function defined by the relation (1.II.25).

2) Let $p \in S^m$. Prove that for all $k \in \mathbb{N}$, there exists $C_{m,k} > 0$, such that for all $\lambda_1 \geq 0$, $\lambda_2 \geq 0$, we have

$$|\mathcal{F}_B(p(., \lambda_1))(\lambda_2)| \leq C_{m,k}(1 + \lambda_1)^m (1 + \lambda_2^{2k})^{-1}.$$

$$|P(\lambda_1, \lambda_2)| \leq C_{m,k}(1 + \lambda_1)^m (1 + \lambda_2^{2k})^{-1}.$$

3) Let $p \in S^m$. Show that the associated pseudo-differential h_p possesses the following representation

$$\forall\, u \in \mathcal{S}_*(\mathbb{R}),\ \forall\, x \geq 0,\ h_p(u)(x) = \int_0^\infty \int_0^\infty P(\lambda_1, \lambda_2)\, \mathcal{F}_B(u)\,(\lambda_1)\, j_\alpha(\lambda_2 x) \times d\mu_\alpha(\lambda_1)\, d\mu_\alpha(\lambda_2).$$

and we have

$$\forall\ \lambda_2 \geq 0,\ \mathcal{F}_B(h_p(u))\,(\lambda_2) = \int_0^\infty P(\lambda_1, \lambda_2)\, \mathcal{F}(u)\,(\lambda_1)\, d\mu_\alpha(\lambda_1).$$

4) The $G^s_*(\mathbb{R})$, $s \in \mathbb{R}$, is defined to be the subspace of all elements $u \in \mathcal{S}_*(\mathbb{R})$, satisfying

$$\|u\|_{G^s_*} = \|(1+\lambda)^s\, \mathcal{F}_B(u)\,(\lambda)\|_{1,\alpha} < +\infty.$$

Let $p \in S^m$ and h_p be the associated pseudo-differential operator. Prove that there exists $C > 0$ such that

$$\|h_p(u)\|_{G^0_*} \leq C\, \|u\|_{G^s_*}.$$

(See [71] [72] and [18]).

Problem 5.VIII.19.

We denote by

• $\mathcal{P}_{-\frac{1}{2}}$ the space of even, bounded and continuous functions f on $\mathbb{R}$, which are Bohr almost-periodic on $\mathbb{R}$, and satisfying

$$\mathcal{M}(f) = \lim_{T \to +\infty} \frac{1}{T} \int_0^T f(x)dx = 0.$$

• $\mathcal{P}_\alpha$ the image of $\mathcal{P}_{-\frac{1}{2}}$ by the Riemann-Liouville integral transform $\mathcal{R}_\alpha$.

If f belons to $\mathcal{P}_\alpha$ we designate by F_f its pre-image in $\mathcal{P}_{-\frac{1}{2}}$:

$$f(x) = \mathcal{R}_\alpha(F_f)\,(x).$$

The elements of $\mathcal{P}_\alpha$ are called almost-periodic functions associated with the Bessel operator.

1) Show that F_f belongs to $\mathcal{P}_{-\frac{1}{2}}$ if and only if the function

$$\Psi_f(x, y) = \sigma_x(F_f)(y),$$

where σ_x, $x \geq 0$, are the translation operators given by (3.I.1), satisfies

$$\mathcal{M}(\Psi_f(x, .)) = \lim_{T \to +\infty} \frac{1}{T} \int_0^T \Psi_f(x,y)\, dy = 0.$$

and belongs to $\mathcal{P}_{-\frac{1}{2}}$ uniformly with respect to x.

2) Prove the following results

i) All function f in $\mathcal{P}_\alpha$ is bounded and we have

$$\|f\|_u \leq \|F_f\|_u.$$

Where

$$\|f\|_u = \sup_{x\geq 0}|f(x)|.$$

ii) All function f of the form

$$f(x) = \sum_{k=0}^{N} \beta_k j_\alpha(\lambda_k x), \ \beta_k \in \mathbb{C}, \ \lambda_k \geq 0,$$

belongs to $\mathcal{P}_\alpha$ and its pre-image is the trigonometric polynomial

$$F_f(x) = \sum_{k=0}^{N} \beta_k \cos \lambda_k x.$$

3) Prove that all f in $\mathcal{P}_\alpha$ can be approximated uniformly on $\mathbb{R}$ by a sequence of functions P_n of the form

$$P_n(x) = \sum_{k=0}^{N_n} \beta_k^{(n)} j_\alpha(\lambda_k x).$$

4) Let f be in $\mathcal{P}_\alpha$ and T_x, $x \geq 0$, the generalized translation operators given by (4.II.3). Show that the functions $T_x(f)$, $x \geq 0$, belong to $\mathcal{P}_\alpha$.

(See [21] and [13]).

Problem 5.VIII.20.

We use in this problem notations and results of Problem 5.VIII.19.

Let f be in $\mathcal{P}_\alpha$ and F_f its pre-image in $\mathcal{P}_{-\frac{1}{2}}$. We denote by $\{\lambda_k\}_{k\in\mathbb{N}}$ and $\{\beta_k\}_{k\in\mathbb{N}}$ respectively the Fourier exponents and coefficients of F_f. From [30] p. 22-27 and p. 65-66, we have the relations

$$\beta_k = \lim_{T\to+\infty} \frac{1}{T}\int_0^T F_f(x) \cos \lambda_k x \, dx.$$

$$\sum_{k=0}^{\infty} |\beta_k|^2 = \lim_{T\to+\infty} \frac{1}{T}\int_0^T |F_f(x)|^2 dx.$$

We write

$$F_f(x) \sim \sum_k \beta_k \cos \lambda_x x.$$

The second member is called Fourier series of F_f.

1) Let $\lambda > 0$ and $\nu > 0$. Using the relations

$$\frac{1}{T}\int_0^T J_\alpha(\lambda x)\, J_\alpha(\nu x)\, x\, dx = \frac{\nu J_\alpha(\lambda T) J_{\alpha-1}(\nu T) - \lambda J_{\alpha-1}(\lambda T) J_\alpha(\nu T)}{\lambda^2 - \nu^2}, \text{ if } \lambda \neq \nu.$$

$$\frac{1}{T}\int_0^T J_\alpha^2(\lambda x)\, x\, dx = \frac{T}{2}\{J_\alpha^2(\lambda T) - J_{\alpha-1}(\lambda T)\, J_{\alpha+1}(\lambda T)\}, \text{ if } \lambda = \nu.$$

deduce that

$$\lim_{T\to+\infty} \frac{1}{T}\int_0^T j_\alpha(\lambda x)\, j_\alpha(\nu x)\, d\mu_\alpha(x) = 0, \text{ if } \lambda \neq \nu.$$

$$\lim_{T\to+\infty} \frac{1}{T}\int_0^T (j_\alpha(\lambda x))^2\, d\mu_\alpha(x) = \frac{2^\alpha\, \Gamma(\alpha+1)}{\pi \lambda^{2\alpha+1}}, \text{ if } \lambda \neq \nu.$$

2) Let f be in $\mathcal{P}_\alpha$ and $\{\lambda_k\}_{k\in\mathbb{N}}$, $\{\beta_k\}_{k\in\mathbb{N}}$ respectively the Fourier exponents and coefficients of its pre-image F_f in $\mathcal{P}_{-\frac{1}{2}}$. prove that

$$\lim_{T\to+\infty} \frac{1}{T}\int_0^T f(x)\, j_\alpha(\nu x)\, d\mu_\alpha(x) = \begin{cases} 0, & \text{if } \nu \neq \lambda_k. \\ \dfrac{2^\alpha\, \Gamma(\alpha+1)}{\pi \lambda_k^{2\alpha+1}}\, \beta_k, & \text{if } \nu = \lambda_k. \end{cases}$$

We call the sequences $\{\lambda_k\}_{k\in\mathbb{N}}$ and $\{\beta_k\}_{k\in\mathbb{N}}$ respectively the generalized Fourier exponents and coefficients of the function f. We write

$$f(x) \sim \sum_k \beta_k\, j_\alpha(\lambda_k x).$$

The second member is called the generalized Fourier series of f.

3) Let $\{\lambda_k\}_{k\in\mathbb{N}}$ and $\{\beta_k\}_{k\in\mathbb{N}}$ the generalized Fourier exponents and coefficients of f in $\mathcal{P}_\alpha$.

i) Show that $\{\lambda_k\}_{k\in\mathbb{N}}$ are also the Fourier exponents of the functions $T_x(f)$, $x \geq 0$.

ii) Prove that the generalized Fourier coefficients of the functions $T_x(f)$, $x \geq 0$, are $\{\beta_k\, j_\alpha(\lambda_k x)\}_{k\in\mathbb{N}}$.

(See [21] and [13]).

Problem 5.VIII.21.

We consider in this problem notations and results of Problem 5.VIII.19 and 5.VIII.20.

We denote by

$\mathcal{P}^0_{-\frac{1}{2}} = \{F \in \mathcal{P}_{-\frac{1}{2}} \,/\, G(y) = \int_0^y F(x)\,dx$ is an odd Bohr almost-periodic function on $\mathbb{R}\}$.

$$H_\alpha = \{f \in \mathcal{P}_\alpha \,/\, \text{the pre-image } F_f \in \mathcal{P}^0_{-\frac{1}{2}}\}.$$

1) Prove the following results

i) the functions

$$P_n(x) = \sum_{k=0}^{N_n} \beta_k^{(n)} j_\alpha(\lambda_k x), \quad n \in \mathbb{N},$$

are in H_α.

ii) Let f be in H_α and F_f its pre-image. If

$$F_f(x) \sim \sum_k \beta_k \cos\lambda_k x,$$

then

$$G_f(y) = \int_0^y F_f(x)dx \sim \sum_k \beta_k \frac{\sin \lambda_k y}{\lambda_k}.$$

iii) If $f \in H_\alpha$, then the functions $T_x(f)$, $x \geq 0$, belong also to H_α.

2) Let f and g be in H_α such that

$$f(x) \sim \sum_k a_k j_\alpha(\lambda_k x) \text{ and } g(x) \sim \sum_k b_k j_\alpha(\lambda_k x).$$

Prove that the limit

$$\lim_{T\to+\infty} \frac{1}{T}\int_0^T f(x)\,\overline{g(x)}\,d\mu_\alpha(x),$$

exists and is finite. It is equal to zero if the two functions f and g have no common generalized Fourier exponents, in the contrary case it is equal to the convergent series

$$\frac{2^\alpha \Gamma(\alpha+1)}{\pi} \sum_k \frac{a_k \bar{b}_k}{\lambda_k^{2\alpha+1}}.$$

The summation is on the common generalized Fourier exponents of f and g.

3) For f and g in H_α we put

$$<f, g>_{H_\alpha} = \lim_{T\to+\infty} \frac{1}{T}\int_0^T f(x)\, \overline{g(x)}\, d\mu_\alpha(x).$$

$$\|f\|_{H_\alpha} = <f, f>_{H_\alpha}.$$

i) Show that $<.,.>_{H_\alpha}$ is a scalar product on H_α.

ii) Prove that H_α provided with the scalar product $<.,.>_{H_\alpha}$ is a prehilbertian space.

iii) Show that the subspace consisting of the functions P_n, $n \in \mathbb{N}$, given by

$$P_n(x) = \sum_{k=0}^{N_n} \beta_k^{(n)}\, j_\alpha(\lambda_x x),$$

is dense in H_α.

iv) Let f be in H_α. Prove that the generalized Fourier series of f converge to f in H_α.

(See [21] and [13]).

Problem 5.VIII.22.

We said that a function f in $\mathcal{E}_*(\mathbb{R})$ is generalized mean-periodic if the closed subspace $V_{\ell_\alpha}(f)$ of $\mathcal{E}_*(\mathbb{R})$ generated by the $T_x(f)$, $x \geq 0$, satisfies

$$V_{\ell_\alpha}(f) \neq \mathcal{E}_*(\mathbb{R}).$$

Where T_x, $x \geq 0$, are the generalized translation operators given by the relation (4.II.3), and ℓ_α the Bessel operator

$$\ell_\alpha = \frac{d^2}{dx^2} + \frac{2\alpha+1}{x}\frac{d}{dx}.$$

1) Show that a function $f \in \mathcal{E}_*(\mathbb{R})$ is generalized mean-periodic if and only if there exists a nonzero $S \in \mathcal{E}'_*(\mathbb{R})$ such that

$$S *_B f(x) = 0, \text{ for all } x \geq 0.$$

2) i) Let $a \in]0, +\infty[$. Prove the following results

- All function $f \in \mathcal{E}_*(\mathbb{R})$ satisfying

$$T_x(f)(a) = 0, \text{ for all } x \geq 0,$$

is generalized mean-periodic.

- All function $f \in \mathcal{E}_*(\mathbb{R})$ satisfying

$$T_x(f)(a) = f(x), \text{ for all } x \geq 0,$$

is generalized mean-periodic.

ii) Show that the function

$$\phi_{\pm\lambda,j}(x) = \left(\frac{d^j}{d\mu^j} j_\alpha(\mu x)\right)_{/\mu=\pm\lambda}, \; x \geq 0,$$

is generalized mean-periodic.

3) Prove that the nonzero functions of $\mathcal{D}_*(\mathbb{R})$ and $\mathcal{E}_*(\mathbb{R}) \cap L^1([0,+\infty[, d\mu_\alpha)$ are not generalized mean-periodic.

(See [86]).

Problem 5.VIII.23.

We use in this problem notations and results of Problem 5.VIII.22.

1) Prove that the functions $\phi_{\pm\lambda,j}$, $0 \leq j \leq \ell$, belong to $V_{\ell_\alpha}(f)$ if and only if for all $S \in \mathcal{E}'_*(\mathbb{R})$ satisfying

$$S *_B f(x) = 0, \text{ for all } x \geq 0.$$

We have

$$\frac{d^j}{d\mu^j}(\mathcal{F}_B(S))(\mu)_{/\mu=\pm\lambda} = 0, \; j = 0, 1, 2, \ldots, \ell.$$

We denote by

$$S_p(f) = \{(\pm\lambda, \ell), \lambda \in \mathbb{C}, \ell \in \mathbb{N} \,/\, \phi_{\pm\lambda,j} \in V_{\ell_\alpha}(f) \text{ for } 0 \leq j \leq \ell, \text{ and } \phi_{\pm\lambda,\ell+1} \notin V_{\ell_\alpha}(f)\},$$

this set is called spectrum of f.

2) Show that

$$\forall\, x \geq 0, \; \phi_{\pm\lambda,j}(x) = \mathcal{R}_\alpha(\varphi_{\pm\lambda,j})(x).$$

Where

$$\forall\, x \geq 0,\ \varphi_{\pm\lambda,j}(x) = \left(\frac{d^j}{d\mu^j}\cos\mu x\right)/_{\mu=\pm\lambda}.$$

3) Let $S \in \mathcal{E}'_*(\mathbb{R})$ and $f \in \mathcal{E}_*(\mathbb{R})$. Prove the following relations

$$W_\alpha^{-1}(S) *_B \mathcal{R}_\alpha(f) = \mathcal{R}_\alpha(S *_0 f).$$

$$W_\alpha(S) *_0 \mathcal{R}_\alpha^{-1}(f) = \mathcal{R}_\alpha^{-1}(S *_0 f),$$

where $\mathcal{R}_\alpha$ is the Riemann-Liouville transform and W_α the Weyl transform studied is chapter 2.

4) We say that a function $f \in \mathcal{E}_*(\mathbb{R})$ is even mean-periodic if the closed subspace $V_{D^2}(f)$ of $\mathcal{E}_*(\mathbb{R})$ generated by the $\sigma_x(f)$, $x \geq 0$, satisfies

$$V_{D^2}(f) \neq \mathcal{E}_*(\mathbb{R}).$$

Where σ_x, $x \geq 0$, are the translation operators given by the relation (3.I.1), and $D = \dfrac{d}{dx}$.

From the results obtained by L. Schwartz [74], we deduce that if

$$Sp(f) = \{(\pm\lambda, \ell),\ \lambda \in \mathbb{C},\ \ell \in \mathbb{N}\ /\ \varphi_{\pm\lambda,j} \in V_{D^2}(f)\ \text{ for }\ 0 \leq j \leq \ell,\ \text{ and }\ \varphi_{\pm\lambda,\ell+1} \notin V_{D^2}(f)\}.$$

then all even mean-periodic function $f \in \mathcal{E}_*(\mathbb{R})$ can be approached in $\mathcal{E}_*(\mathbb{R})$ by finite linear combinations of the function $\varphi_{\pm\lambda,j}$, $0 \leq j \leq \ell$, such that $(\pm\lambda, \ell) \in Sp(f)$.

i) Using the precedent question deduce that a function $f \in \mathcal{E}_*(\mathbb{R})$ is generalized mean-periodic if and only if the function $\mathcal{R}_\alpha^{-1}(f)$ is even mean-periodic.

ii) Prove that all generalized mean-periodic function $f \in \mathcal{E}_*(\mathbb{R})$ can be approached in $\mathcal{E}_*(\mathbb{R})$ by finite linear combinations of the functions $\phi_{\pm\lambda,j}$, $0 \leq j \leq \ell$, such that $(\pm\lambda, \ell) \in Sp(f)$.

5) Let $f \in \mathcal{E}_*(\mathbb{R})$ be a generalized mean-periodic function. Deduce that if $Sp(f) = \emptyset$, then $f \equiv 0$.

6) Let $a \in]0, +\infty[$, and $f \in \mathcal{E}_*(\mathbb{R})$ satisfying

$$T_x(f)(a) = 0,\ \text{ for all }\ x \geq 0.$$

Prove that f can be approached in $\mathcal{E}_*(\mathbb{R})$ by finite linear combinations of the functions $j_\alpha(\frac{k\pi}{a}x)$, $k \in \mathbb{N}^*$.

(See [86]).

Problem 5.VIII.24.

We use here notations of Problem 5.VIII.22.

We denote by

• $\mathcal{G}_{\ell_\alpha}$ (resp. $\mathcal{G}_{D^2}$) the group of isomorphisms Y of $\mathcal{E}_*(\mathbb{R})$ onto itself such that

$$Y\ell_\alpha = \ell_\alpha Y \text{ (resp. } YD^2 = D^2Y).$$

• $\mathcal{V}_{\ell_\alpha}(f)$(resp. $\mathcal{V}_{D^2}(f)$) the closed subspace of $\mathcal{E}_*(\mathbb{R})$ generated by Y(f) with $Y \in \mathcal{G}_{\ell_\alpha}$ (resp. $\mathcal{G}_{D^2}$).

1°) Prove that all element of $\mathcal{G}_{D^2}$ is of the form

$$Y(f)(x) = a\,\sigma_y(f)(x), \quad a \in \mathbb{C},\ a \neq 0,\ y \geq 0.$$

Where σ_y, $y \geq 0$, are the translation operators given by the relation (3.I.1).

2) Deduce that the space of functions f in $\mathcal{E}_*(\mathbb{R})$ which satisfy

$$\mathcal{V}_{D^2}(f) \neq \mathcal{E}_*(\mathbb{R}),$$

is equal to the space of even mean-periodic functions f in $\mathcal{E}_*(\mathbb{R})$.

3) Prove the following results

i) The group $\mathcal{G}_{\ell_\alpha}$ is isomorphic to the group $\mathcal{G}_{D^2}$.

ii) We have the relation

$$\mathcal{V}_{\ell_\alpha}(f) = \mathcal{R}_\alpha\, \mathcal{V}_{D^2}(\mathcal{R}_\alpha^{-1}(f)), \quad f \in \mathcal{E}_*(\mathbb{R}).$$

4) Deduce that the space of functions f in $\mathcal{E}_*(\mathbb{R})$ which satisfy

$$\mathcal{V}_{\ell_\alpha}(f) \neq \mathcal{E}_*(\mathbb{R}),$$

is equal to the space of generalized mean-periodic functions f in $\mathcal{E}_*(\mathbb{R})$.

(See [86]).

6 Infinitely Divisible Probabilities And Central Limit Theorem Associated With The Bessel Operator

Introduction.

In this chapter we use the Harmonic Analysis associated with the Bessel operator given in the precedent chapters, to study some applications to the probability theory. We begin by showing that the functions

$$Q(x) = cx^2, \quad c \in \mathbb{R},$$

are all the generalized quadratic forms associated with the Bessel operator i.e. the even C^∞-functions on $\mathbb{R}$ satisfying

$$\forall\, x \geq 0, \quad y \geq 0, \quad T_x(Q)(y) = Q(x) + Q(y).$$

Where T_x, $x \geq 0$, are the generalized translation operators given by the relation (4.II.3).

Using the generalized quadratic form

$$q(x) = \frac{x^2}{4(\alpha + 1)},$$

we define the dispersion (or variance) of a probability measure ν on $[0, +\infty[$ by

$$D(\nu) = \int_0^\infty q(x)\, d\nu(x).$$

We study the properties of the mapping D and we prove the main relation

$$D(\nu *_B \tau) = D(\nu) + D(\tau),$$

where ν and τ are probability measures on $[0, +\infty[$ and $*_B$ the generalized convolution product given by the relation (4.III.20).

Next we define the continuous negative definite functions for the Bessel operator, and we prove for them a Levy-Khintchine's formula (see also [44]). This formula is useful to characterize the convolution semigroup and the infinitely divisible probabilities associated with the Bessel operator.

Finally we define the Gaussian (or normal) distribution associated with the Bessel operator and we prove a central limit theorem.

Analogous resuls are obtained in [82] in the case of a Harmonic Analysis associated with a singular differential operator more general than the Bessel operator (see also [6] [8] [94] [95]).

6.I. Dispersion of a probability measure on $[0, +\infty[$.

Notations. We denote by

- $M_+^b([0, +\infty[)$ the set of positive bounded Radon measures on $[0, +\infty[$.
- $M^1([0, +\infty[)$ the set of probability measures on $[0, +\infty[$.

6.I.1. Generalized quadratic form.

Definition 6.I.1. A generalized quadratic form associated with the Bessel operator is a function Q in $\mathcal{E}_*(\mathbb{R})$ satisfying

(6.I.1) $$\forall\ x \geq 0,\ y \geq 0,\ T_x(Q)(y) = Q(x) + Q(y)\ ,$$

where T_x, $x \geq 0$, are the generalized translation operators associated with the Bessel operator defined by the relation (4.II.3).

Remarks.

i) The relation (6.I.1) generalizes that which is satisfied by the classical quadratic forms on $[0, +\infty[$:

(6.I.2) $$\forall\ x \geq 0,\ y \geq 0,\ \sigma_x(Q)(y) = Q(x) + Q(y)\ ,$$

where σ_x, $x \geq 0$, the translation operators given by the relation (3.I.1).

ii) Using the relation (6.I.1) and the fact that

$$T_0 = \mathrm{id}\ ,$$

we obtain

$$Q(0) = 0\ .$$

Theorem 6.I.1. The generalized quadratic forms associated with the Bessel operator are

(6.I.3) $$\forall\, x \geq 0,\ Q(x) = c\, x^2,$$

where c is a real constant.

Proof.

a) We consider the function

$$\forall\, x \geq 0,\ Q(x) = c\, x^2$$

where c is a real constant.

From the relation (4.II.3) we have

$$\forall\, x \geq 0, y \geq 0,\ T_x(Q)(y) = \frac{c\,\Gamma(\alpha+1)}{\sqrt{\pi}\,\Gamma(\alpha+\frac{1}{2})}\int_0^{\pi} (x^2 + y^2 - 2xy\cos\theta)(\sin\theta)^{2\alpha}\, d\theta\,,$$
$$= c(x^2 + y^2).$$

Then

$$\forall\, x \geq 0, y \geq 0,\ T_x(Q)(y) = Q(x) + Q(y)\,.$$

b) Let Q be a function in $\mathcal{E}_*(\mathbb{R})$ satisfying

(6.I.4) $$\forall\, x \geq 0, y \geq 0, T_x(Q)(y) = Q(x) + Q(y)\,.$$

From Proposition 4.II.7 we have

$$\forall\, x \geq 0, y \geq 0,\ \ell_{\alpha,x}(T_x(Q))(y) = T_x(\ell_\alpha(Q))\,(y)\,.$$

Thus using (6.I.4) we obtain

$$\forall\, x \geq 0, y \geq 0,\ T_x\,(\ell_\alpha(Q))(y) = \ell_\alpha(Q)\,(y)\,.$$

We take $x = 0$. Then

$$\forall\, y \geq 0,\ \ell_\alpha(Q))(y) = \ell_\alpha(Q)\,(0)..$$

Using the fact that

$$\ell_\alpha = \frac{1}{y^{2\alpha+1}}\frac{d}{dy}\,[y^{2\alpha+1}\frac{d}{dy}],$$

and that the function Q belongs to $\mathcal{E}_*(\mathbb{R})$ we obtain

$$\forall\, y \geq 0,\ y^{2\alpha+1}\frac{d}{dy}Q(y) = \ell_\alpha(Q)\,(0)\int_0^y t^{2\alpha+1}dt,$$
$$= \ell_\alpha(Q)\,(0)\,\frac{y^{2\alpha+2}}{2\alpha+2}.$$

Thus

$$\forall\, y \geq 0,\ Q(y) = \ell_\alpha(Q)(0)\ \frac{y^2}{2(2\alpha+2)}.$$

We put

$$c = \frac{\ell_\alpha(Q)(0)}{4(\alpha+1)}.$$

Then

$$\forall\, y \geq 0,\ Q(y) = cy^2.$$

Proposition 6.I.1. The generalized quadratic form $q(x) = \dfrac{x^2}{4(\alpha+1)}$ satisfies

(6.I.5) $$\forall\, x \geq 0,\ q(x) = -\ \frac{d}{d\lambda^2}\, j_\alpha(\lambda x)/_{\lambda=0}\,.$$

Proof.

For all $x \geq 0$, the function $\lambda \rightarrow j_\alpha(\lambda x)$ verifies the equation

$$\frac{d^2}{d\lambda^2}\, j_\alpha(\lambda x) + \frac{2\alpha+1}{\lambda}\ \frac{d}{d\lambda}\, j_\alpha(\lambda x) = -\, x^2\, j_\alpha(\lambda x).$$

Using the fact that

$$j_\alpha(0) = 1,$$

we obtain

$$\frac{d^2}{d\lambda^2}\, j_\alpha(\lambda x)\ /_{\lambda=0} = -\ \frac{x^2}{2(\alpha+1)}\,.$$

Thus

$$\frac{d}{d\lambda^2}\, j_\alpha(\lambda x)/_{\lambda=0} = -\ \frac{x^2}{4(\alpha+1)} = -\ q(x).$$

6.I.2. Dispersion of a probability measure on $[0, +\infty[$.

Definition 6.I.2. The dispersion (or variance) of a mesure ν in $M^1([0, +\infty[)$ is given by

(6.I.6) $$D(\mu) = \int_0^\infty q(x)\, d\nu(x).$$

Lemma 6.I.1. We have

(6.I.7) $$\forall\, x \geq 0, \lambda \geq 0, j_\alpha(\lambda x) = 1 - \lambda^2 \int_0^x v(x,t)\, j_\alpha(\lambda t)\, t^{2\alpha+1}\, dt,$$

with

(6.I.8)
$$v(x,t) = \int_t^x \frac{dr}{r^{2\alpha+1}}.$$

Proof.

For all $\lambda \geq 0$, the function $x \to j_\alpha(\lambda x)$ satisfies the equation

$$\ell_\alpha (j_\alpha(\lambda .))(x) = - \lambda^2 j_\alpha(\lambda x).$$

Using the fact that

$$\ell_\alpha = \frac{1}{x^{2\alpha+1}} \frac{d}{dx} [x^{2\alpha+1} \frac{d}{dx}],$$

we obtain

$$\forall x \geq 0, x^{2\alpha+1} \frac{d}{dx} j_\alpha(\lambda x) = - \lambda^2 \int_0^x j_\alpha(\lambda t) t^{2\alpha+1} dt .$$

Thus

$$\forall x \geq 0, j_\alpha(\lambda x) = 1 - \lambda^2 \int_0^x \frac{1}{r^{2\alpha+1}} (\int_0^r j_\alpha(\lambda t) t^{2\alpha+1} dt) dr .$$

By applying Fubini's theorem we deduce that

$$\forall x \geq 0, j_\alpha(\lambda x) = 1 - \lambda^2 \int_0^x (\int_t^x \frac{dr}{r^{2\alpha+1}}) j_\alpha(\lambda t) t^{2\alpha+1} dt.$$

This completes the proof.

Theorem 6.I.2. Let ν be in $M^1([0, +\infty[)$. The function $f(x) = x^2$ is ν-integrable if and only if the function $\mathfrak{F}_B(\nu)(\lambda)$ is derivable with respect to λ^2 at the point $\lambda = 0$, and we have

(6.I.9)
$$D(\nu) = - \frac{d}{d\lambda^2} \mathfrak{F}_B(\nu)(\lambda) /_{\lambda=0}.$$

Proof.

a) We suppose that the function $\mathfrak{F}_B(\nu)(\lambda)$ is derivable with respect to λ^2. From the relation (1.II.1) we have

(6.I.10)
$$\int_0^\infty q(x)\, d\nu(x) = \int_0^\infty \lim_{\substack{\lambda \to 0 \\ \lambda > 0}} \frac{1 - j_\alpha(\lambda x)}{\lambda^2} d\nu(x) .$$

On the other hand we have

(6.I.11)
$$\frac{d}{d\lambda^2} \mathfrak{F}_B(\nu)(\lambda) /_{\lambda=0} = \lim_{\substack{\lambda \to 0 \\ \lambda > 0}} \int_0^\infty \frac{1 - j_\alpha(\lambda x)}{\lambda^2} d\nu(x) .$$

Using Fatou's Lemma we obtain

$$\int_0^\infty \lim_{\substack{\lambda\to 0\\ \lambda>0}} \frac{1-j_\alpha(\lambda x)}{\lambda^2}\, d\nu(x) \leq \lim_{\substack{\lambda\to 0\\ \lambda>0}} \int_0^\infty \frac{1-j_\alpha(\lambda x)}{\lambda^2}\, d\nu(x)\ .$$

From the relations (6.I.10) and (6.I.11) we deduce that the function $f(x) = x^2$ is ν-integrable.

b) We suppose that the function $f(x) = x^2$ is ν-integrable. From Lemma 6.I.1 we have

$$\forall\, \lambda \geq 0,\ \mathfrak{F}_B(\nu)(\lambda) = 1 - \lambda^2 \int_0^\infty \Big(\int_0^x v(x,t)\, j_\alpha(\lambda t)\, t^{2\alpha+1}\, dt\Big)\, d\nu(x).$$

Thus for $\lambda > 0$:

$$\frac{\mathfrak{F}_B(\lambda) - 1}{\lambda^2} = -\int_0^\infty \Big(\int_0^x v(x,t)\, j_\alpha(\lambda t)\, t^{2\alpha+1}\, dt\Big)\, d\nu(x).$$

We have

- $\forall\, t > 0,\ \lim\limits_{\lambda\to 0} j_\alpha(\lambda t) = 1.$

- From the fact that

$$\forall\, t \geq 0,\ \lambda \geq 0,\ |j_\alpha(\lambda t)| \leq 1,$$

we deduce that for all $x \geq 0$:

$$\Big|\int_0^x v(x,t)\, j_\alpha(\lambda t)\, t^{2\alpha+1}\, dt\,\Big| \leq \int_0^x v(x,t)\, t^{2\alpha+1}\, dt = q(x)\ .$$

Then from dominated convergence theorem we have

$$\lim_{\substack{\lambda\to 0\\ \lambda>0}} \frac{\mathfrak{F}_B(\lambda) - 1}{\lambda^2} = -\int_0^\infty \Big(\int_0^x v(x,t)\, t^{2\alpha+1}\, dt\Big)\, d\nu(x),$$

$$= -\int_0^\infty q(x)\, d\nu(x),$$

$$= -D(\nu).$$

hence the function $\mathfrak{F}_B(\nu)(\lambda)$ is derivable with respect to the variable λ^2 at the point $\lambda = 0$, and we have

$$D(\nu) = -\frac{d}{d\lambda^2}\, \mathfrak{F}_B(\nu)(\lambda)\,/_{\lambda=0}\, .$$

Corollary 6.I.1. For ν and τ in $M^1([0, +\infty[)$ such that the function $f(x) = x^2$ is ν and τ-integrable. We have

(6.I.12) $D(\nu *_B \tau) = D(\nu) + D(\tau)$.

Proof.

From the relation (5.VI.7) we have

$$\forall \lambda \geq 0, \mathfrak{F}_B(\nu *_B \tau)(\lambda) = \mathfrak{F}_B(\nu)(\lambda).\mathfrak{F}_B(\tau)(\lambda),$$

we deduce the result from Theorem 6.I.2 and this relation.

6.I.3. Levy's theorem.

Proposition 6.I.2. Let $\{\nu_n\}_{n\in\mathbb{N}}$ be a sequence in $M^1([0, +\infty[)$ which converges vaguely to a measure ν in $M_+^b([0, +\infty[)$. Then

(6.I.13) $$\forall \lambda > 0, \lim_{n\to+\infty} \mathfrak{F}_B(\nu_n)(\lambda) = \mathfrak{F}_B(\nu)(\lambda).$$

Proof.

From Lemma 5.II.1 we have

(6.I.14) $$\forall \lambda > 0, \lim_{x\to+\infty} j_\alpha(\lambda x) = 0.$$

Thus

(6.I.15) $$\forall \varepsilon \geq 0, \exists A \geq 0 / x > A \Rightarrow |j_\alpha(\lambda x)| < \varepsilon.$$

We consider an even continuous function Ψ on $\mathbb{R}$ with compact support such that

$$\Psi(x) = 1, \text{ for all } x \in [-A, A],$$

$$|\Psi(x)| \leq 1, \text{ for all } x \notin [-A, A].$$

Then for all $\lambda > 0$:

$$|\mathfrak{F}_B(\nu_n)(\lambda) - \mathfrak{F}_B(\nu)(\lambda)| \leq \left|\int_0^\infty \Psi(x)\, j_\alpha(\lambda x)\, d\nu_n(x) - \int_0^\infty \Psi(x)\, j_\alpha(\lambda x) d\nu(x)\right|$$
$$+ \int_A^\infty |(1-\Psi(x))\, j_\alpha(\lambda x)|\, d\nu_n(x) + \int_A^\infty |(1 - \Psi(x))\, j_\alpha(\lambda x)\, |d\nu(x).$$

using the relation (6.I.15) we obtain for all $\lambda > 0$:

$$|\mathfrak{F}_B(\nu_n)(\lambda) - \mathfrak{F}_B(\nu)(\lambda)| \leq \left|\int_0^\infty \Psi(x)\, j_\alpha(\lambda x)\, d\nu_n(x) - \int_0^\infty \Psi(x)\, j_\alpha(\lambda x) d\nu(x)\right|$$
$$+ 2\varepsilon\left(1 + \int_0^\infty d\nu(x)\right).$$

We deduce the result from the vague convergence of $\{\nu_n\}_{n\in\mathbb{N}}$ to ν when $n \to +\infty$.

Theorem 6.I.3. (Levy's theorem). Let $\{\nu_n\}_{n\in\mathbb{N}}$ be a sequence in $M^1([0, +\infty[)$ such that

(6.I.16) $$\forall\, \lambda \geq 0, \quad \lim_{n\to+\infty} \mathfrak{F}_B(\nu_n)(\lambda) = f(\lambda),$$

where $f : [0, +\infty[\to \mathbb{R}$, is a right continuous function at zero. Then the sequence $\{\nu_n\}_{n\in\mathbb{N}}$ converges vaguely when n tends to infinity, to a measure ν in $M_+^b([0, +\infty[)$ such that

(6.I.17) • $\|\nu\| \leq 1$.

(6.I.18) • $\forall\, \lambda \geq 0,\ \mathfrak{F}_B(\nu)(\lambda) = f(\lambda)$.

If morever the measure ν belongs to $M^1([0, +\infty[)$, the vague convergence becomes weak convergence.

Proof.

We can extract from the sequence $\{\nu_n\}_{n\in\mathbb{N}}$ a subsequence $\{\nu_{n_k}\}_{k\in\mathbb{N}}$ which converges vaguely when n tends to infinity, to a measure ν in $M_+^b([0, +\infty[)$ such that

$$\|\nu\| \leq 1.$$

From Proposition 6.I.2 we have

$$\forall\, \lambda > 0, \quad \lim_{k\to+\infty} \mathfrak{F}_B(\nu_{n_k})(\lambda) = \mathfrak{F}_B(\nu)(\lambda).$$

But from (6.I.16) we have

$$\forall\, \lambda \geq 0, \quad \lim_{k\to+\infty} \mathfrak{F}_B(\nu_{n_k})(\lambda) = f(\lambda).$$

Thus

(6.I.19) $$\forall\, \lambda \geq 0,\ \mathfrak{F}_B(\nu)(\lambda) = f(\lambda).$$

If $\{\nu_{n'_k}\}_{k\in\mathbb{N}}$ is another subsequence of $\{\nu_n\}_{n\in\mathbb{N}}$ which convvergeces vaguely when n tends to infinity, to a measure ν' in $M_+^b([0, +\infty[)$, then

$$\forall\, \lambda \geq 0,\ \mathfrak{F}_B(\nu)(\lambda) = f(\lambda) = \mathfrak{F}_B(\nu')(\lambda).$$

From Proposition 5.VI.3 we deduce that

$$\nu = \nu'.$$

Then the sequence $\{\nu_n\}_{n\in\mathbb{N}}$ converges vaguely when n tends to infinity, to the measure ν, and we have the relation (6.I.19).

If we suppose moreover that

$$\|\nu\| = 1,$$

then the vague convergence becomes weak convergence.

Proposition 6.I.3. Let ν be a measure in $M_+^b([0, +\infty[)$. Then the function $f(\lambda)$ defined by

$$\forall\, \lambda \geq 0,\ f(\lambda) = \exp\{-t[\mathcal{F}_B(\nu)(\lambda) - \mathcal{F}_B(\nu)(0)]\},\ t \geq 0,$$

is the Fourier-Bessel transform of a measure in $M^1([0, +\infty[)$.

Proof.

We put

$$\nu_n = e^{-at} \sum_{k=0}^{n} \frac{t^k}{k!} \nu^{*_B k},$$

where

$$a = \mathcal{F}_B(\nu)(0).$$

We have

$$\forall\, n \in \mathbb{N},\ \|\nu\| = \mathcal{F}_B(\nu_n)(0) \leq 1.$$

On the other hand we have

$$\forall\, \lambda \geq 0,\quad \lim_{n\to+\infty} \mathcal{F}_B(\nu_n)(\lambda) = f(\lambda).$$

The same proof as for Theorem 6.I.3 shows that there exists a measure τ in $M_+^b([0, +\infty[)$ such that

- $\|\tau\| \leq 1$.
- $\forall\, \lambda \geq 0,\ \mathcal{F}_B(\tau)(\lambda) = f(\lambda)$.

From this equality we deduce that

$$\|\tau\| = \mathcal{F}_B(\tau)(0) = f(0) = 1.$$

This completes the proof.

6.II.Levy-Khintchine's formula.

Definition 6.II.1. A continuous function $\Psi : [0, +\infty[\to \mathbb{R}$ is said to be negative definite for the Bessel operator if for every $t \in [0, +\infty[$, the function $\lambda \to \exp(-t\,\Psi(\lambda))$ is a bounded positive definite function associated with the Bessel operator (see Problem 5.VIII.12).

The following proposition gives properties of negative definite functions for the Bessel operator.

Proposition 6.II.1.

i) For all negative definite function Ψ for the Bessel operator, we have

(6.II.1) $$\forall\, \lambda \geq 0,\ \ \Psi(\lambda) \geq \Psi(0).$$

ii) For all negative definite functions f, g for the Bessel operator and a positive constant c, the functions cf, f + g and f - f(0) are negative definite functions for the Bessel operator.

iii) Let f be a continuous positive definite function associated with the Bessel operator. Then the function f(0) - f is negative definite function for the Bessel operator.

Theorem 6.II.1 (Levy-Khintchine's formula). Let Ψ be a negative definite function for the Bessel operator. Then there exist nonnegative constants c, d and a measure τ in $M_+^b([0, +\infty[)$, such that

(6.II.2) $$\forall\, \lambda \geq 0,\ \ \Psi(\lambda) = c\lambda^2 + d + \int_0^\infty (1 - j_\alpha(\lambda x))\,d\tau(x).$$

The measure τ satisfies

(6.II.3) $$\int_0^\infty \frac{x^2}{1+x^2}\,d\tau(x) < +\infty.$$

The proof of this theorem results from the following four Lemmas. Without loss of generality we can suppose that

$$\Psi(0) = 0.$$

Lemma 6.II.1. Let Ψ be a negative definite function for the Bessel operator and $\lambda_1 \in [0, +\infty[$. Then we have

(6.II.4) $$\forall \lambda_2 \geq 0, \; T_{\lambda_1}(\Psi)(\lambda_2) - \Psi(\lambda_2) = \int_0^\infty j_\alpha(\lambda_2 x) d\, \nu_{\lambda_1}(x).$$

Where T_{λ_1} is the generalized translation operator given by the relation (4.II.3) and ν_{λ_1} is a unique measure in $M_+^b([0, +\infty[)$.

Proof.

i) Assume that

$$\forall \lambda_2 \geq 0, \; \Psi(\lambda_2) = h(0) - h(\lambda_2),$$

where h is a continuous bounded positive definite function associated with the Bessel operator. Then from Problem 5.VIII.12, we have

$$\forall \lambda_2 \geq 0, \; h(\lambda_2) = \mathfrak{F}_B(\nu)(\lambda_2),$$

where ν is a measure in $M_+^b([0, +\infty[)$.

An easy computation gives

$$\forall \lambda_2 \geq 0, \; T_{\lambda_1}(\Psi)(\lambda_2) - \Psi(\lambda_2) = \int_0^\infty j_\alpha(\lambda_2 x)\,(1 - j_\alpha(\lambda_1 x))\, d\nu(x).$$

ii) In the general case, we write for every $\lambda_2 \geq 0$:

$$\Psi(\lambda_2) = \lim_{n \to +\infty} [h_n(0) - h_n(\lambda_2)],$$

where

$$h_n(\lambda_2) = n \exp(-\frac{1}{n}\Psi(\lambda_2)), \; n \in \mathbb{N}^*.$$

Put

$$\Psi_n(\lambda_2) = h_n(0) - h_n(\lambda_2).$$

From i) we have

$$T_{\lambda_1}(\Psi_n)(\lambda_2) - \Psi_n(\lambda_2) = \int_0^\infty j_\alpha(\lambda_2 x)\,(1 - j_\alpha(\lambda_1 x))\, d\nu_n(x).$$

Where ν_n is the measure associated with the continuous bounded positive definite function h_n associated with the Bessel operator. We consider the sequence $\{\tau_n\}_{n\in\mathbb{N}}$ of measures in $M_+^b([0, +\infty[)$ defined by

$$d\tau_n(x) = (1 - j_\alpha(\lambda_1 x))d\nu_n(x).$$

We extract from this sequence a subsequence $\{\tau_{n_p}\}_{p\in\mathbb{N}}$ which converges vaguely when p tends to infinity, to a measure τ_{λ_1} in $M_+^b([0, +\infty[)$.

We have

$$T_{\lambda_1}(\Psi)(\lambda_2) - \Psi(\lambda_2) = \lim_{p\to+\infty} (T_{\lambda_1}(\Psi_{n_p})(\lambda_2) - \Psi_{n_p}(\lambda_2)),$$

$$= \lim_{p\to+\infty} \int_0^\infty j_\alpha(\lambda_2 x) d\tau_{n_p}(x).$$

We can see that

$$\lim_{p\to+\infty} \int_0^\infty j_\alpha(\lambda_2 x) d\tau_{n_p}(x) = \int_0^\infty j_\alpha(\lambda_2 x) d\tau_{\lambda_1}(x).$$

We repeat the reasoning for any other convergent subsequence, we obtain that the full sequence converges to τ_{λ_1}, and Proposition 5.VI.3 shows that τ_{λ_1} is unique.

Lemma 6.II.2. The measure τ_{λ_1}, $\lambda_1 \geq 0$, defined in Lemma 6.II.1, can be written

(6.II.5) $$d\tau_{\lambda_1}(x) = Q(\lambda_1)\delta + (1 - j_\alpha(\lambda_1 x))\, d\tau(x),$$

where $Q(\lambda_1) \geq 0$, and τ is a measure in $M_+^b([0, +\infty[)$ which does not depend on λ_1.

Proof.

By decomposing the measure τ_{λ_1} we have

(6.II.6) $$\tau_{\lambda_1} = Q(\lambda_1)\,\delta + \tau'_{\lambda_1},$$

with $Q(\lambda_1) \geq 0$ and $\tau'_{\lambda_1}(\{0\}) = 0$, (here δ denotes the Dirac measure at zero). We have seen in the proof of Lemma 6.II.1 that

$$\tau_{\lambda_1} = \lim_{n\to+\infty} (1 - j_\alpha(\lambda_1 x))\nu_n.$$

We deduce that for every $\lambda_1 > 0$ and $\lambda_2 > 0$:

$$(1 - j_\alpha(\lambda_2 x))\, \nu_{\lambda_1} = (1 - j_\alpha(\lambda_1 x))\, \nu_{\lambda_2},$$

then we conclude that the measure defined by

(6.II.7) $$\frac{1}{1 - j_\alpha(\lambda_1 x)}\, d\tau'_{\lambda_1}(x),$$

does not depend on λ_1 (we recall that $(1 - j_\alpha(\lambda_1 x))$ does not vanish for $\lambda_1 > 0$). We denote it by $d\tau(x)$.

Hence from the relation (6.II.6) the formula (6.II.4) can be written

(6.II.8) $$\forall \lambda_2 \geq 0,\ T_{\lambda_1}(\Psi)(\lambda_2) - \Psi(\lambda_2) = Q(\lambda_1) + \int_0^\infty j_\alpha(\lambda_2 x)\,(1 - j_\alpha(\lambda_1 x))\, d\tau(x).$$

Lemma 6.II.3. The measure τ defined by the relation (6.II.7) satisfies

(6.II.9) $$\int_0^\infty \frac{x^2}{1+x^2}\, d\tau(x) < +\infty.$$

Proof.

Let $\lambda_1 > 0$ be fixed. In the complement of any neighborhood of zero the measure τ defined by (6.II.7) is bounded. We deduce the result from the fact that the function

$$\frac{x^2}{(1+x^2)(1-j_\alpha(\lambda_1 x))},$$

is bounded on $[0, +\infty[$, and the property of the measure τ'_{λ_1}.

Lemma 6.II.4. Put

$$\forall\, \lambda \geq 0,\ \Psi_1(\lambda) = \int_0^\infty (1 - j_\alpha(\lambda x))\, d\tau(x).$$

Then we have

$$\forall\, \lambda \geq 0,\ \Psi(\lambda) - \Psi_1(\lambda) = c\,\lambda^2,$$

where c is a nonnegative constant.

Proof.

It is clear that the function Ψ_1 is continuous on $[0, +\infty[$, then the function

$$Q = \Psi - \Psi_1,$$

is continuous on $[0, +\infty[$.

On the other hand for every $\lambda_1 > 0$ and $\lambda_2 > 0$, we have

$$(\delta_{\lambda_2} - \delta) *_B [(\delta_{\lambda_1} - \delta) *_B \Psi_1] = (\delta_{\lambda_2} - \delta) *_B [(\delta_{\lambda_1} - \delta) *_B \Psi].$$

Then the function Q satisfies

$$(\delta_{\lambda_2} - \delta) *_B [(\delta_{\lambda_1} - \delta) *_B Q](\lambda) = 0, \text{ for all } \lambda \geq 0.$$

We put $\lambda = 0$. Then we obtain

$$T_{\lambda_1}(Q)(\lambda_2) = Q(\lambda_1) + Q(\lambda_2), \text{ for all } \lambda_1 > 0,\ \lambda_2 > 0.$$

The continuity of the function Q on $[0, +\infty[$ and Proposition 4.II.3 imply that this relation is also true for $\lambda_1 = 0$ and $\lambda_2 = 0$.

From Theorem 6.I.1 there exists a constant c such that

$$\forall\, \lambda \geq 0,\ Q(\lambda) = c\lambda^2.$$

As

$$\forall \lambda \geq 0, Q(\lambda) \geq 0,$$

then c is nonnegative.

Hence the proof of Theorem 6.II.1 is completed.

6.III. Convolution semigroups and infinitely divisible probabilities associated with the Bessel operator.

6.III.1. Convolution semigroups.

Definition 6.III.1. A net $\{\nu_t\}_{t\geq 0}$ of measure in $M_+^b([0, +\infty[)$ is called a convolution semigroup associated with the Bessel operator if it satisfies

i) $\nu_t *_B \nu_s = \nu_{t+s}$, $t \geq 0, s \geq 0$.

ii) $\nu_0 = \delta$.

iv) $t \to \nu_t$ is vaguely continuous (i.e that ν_t converges vaguely to ν_0 as $t \to 0$).

Theorem 6.III.1. The net $\{\nu_t\}_{t\geq 0}$ is a convolution semigroup associated with the Bessel operator if and only if

(6.III.1) $$\forall \lambda \geq 0, \mathcal{F}_B(\nu_t)(\lambda) = \exp(-t\Psi(\lambda)),$$

where

(6.III.2) $$\Psi(\lambda) = c\lambda^2 + d + \int_0^\infty (1 - j_\alpha(\lambda x))\, d\tau(x),$$

with c, d nonnegative constants and τ a measure in $M_+^b([0, +\infty[)$ satisfying

(6.III.3) $$\int_0^\infty \frac{x^2}{1+x^2}\, d\tau(x) < +\infty.$$

Proof.

a) If $\{\nu_t\}_{t\geq 0}$ is a convolution semigroup associated with the Bessel operator, then for fixed $\lambda \geq 0$ the function $t \to \mathcal{F}_B(\nu_t)(\lambda)$ satisfies the following properties

- $\mathcal{F}_B(\nu_{t+s})(\lambda) = \mathcal{F}_B(\nu_t)(\lambda).\mathcal{F}_B(\nu_s)(\lambda)$, for all $t \geq 0, s \geq 0$,
- $|\mathcal{F}_B(\nu_t)(\lambda)| \leq \|\nu_t\|$, for all $t \geq 0$,
- $\mathcal{F}_B(\nu_0)(\lambda) = 1$,
- $\mathcal{F}_B(\nu_t)(\lambda)$ is continuous on $[0, +\infty[$.

Then there exists a continuous function Ψ on $[0, +\infty[$ such that

$$\Psi(0) \geq 0$$

and

$$\mathfrak{F}_B(\nu_t)(\lambda) = \exp(-\, t\,\Psi(\lambda)).$$

Since $\mathfrak{F}_B(\nu_t)$ is a continuous bounded positive definite function associated with the Bessel operator, then Ψ is a negative definite function for the Bessel operator and the Levy-Khintchine formula (Theorem 6.II.1) gives the result.

b) The converse is immediate.

6.III.2. Infinitely divisible probabilities associated with the Bessel operator.

Definition 6.III.2. A measure μ in $M^1([0, +\infty[)$ is said to be infinitely divisible if for every integer $n \geq 1$, there exists ν_n in $M^1([0, +\infty[)$ such that

$$\mu = \nu_n *_B \nu_n *_B \dots *_B \nu_n \text{ , n times,} \tag{6.III.4}$$
$$= \nu_n^{*_B n}.$$

Proposition 6.III.1. Let μ in $M^1([0, +\infty[)$ be infinitely divisible. Then we have

i) $\forall\, \lambda \geq 0,\ \mathfrak{F}_B(\mu)(\lambda) > 0,$

ii) $\forall\, \lambda \geq 0,\ \lim\limits_{n \to +\infty} \mathfrak{F}_B(\nu_n)(\lambda) = 1.$

Proof.

i) As for $\lambda \geq 0$, the functions $\mathfrak{F}_B(\mu)(\lambda)$ and $\mathfrak{F}_B(\nu_n)(\lambda)$ are real and

$$\forall\, \lambda \geq 0,\ |\mathfrak{F}_B(\mu)(\lambda)| \leq 1,$$

then

$$\forall\, \lambda \geq 0,\ (\mathfrak{F}_B(\nu_n)(\lambda))^2 = [(\mathfrak{F}_B(\mu)(\lambda))^2]^{1/n} \xrightarrow[n \to +\infty]{} f(\lambda),$$

with

$$f(\lambda) = \begin{cases} 0, & \text{if } \mathfrak{F}_B(\mu)(\lambda) = 0, \\ 1, & \text{if } \mathfrak{F}_B(\mu)(\lambda) \neq 0. \end{cases}$$

But

$$\mathfrak{F}_B(\mu)(0) = 1.$$

Then

$$f(0) = 1.$$

On the other hand the function $\mathfrak{F}_B(\mu)(\lambda)$ is continuous on $[0, +\infty[$. Then

$$\mathfrak{F}_B(\mu)(0) \neq 0.$$

in a neighborhood V of zero.

Hence

$$f(\lambda) = 1, \text{ for all } \lambda \in V.$$

We deduce that the function f is continuous at the point zero.

As

$$\forall \lambda \geq 0, \ (\mathfrak{F}_B(\nu_n)(\lambda))^2 = \mathfrak{F}_B(\nu_n^{*2}_B)(\lambda),$$

then from Theorem 6.I.3 there exists a measure θ in $M_+^b([0, +\infty[)$ such that

$$\forall \lambda \geq 0, \ \mathfrak{F}_B(\theta)(\lambda) = f(\lambda).$$

Then f is continuous on $[0, +\infty[$ and hence

$$\forall \lambda \geq 0, \quad f(\lambda) = 1.$$

Thus

$$\forall \lambda \geq 0, \ \mathfrak{F}_B(\mu)(\lambda) > 0.$$

ii) We deduce the result from the fact that

$$\forall \lambda \geq 0, \ \mathfrak{F}_B(\nu_n)(\lambda) = \exp(\frac{1}{n} \operatorname{Log} \mathfrak{F}_B(\mu)(\lambda)).$$

Proposition 6.III.2. Let μ be in $M^1([0, +\infty[)$ infinitely divisible and $t \geq 0$. Then there exists μ_t in $M^1([0, +\infty[)$ such that

$$\forall \lambda \geq 0, \ (\mathfrak{F}_B(\mu)(\lambda))^t = \mathfrak{F}_B(\mu_t)(\lambda). \tag{6.III.5}$$

Proof.

Let μ be in $M^1([0, +\infty[)$ infinitely divisible. From Definition 6.III.2, for every integer $n \geq 1$, there exists ν_n in $M^1([0, +\infty[)$ such that

$$\mu = \nu_n^{*n}_B.$$

We put

$$\forall \lambda \geq 0, \ a_n(\lambda) = 1 - \mathfrak{F}_B(\nu_n)(\lambda) = \int_0^\infty (1 - j_\alpha(\lambda t)) \, d\nu_n(t).$$

From Proposition 6.III.1, for all $\lambda \geq 0$, the function $\mathrm{Log}\, \mathfrak{F}_B(\mu)(\lambda)$ is well defined.

Moreover we have

$$\forall \lambda \geq 0, \quad \lim_{n\to+\infty} a_n(\lambda) = 0.$$

Thus

$$\begin{aligned}\forall \lambda \geq 0, \mathrm{Log}\, \mathfrak{F}_B(\mu)(\lambda) &= n\, \mathrm{Log}\, \mathfrak{F}_B(\nu_n)(\lambda), \\ &= -n \sum_{k=1}^{\infty} \frac{(a_n(\lambda))^k}{k}.\end{aligned}$$

We have

$$\forall \lambda \geq 0, \mathrm{Log}\, \mathfrak{F}_B(\mu)(\lambda) = -n\, a_n(\lambda) - n \sum_{k=2}^{\infty} \frac{(a_n(\lambda))^k}{k}.$$

As the $a_n(\lambda)$, $n \in \mathbb{N}$, are positive, then

$$\forall \lambda \geq 0, \ 0 \leq n \sum_{k=2}^{\infty} \frac{(a_n(\lambda))^k}{k} \leq a_n(\lambda)\, (-\mathrm{Log}\, \mathfrak{F}_B(\mu)(\lambda)) \xrightarrow[n\to+\infty]{} 0\,.$$

By putting

$$\theta_n = n\, \nu_n,$$

we obtain

$$\forall \lambda \geq 0, t\, \mathrm{Log}\, \mathfrak{F}_B(\mu)(\lambda) = \lim_{n\to+\infty} \{-t \int_0^{\infty} (1 - j_\alpha(\lambda t))d\theta_n(t)\}.$$

Thus

$$\forall \lambda \geq 0, (\mathfrak{F}_B(\mu)(\lambda))^t = \lim_{n\to+\infty} \exp[-t\, (\mathfrak{F}_B(\theta_n)(\lambda) - \mathfrak{F}_B(\theta_n)(0))].$$

From Proposition 6.I.3 there exists a measure $\gamma_{t,n}$ in $M^1([0, +\infty[)$ such that

$$\forall \lambda \geq 0, \mathfrak{F}_B(\gamma_{t,n})(\lambda) = \exp\, [-t(\mathfrak{F}_B(\theta_n)(\lambda) - \mathfrak{F}_B(\theta_n)(0))].$$

Thus

$$\forall \lambda \geq 0, (\mathfrak{F}_B(\mu)(\lambda))^t = \lim_{n\to+\infty} \mathfrak{F}_B(\gamma_{t,n})(\lambda).$$

Hence, from Theorem 6.I.3 there exists a measure μ_t in $M^1([0, +\infty[)$ such that

$$\forall \lambda \geq 0, (\mathfrak{F}_B(\mu)(\lambda))^t = \mathfrak{F}_B(\mu_t)(\lambda).$$

Theorem 6.III.1. A measure μ in $M^1([0, +\infty[)$ is infinitely divisible if and only if

(6.III.6) $$\forall \lambda \geq 0, \mathfrak{F}_B(\mu)(\lambda) = \exp(-\Psi(\lambda)).$$

Where Ψ is a negative definite function for the Bessel operator defined by the formula (6.II.2) with

(6.III.7) $$\Psi(0) = 0.$$

Proof.

a) Let μ be in $M^1([0, +\infty[)$ infinitely divisible, the net $\{\mu_t\}_{t\geq 0}$ defined by

$$\forall \lambda \geq 0, \mathcal{F}_B(\mu_t)(\lambda) = (\mathcal{F}_B(\mu)(\lambda))^t,$$

is a convolution semigroup associated with the Bessel operator. Then from Theorem 6.III.1 we have

$$\forall \lambda \geq 0, \mathcal{F}_B(\mu)(\lambda) = \exp(-\Psi(\lambda)).$$

with Ψ defined by the formula (6.II.2).

From the relation

$$\mathcal{F}_B(\mu)(0) = 0,$$

we deduce that

$$\Psi(0) = 0.$$

b) Conversely, let μ be in $M^1([0, +\infty[)$ satisfying

$$\forall \lambda \geq 0, \mathcal{F}_B(\mu)(\lambda) = \exp(-\Psi(\lambda)).$$

Let n be in $\mathbb{N}^*$. From Theorem 6.III.1 there exists a measure ν_n in $M_+^b([0, +\infty[)$ such that

$$\forall \lambda \geq 0, \mathcal{F}_B(\nu_n)(\lambda) = \exp(-\frac{1}{n}\Psi(\lambda).$$

As

$$\Psi(0) = 0.$$

Then the measure ν_n belongs to $M^1([0, +\infty[)$.

We have

$$\forall \lambda \geq 0, \mathcal{F}_B(\mu)(\lambda) = (\mathcal{F}_B(\nu_n)(\lambda))^n = \mathcal{F}_B(\nu_n^{*n_B})(\lambda).$$

From Proposition 5.VI.3 we deduce that

$$\mu = \nu_n^{*n_B}.$$

6.IV. Central limit theorem associated with the Bessel operator.

Definition 6.IV.1. For $t \geq 0$, the measure μ_t in $M^1([0, +\infty[)$ given by

$$\mu_t = \begin{cases} k_t \, \mu_\alpha, & \text{if } t > 0, \\ \delta, & \text{if } t = 0. \end{cases} \tag{6.IV.1}$$

where

$$\forall\, x \geq 0, \quad k_t(x) = \frac{e^{-\frac{x^2}{4t}}}{(2t)^{\alpha+1}}, \tag{6.IV.2}$$

is called Gaussian (or normal) distribution associated with the Bessel operator.

Proposition 6.IV.1. The dispersion of the normal distribution μ_t, $t \geq 0$, associated with the Bessel operator, satisfies

$$D(\mu_t) = t. \tag{6.IV.3}$$

Proof.

From the relation (5.II.2) we have

$$\forall\, \lambda \geq 0, \quad \mathcal{F}_B(\mu_t)(\lambda) = e^{-t\lambda^2}.$$

We deduce the result from the relation (6.I.9).

Lemma 6.IV.1. We have

$$\forall\, x \geq 0, \quad \lambda \geq 0, \quad j_\alpha(\lambda x) = 1 - \lambda^2 q(x) + K(\lambda, x), \tag{6.IV.4}$$

with

$$q(x) = \frac{x^2}{4(\alpha+1)}, \tag{6.IV.5}$$

and K a continuous function on $[0, +\infty[\times [0, +\infty[$ satisfying

$$\forall\, x \geq 0, \quad \lambda \geq 0, \quad |K(\lambda, x)| \leq C(\lambda)\, x^2, \tag{6.IV.6}$$

with

$$C(\lambda) = \frac{\lambda^2}{2(\alpha+1)}. \tag{6.IV.7}$$

Proof.

Using Lemma 6.I.1 we have

$$\forall\, x \geq 0, \quad \lambda \geq 0, \; j_\alpha(\lambda x) = 1 - \lambda^2 \int_0^x v(x,t)\, j_\alpha(\lambda t) t^{2\alpha+1} dt, \tag{6.IV.8}$$

with

$$v(x, t) = \int_t^x \frac{dr}{r^{2\alpha+1}}.$$

The relation (6.IV.8) can also be written in the form

$$\forall\, x \geq 0,\quad \lambda \geq 0, j_\alpha(\lambda x) = 1 - \lambda^2 \int_0^x v(x,t)\, t^{2\alpha+1} dt,$$
$$+ \lambda^2 \int_0^x v(x,t)\, (1 - j_\alpha(\lambda t)) t^{2\alpha+1} dt.$$

But

$$\int_0^x v(x,t)\, t^{2\alpha+1} dt = q(x).$$

Thus

$$\forall\, x \geq 0,\quad \lambda \geq 0, j_\alpha(\lambda x) = 1 - \lambda^2 q(x) + K(\lambda, x)$$

with

$$K(\lambda,x) = \lambda^2 \int_0^x v(x,t)\, (1-j_\alpha(\lambda t))\, t^{2\alpha+1} dt.$$

Using the fact that

$$\forall\, t \geq 0,\quad \lambda \geq 0, |j_\alpha(\lambda t)| \leq 1,$$

we obtain

$$\forall\, x \geq 0,\quad \lambda \geq 0,\ |K(\lambda,x)| \leq 2\lambda^2 q(x).$$

Thus

$$\forall\, x \geq 0,\quad \lambda \geq 0,\ |K(\lambda,x)| \leq \frac{\lambda^2}{2(\alpha+1)} x^2.$$

We put

$$C(\lambda) = \frac{\lambda^2}{2(\alpha+1)}.$$

Notation. We denote by $M_0^1([0, +\infty[)$ the set of probability measures μ such that

$$\int_0^\infty x^2\, d\mu(x) < +\infty.$$

Let $\mu_{n,j}$, $1 \leq j \leq k_n$, $n \in \mathbb{N}^*$, $k_n \in \mathbb{N}$, be a sequence of measures in $M_0^1([0, +\infty[)$.

We put

$$\mu_n = \mu_{n,1} *_B \mu_{n,2} *_B \cdots *_B \mu_{n,k_n}. \tag{6.IV.9}$$

Theorem 6.IV.1. (Central limit theorem). We make the following assumptions

(6.IV.10) i) $\lim_{n\to+\infty} \sum_{j=1}^{k_n} \int_0^\infty x^2\, d\mu_{n,j}(x) = 0,$

(6.IV.11) ii) $\lim_{n\to+\infty} \sup_{1\le j\le k_n} D(\mu_{n,j}) = 0,$

(6.IV.12) iii) $\lim_{n\to+\infty} D(\mu_n) = t \ge 0.$

Then the sequence $\{\mu_n\}_{n\in\mathbb{N}}$ converges weakly to the Gaussian distribution μ_t .

Proof.

From Lemma 6.IV.1 we have for all $\lambda \ge 0$:

(6.IV.13) $$1 - \mathfrak{F}_B(\mu_{n,j})(\lambda) = \lambda^2 D(\mu_{n,j}) + \int_0^\infty K(\lambda,x)\,d\mu_{n,j}(x).$$

Then

(6.IV.14) $$|1 - \mathfrak{F}_B(\mu_{n,j})(\lambda)| \le \lambda^2 D(\mu_{n,j}) + C(\lambda) \int_0^\infty x^2\, d\mu_{n,j}(x),$$

$$|1 - \mathfrak{F}_B(\mu_{n,j})(\lambda)| \le \lambda^2 \sup_{1\le j\le k_n} D(\mu_{n,j}) + C(\lambda) \sum_{j=1}^{k_n} \int_0^\infty x^2\, d\mu_{n,j}(x).$$

Hence

$$\sup_{1\le j\le k_n} |1 - \mathfrak{F}_B(\mu_{n,j})(\lambda)| \le \lambda^2 \sup_{1\le j\le k_n} D(\mu_{n,j}) + C(\lambda) \sum_{j=1}^{k_n} \int_0^\infty x^2\, d\mu_{n,j}(x).$$

From the relations (6.IV.10) and (6.IV.11) we deduce that

(6.IV.15) $$\lim_{n\to+\infty} \sup_{1\le j\le k_n} |1 - \mathfrak{F}_B(\mu_{n,j})(\lambda)| = 0.$$

This relation implies for n large enough

$$\mathfrak{F}_B(\mu_{n,j})(\lambda) \ne 0, \text{ for all } j \text{ such that } 1 \le j \le k_n .$$

Thus

$$\mathfrak{F}_B(\mu_n)(\lambda) = \prod_{j=1}^{k_n} \mathfrak{F}_B(\mu_{n,j})(\lambda) \ne 0.$$

For such n, the function $\mathrm{Log}\mathfrak{F}_B(\mu_n)(\lambda)$ is well defined and we have

(6.IV.16) $$\mathrm{Log}\mathfrak{F}_B(\mu_n)(\lambda) = \sum_{j=1}^{k_n} \mathrm{Log}\mathfrak{F}_B(\mu_{n,j})(\lambda) = \sum_{j=1}^{k_n} \mathrm{Log}(1 - \beta_{n,j}(\lambda)),$$

$$= - \sum_{j=1}^{k_n} \sum_{m=1}^{\infty} \frac{(\beta_{n,j}(\lambda))^m}{m},$$

where

$$\beta_{n,j}(\lambda) = 1 - \mathfrak{F}_B(\mu_{n,j})(\lambda).$$

From the relation (6.IV.13) we have

$$\sum_{j=1}^{k_n} \beta_{n,j}(\lambda) = \lambda^2 D(\mu_n) + \sum_{j=1}^{k_n} \int_0^\infty K(\lambda,x)\, d\mu_{n,j}(x).$$

Then from the relations (6.IV.10) and (6.IV.12) we have

$$\lim_{n\to+\infty} \sum_{j=1}^{k_n} \beta_{n,j}(\lambda) = \lambda^2 t. \tag{6.IV.17}$$

On the other hand we have

$$0 \le \sum_{j=1}^{k_n} \sum_{m=2}^{\infty} \frac{(\beta_{n,j}(\lambda))^m}{m} \le \frac{1}{2} \sum_{j=1}^{k_n} \sum_{m=2}^{\infty} (\beta_{n,j}(\lambda))^m .$$

We have the inequality

$$\sum_{j=1}^{k_n} (\beta_{n,j}(\lambda))^2 \le \left(\sup_{1\le j\le k_n} \beta_{n,j}(\lambda)\right) \sum_{j=1}^{k_n} \beta_{n,j}(\lambda).$$

Thus from the relations (6.IV.15) and (6.IV.17) we deduce that

$$\lim_{n\to+\infty} \sum_{j=1}^{k_n} (\beta_{n,j}(\lambda))^2 = 0.$$

Hence

$$\lim_{n\to+\infty} \sum_{j=1}^{k_n} (\beta_{n,j}(\lambda))^m = 0, \text{ for all } m \ge 2.$$

Then

$$\lim_{n\to+\infty} \sum_{j=1}^{k_n} \sum_{m=2}^{\infty} \frac{(\beta_{n,j}(\lambda))^m}{m} = 0. \tag{6.IV.18}$$

Using the relation (6.IV.16) we obtain

$$\mathfrak{F}_B(\mu_n)(\lambda) = \exp\Big(- \sum_{j=1}^{k_n} \beta_{n,j}(\lambda)\Big) \exp\Big(- \sum_{j=1}^{k_n} \sum_{m=2}^{\infty} \frac{(\beta_{n,j}(\lambda))^m}{m}\Big).$$

Then from the relations (6.IV.17) and (6.IV.18) we have

$$\lim_{n\to+\infty} \mathfrak{F}_B(\mu_n)(\lambda) = \exp(-t\lambda^2).$$

From Theorem 6.I.3, Proposition 5.VI.3, Definition 6.IV.1 and the relation (5.II.2) we deduce that the sequence $\{\mu_n\}_{n\in\mathbb{N}}$ converges weakly to the Gaussian distribution μ_t.

7 Continuous Wavelet Transform Associated With The Bessel Operator

Introduction

The classical continuous wavelet transform for functions on $\mathbb{R}$ is an integral transform for which the kernel is the dilated translate of a so-called wavelet g, a quite arbitrary square integrable function on $\mathbb{R}$. A Plancherel formula for this transform is obtained, and in the course of its derivation we naturally arrive at an admissibility assumption for the wavelet g. There follow a Parseval and a pointwise reconstruction formula.

In this chapter, using the paper [52] we give whithout proofs the main results on the classical wavelets on $[0, +\infty[$, and on the classical wavelet transform on $[0, \infty[$.

Next, we define wavelets associated with the Bessel operator, and using the generalized translation operators T_x, $x \geq 0$, defined by the relation (4.II.3) and the dilation $g \to g_a$ given by

$$g_a(x) = \frac{1}{a^{2\alpha+2}} g(\frac{x}{a}), a > 0, x \geq 0,$$

we define a continuous wavelet transform associated with the Bessel operator. We prove for this transform a Plancherel formula, a Calderon's reproducing formula and a pointwise reconstruction formula, and we discuss it in the framework of coherent states. Also a generalization of the Calderon's reproducing formula associated with the Bessel operator, to the measures has been given. (See Problems 7.V.4, 7.V.5, 7.V.6, 7.V.7 and [65]).

Using the precedent results and the Fourier-Bessel transform studied in Chapter 5, we establish for the Riemann-Liouville and Weyl integral transforms $\mathcal{R}_\alpha$ and W_α the following results.

- We define and characterize other spaces of functions than $\mathcal{E}_*(\mathbb{R})$ and $\mathcal{D}_*(\mathbb{R})$ on which the transforms $\mathcal{R}_\alpha$ and W_α are bijective.

- We give the following inversion formula for $\mathcal{R}_\alpha$ and W_α when applied to some spaces of functions

$$\mathcal{R}_\alpha K_0 W_\alpha(f) = 2^\alpha \Gamma(\alpha+1)f,$$

$$\mathcal{R}_\alpha W_\alpha K_B(f) = 2^\alpha \Gamma(\alpha+1)f,$$

$$W_\alpha \mathcal{R}_\alpha K_0(f) = 2^\alpha \Gamma(\alpha+1)f,$$

$$W_\alpha K_B \mathcal{R}_\alpha(f) = 2^\alpha \Gamma(\alpha+1)f,$$

where K_0 and K_B are integro-differential operators.

- Using the transforms $\mathcal{R}_\alpha$ and W_α we give relations between the continuous wavelet transform associated with the Bessel operator and the classical continuous wavelet transform on $[0, +\infty[$, and we deduce other formulas which give the inverse operators $\mathcal{R}_\alpha^{-1}$ and W_α^{-1}. These formulas are better than those obtained in Chapter 2 because they are simple and we have a large choice of wavelets associated with the Bessel operator , that we can use in these formulas.

We use these results to determine the inverse of the Radon transform on $\mathbb{R}^n$, $n \geq 2$. (See Problems 7.V.1, 7.V.2, 7.V.3 and [66]).

Wavelets and the continuous wavelet transform associated with a singular differential operator more general than the Bessel operator are studied in [35] [87] [88] with some applications.

7.I. Classical continuous wavelet transform on $[0, +\infty[$.

7.I.1. Classical wavelets on $[0, +\infty[$.

Definition 7.I.1. A classical wavelet on $[0, +\infty[$, is a measurable function g on $[0, +\infty[$ satisfying the following admissibility condition

(7.I.1) $$0 < C_g^0 = \int_0^\infty |\mathfrak{F}_0(g)(\lambda)|^2 \frac{d\lambda}{\lambda} < +\infty.$$

Where $\mathfrak{F}_0$ is the Fourier-cosine transform studied in chapter 3.

Proposition 7.I.1. Let g be a nonzero function in $L^2([0, +\infty[, dx)$ verifying

i) $\mathfrak{F}_0(g)$ is right continuous at zero.

ii) $\exists\, \gamma > 0$ such that $\mathfrak{F}_0(g)(\lambda) - \mathfrak{F}_0(g)(0) = O(\lambda^\gamma)$, $\lambda \to 0^+$.

Then the condition (7.I.1) is equivalent to

(7.I.2) $$\mathfrak{F}_0(g)(0) = 0.$$

Remark.

The continuity hypothesis in Proposition 7.I.1 holds if g is moreover in $L^1([0, +\infty[, dx)$.

Corollary 7.I.1. Let g be a nonzero function in $\mathcal{S}_*(\mathbb{R})$ (resp. $\mathcal{D}_*(\mathbb{R})$). Then $D^2 g$ is a classical wavelet on $[0, +\infty[$, in $\mathcal{S}_*(\mathbb{R})$ (resp. $\mathcal{D}_*(\mathbb{R})$) and we have

(7.I.3) $$C_{D_g^2}^0 = \int_0^\infty \lambda^3 |\mathfrak{F}_0(g)(\lambda)|^2 d\lambda,$$

where

$$D = \frac{d}{dx}.$$

Example.

Let $t \in]0, +\infty[$. The function $x \to e^{-tx^2}$ belongs to $\mathcal{S}_*(\mathbb{R})$ and we have

(7.I.4) $$\forall\, \lambda \geq 0, \mathfrak{F}_0(e^{-tx^2})(\lambda) = \frac{\sqrt{\pi}}{2\sqrt{t}} e^{-\lambda^2/4t}.$$

Using Corollary 7.I.1 we deduce that the function

(7.I.5) $$g(x) = -\frac{1}{2t} D_x^2 e^{-tx^2} = (1-2tx^2)\, e^{-tx^2},$$

is a classical wavelet on $[0, +\infty[$ in $\mathcal{S}_*(\mathbb{R})$ and we have

(7.I.6) $$C_g^0 = \frac{\pi t}{2}.$$

Definition 7.I.2. Let $a \in]0, +\infty[$. The dilation operator H_a is defined for a measurable function f on $[0, +\infty[$ by

(7.I.7) $$\forall\, x \geq 0,\ H_a(f)(x) = \frac{1}{\sqrt{a}} f(\frac{x}{a}).$$

The properties of this operator are given by the following propositions.

Proposition 7.I.2.

i) We have

(7.I.8) $$H_1 = \mathrm{id}.$$

(7.I.9) $$H_a \circ H_b = H_{ab},\ a, b \in]0, +\infty[.$$

(7.I.10) $$H_a^{-1} = H_{a^{-1}},\ a \in]0, +\infty[.$$

ii) For all $a \in]0, +\infty[$, the operator H_a is linear and isometric from $L^2([0, +\infty[, dx)$ onto itself.

iii) For all $a \in]0, +\infty[$, the operator H_a is a topological automorphism of $\mathcal{S}_*(\mathbb{R})$ (resp. $\mathcal{D}_*(\mathbb{R})$).

Proposition 7.I.3. For all $a \in]0, +\infty[$ and f in $L^2([0, +\infty[, dx)$ we have

(7.I.11) $$\mathfrak{F}_0 \circ H_a(f) = H_{a^{-1}} \circ \mathfrak{F}_0(f).$$

Using a classical wavelet on $[0, +\infty[$, the dilation operators H_a, $a > 0$, and the translation operators σ_b, $b \geq 0$, we can construct a new family of classical wavelets on $[0, +\infty[$.

Definition 7.I.3. Let g be a classical wavelet on $[0, +\infty[$ in $L^2([0, +\infty[, dx)$. Then for all $a \in]0, +\infty[$ and $b \in [0, +\infty[$, the function

(7.I.12) $$g^0_{a,b} = \sigma_b \circ H_a(g),$$

is a classical wavelet on $[0, +\infty[$ in $L^2([0, +\infty[, dx)$ and we have

(7.I.13) $$C^0_{g^0_{a,b}} = a \int_0^\infty \cos^2 \frac{\lambda b}{a} |\mathfrak{F}_0(g)(\lambda)|^2 \frac{d\lambda}{\lambda}.$$

Proposition 7.I.4. Let g be a classical wavelet on $[0, +\infty[$ in $L^2([0, +\infty[, dx)$. Then the mapping $(a, b) \rightarrow g^0_{a,b}$ is continuous from $]0, +\infty[\times [0, +\infty[$ into $L^2([0, +\infty[, dx)$.

7.I.2. Classical continuous wavelet transform on $[0, +\infty[$.

In this subsection we consider a classical wavelet g on $[0, +\infty[$ in $L^2([0, +\infty[, dx)$.

Definition 7.I.4. The classical continuous wavelet transform S_g on $[0, +\infty[$ is defined for regular functions f on $[0, +\infty[$ by

(7.I.14) $$S_g(f)(a,b) = \int_0^\infty f(x)\overline{g^0_{a,b}(x)}\, dx, \quad a > 0, \quad b \geq 0,$$

where $g^0_{a,b}$ is the function defined by the relation (7.I.14).

We can write the relation (6.I.16) in the form

(7.I.15) $$S_g(f)(a,b) = f *_0 \overline{H_a(g)(b)}.$$

Where $*_0$ is the convolution product defined by the relation (3.I.12).

Some properties of the transform S_g are given in the following propositions.

Proposition 7.I.5.

i) For all $a \in]0, +\infty[$ and f in $L^q([0, +\infty[, dx)$, $q \in [1,2]$, the function $b \to S_g(f)(a,b)$ is defined almost everywhere on $[0, +\infty[$, belongs to $L^r([0, +\infty[, dx)$, with $r \in [1, +\infty]$ such that $\frac{1}{r} = \frac{1}{q} - \frac{1}{2}$ and we have

(7.I.16) $$\|S_g(f)(a,.)\|_r \leq \|f\|_q \|g\|_2 .$$

ii) For all $a \in]0, +\infty[$ and f in $L^2([0, +\infty[, dx)$, the function $b \to S_g(f)(a,b)$ is continuous on $[0, +\infty[$, tends to zero at the infinity and we have

(7.I.17) $$\sup_{b\geq 0} |S_g(f)(a,b)| \leq \|f\|_2 \|g\|_2 .$$

iii) If g is in $\mathcal{S}_*(\mathbb{R})$ (resp. $\mathcal{D}_*(\mathbb{R})$), then for all $a \in]0, +\infty[$ and f in $\mathcal{S}_*(\mathbb{R})$ (resp. $\mathcal{D}_*(\mathbb{R})$), the function $b \to S_g(f)(a,b)$ belongs to $\mathcal{S}_*(\mathbb{R})$ (resp. $\mathcal{D}_*(\mathbb{R})$).

Theorem 7.I.1.

i) Plancherel formula for S_g.

For all f in $L^2([0, +\infty[, dx)$ we have

(7.I.18) $$\int_0^\infty |f(x)|^2\, dx = \frac{1}{C^0_g}\int_0^\infty \int_0^\infty |S_g(f)(a,b)|^2 \frac{da}{a^2}\, db.$$

ii) Parseval formula for S_g.

For all f and g in $L^2([0,+\infty[, dx)$ we have

$$\int_0^\infty f_1(x)\,\overline{f_2(x)}\,dx = \frac{1}{C_g^0}\int_0^\infty\int_0^\infty S_g(f_1)(x)\,\overline{S_g(f_2)(x)}\,\frac{da}{a^2}\,db. \tag{7.I.19}$$

The following result is the so-called Calderon's reproducing formula.

Theorem 7.I.2. We suppose that the classical wavelet g on $[0,+\infty[$, is such that the function $\mathfrak{F}_0(g)$ belongs to $L^\infty([0,+\infty[, dx)$. Then for f in $L^2([0,+\infty[, dx)$ and $0<\varepsilon<\delta<+\infty$.

The function

$$f_{\varepsilon,\delta}(x) = \frac{1}{C_g^0}\int_\varepsilon^\delta\int_0^\infty S_g(f)(a,b)\; g_{a,b}^0(x)\,db\,\frac{da}{a^2}, \tag{7.I.20}$$

belongs to $L^2([0,+\infty[, dx)$ and satisfies

$$\lim_{\substack{\varepsilon\to 0^+\\ \delta\to+\infty}} \|f_{\varepsilon,\delta} - f\|_2 = 0. \tag{7.I.21}$$

Next we formulate a pointwise reconstruction formula for the transform S_g.

Theorem 7.I.3. For f in $L^1([0,+\infty[, dx)$ such that $\mathfrak{F}_0(f)$ belongs to $L^1([0,+\infty[, \frac{2}{\pi}\,d\lambda)$, we have

$$f(x) = \frac{1}{C_g^0}\int_0^\infty \left(\int_0^\infty S_g(f)(a,b)\; g_{a,b}^0(x)\,db\right)\frac{da}{a^2},\quad \text{a.e. on } [0,+\infty[, \tag{7.I.22}$$

where, for each $x \ge 0$, both the inner integral and the outer integral are absolutely convergent, but possible not the double integral.

The following theorem gives a characterization of the image of S_g.

Theorem 7.I.4. Let F be in $L^2(]0,+\infty[\times[0,+\infty[, db\,\frac{da}{a^2C_g^0})$ (the space of square integrable functions on $]0,+\infty[\times[0,+\infty[$ with respect to the measure $db\,\frac{da}{a^2C_g^0}$).

Then F belongs to the image of S_g if and only if

$$F(a,b) = \frac{1}{C_g^0}\int_0^\infty\int_0^\infty F(a',b')\left(\int_0^\infty g_{a',b'}^0(x)\,\overline{g_{a,b}^0(x)}\,dx\right)db'\,\frac{da'}{(a')^2}. \tag{7.I.23}$$

7.II. Continuous wavelet transform associated with the Bessel operator.

7.II.1. Wavelets associated with the Bessel opeerator.

Definition 7.II.1. A wavelet associated with the Bessel operator is a measurable function g on $[0, +\infty[$ satisfying the following admissibility condition

(7.II.1)
$$0 < C_g = \int_0^\infty |\mathfrak{F}_B(g)(\lambda)|^2 \frac{d\lambda}{\lambda} < +\infty.$$

Proposition 7.II.1. Let g be a nonzero function in $L^2([0, +\infty[, d\mu_\alpha)$ satisfying

i) $\mathfrak{F}_B(g)$ is right continuous at zero.

ii) $\exists\, \gamma > 0$ such that $\mathfrak{F}_B(g)(\lambda) - \mathfrak{F}_B(g)(0) = O(\lambda^\gamma),\ \lambda \to 0^+$.

Then the condition (7.II.1) is equivalent to

(7.II.2)
$$\mathfrak{F}_B(g)(0) = \int_0^\infty g(x)d\mu_\alpha(x) = 0.$$

Proof.

a) We suppose that the condition (7.II.1) is true. If $\mathfrak{F}_B(g)(0) \neq 0$ then from the i) there exists $\eta > 0$ and $\varepsilon > 0$ such that for all $\lambda \in [0, \eta]$ we have

$$|\mathfrak{F}_B(g)(\lambda)| \geq \varepsilon.$$

Then

$$\int_0^\infty |\mathfrak{F}_B(g)(\lambda)|^2 \frac{d\lambda}{\lambda} \geq \varepsilon^2 \int_0^\eta \frac{d\lambda}{\lambda} = +\infty.$$

b) Conversely, we suppose that $\mathfrak{F}_B(g)(0) = 0$. As $g \not\equiv 0$, Theorems 5.IV.2, 5.IV.3. Show that the first inequality of (7.II.1) is satisfied. On the other hand from ii), there exist $\delta > 0$ and $M > 0$ such that for all $\lambda \in [0, \delta]$, we have

$$|\mathfrak{F}_B(g)(\lambda)| \leq M\lambda^\gamma.$$

Then

$$\int_0^\infty |\mathfrak{F}_B(g)(\lambda)|^2 \frac{d\lambda}{\lambda} \leq M^2 \int_0^\delta \lambda^{2\gamma-1}\, d\lambda + \int_\delta^\infty |\mathfrak{F}_B(g)(\lambda)|^2 \frac{d\lambda}{\lambda}.$$

But

$$\int_\delta^\infty |\mathfrak{F}_B(g)(\lambda)|^2 \frac{d\lambda}{\lambda} \leq \frac{2^\alpha \Gamma(\alpha+1)}{\delta^{2(\alpha+1)}} \int_0^\infty |\mathfrak{F}_B(g)(\lambda)|^2\, d\mu_\alpha(\lambda).$$

Thus from Theorem 5.IV.3 we have

$$\int_0^\infty |\mathfrak{F}_B(g)(\lambda)|^2 \frac{d\lambda}{\lambda} \le \frac{M^2\delta^{2\gamma}}{2\gamma} + \frac{2^\alpha\Gamma(\alpha+1)}{\delta^{2(\alpha+1)}} \|g\|_{2,\alpha}.$$

This proves the second inequality of (7.II.1).

Remark.

The right continuity assumption in Proposition 7.II.1 holds if g is moreover in $L^1([0, +\infty[, d\mu_\alpha)$.

Corollary 7.II.1. Let g be a nonzero function in $\mathcal{S}_*(\mathbb{R})$ (resp. $\mathcal{D}_*(\mathbb{R})$). Then $\ell_\alpha(g)$ is a wavelet associated with the Bessel operator in $\mathcal{S}_*(\mathbb{R})$ (resp. $\mathcal{D}_*(\mathbb{R})$), and we have

(7.II.3) $$C_{\ell_\alpha(g)} = \int_0^\infty \lambda^3 |\mathfrak{F}_B(g)(\lambda)|^2 d\lambda.$$

Where ℓ_α is the Bessel operator given by

(7.II.4) $$\ell_\alpha = \frac{d^2}{dx^2} + \frac{2\alpha+1}{x}\frac{d}{dx}.$$

Proof.

We deduce the result from Proposition 7.II.1 and the relation

$$\forall\, \lambda \ge 0,\ \mathfrak{F}_B(\ell_\alpha(g))(\lambda) = -\lambda^2 \mathfrak{F}_B(g)(\lambda).$$

Example.

Let $t \in]0, +\infty[$. From the relation (5.II.2) the function

(7.II.5) $$k_t(x) = \int_0^\infty e^{-t\lambda^2} j_\alpha(\lambda x) d\mu_\alpha(\lambda) = \frac{e^{-\frac{x^2}{4t}}}{(2t)^{\alpha+1}},\ x \ge 0,$$

belongs to $\mathcal{S}_*(\mathbb{R})$. Using Corollary 7.II.1 we deduce that the function

(7.II.6) $$g(x) = -\frac{d}{dt}k_t(x) = -\ell_{\alpha,x}(k_t)(x) = (x^2 - 4(\alpha+1)t)\,\frac{e^{-\frac{x^2}{4t}}}{(2t)^{\alpha+3}},\ x \ge 0,$$

is a wavelet associated with the Bessel operator in $\mathcal{S}_*(\mathbb{R})$ and we have

(7.II.7) $$C_g = \frac{1}{8t^2}.$$

Proposition 7.II.2. Let g be an even function on $\mathbb{R}$. Then g is a wavelet associated with the Bessel operator, in $\mathcal{S}_*(\mathbb{R})$ (resp. $\mathcal{D}_*(\mathbb{R})$) if and only if $W_\alpha(g)$ is a classical wavelet on $[0, +\infty[$, in $\mathcal{S}_*(\mathbb{R})$ (resp. $\mathcal{D}_*(\mathbb{R})$) and we have

(7.II.8) $$C^0_{W_\alpha(g)} = (2^\alpha\Gamma(\alpha+1))^2 C_g.$$

Proof.

We deduce the results from Theorems 2.II.2, 2.III.1 and the relations (5.II.14), (5.III.1).

Theorem 7.II.1. Let $a \in]0, +\infty[$ and g be in $L^2([0, +\infty[, d\mu_\alpha)$. Then the function g_a defined for almost all $x \geq 0$, by

(7.II.9)
$$g_a(x) = \frac{1}{a^{2\alpha+2}} g\left(\frac{x}{a}\right) = \frac{1}{a^{2\alpha+\frac{3}{2}}} H_a(g)(x),$$

satisfies

i) the function g_a belongs to $L^2([0, +\infty[, d\mu_\alpha)$ and we have

(7.II.10)
$$\|g_a\|_{2,\alpha} = \frac{1}{a^{\alpha+1}} \|g\|_{2,\alpha}.$$

ii) We have for almost all $\lambda \geq 0$:

(7.II.11)
$$\mathfrak{F}_B(g_a)(\lambda) = \mathfrak{F}_B(g)(a\lambda).$$

iii) We have

(7.II.12)
$$g_a = \frac{1}{\sqrt{a}} \mathfrak{F}_B \circ H_{a^{-1}} \circ \mathfrak{F}_B(g).$$

Proof.

The i) and ii) are clear.

The iii) can be deduced from the ii) and Proposition 7.I.2.

Proposition 7.II.3. Let g be in $\mathcal{S}_*(\mathbb{R})$ (resp. $\mathcal{D}_*(\mathbb{R})$). Then for all $a \in]0, +\infty[$, the function g_a given by the relation (7.II.9) belongs to $\mathcal{S}_*(\mathbb{R})$ (resp. $\mathcal{D}_*(\mathbb{R})$) and we have

(7.II.13)
$$g_a = \frac{1}{\sqrt{a}} W_\alpha^{-1} \circ H_a \circ W_\alpha(g).$$

Proof.

We obtain these results from the relation (7.II.12), Theorem 5.III.1, Propositions 7.I.2, 7.I.3 and the relations (5.II.14), (5.III.1).

Theorem 7.II.2. Let g be a wavelet associated with the Bessel operator in $L^2([0, +\infty[, d\mu_\alpha)$. Then for all $a \in]0, +\infty[$ and $b \in [0, +\infty[$ the function

(7.II.14)
$$g_{a,b} = \sqrt{a}\, T_b(g_a),$$

is a wavelet associated with the Bessel operator, in $L^2([0,+\infty[, d\mu_\alpha)$ and we have

(7.II.15) $$C_{g_{a,b}} = a\int_0^\infty (j_\alpha(\frac{\lambda b}{a}))^2 |\mathcal{F}_B(g)(\lambda)|^2 \frac{d\lambda}{\lambda}.$$

Where T_b, $b \geq 0$, are the generalized translation operators defined by the relation (4.II.3).

Proof.

As the function g_a is in $L^2([0,+\infty[, d\mu_\alpha)$, Proposition 4.II.2 shows that the relation (7.II.14) defines an element of $L^2([0,+\infty[, d\mu_\alpha)$. On the other hand from the relations (5.II.8) and (7.II.11) we have for almost all $\lambda \geq 0$:

(7.II.16) $$\mathcal{F}_B(g_{a,b})(\lambda) = \sqrt{a}\; j_\alpha(\lambda b)\, \mathcal{F}_B(g)(a\lambda).$$

This relation implies (7.II.15).

Now, we must prove that the function $g_{a,b}$ satisfies the admissibility condition (7.II.1). As $g \not\equiv 0$, we deduce from the relation (7.II.15) and Theorem 5.IV.2 that $C_{g_{a,b}}$ is nonzero. On the other hand from the relation (7.II.15) and the fact that

$$\forall\, b \geq 0,\ a > 0,\ \lambda \geq 0,\ |j_\alpha(\frac{\lambda b}{a})| \leq 1,$$

we deduce that

$$C_{g_{a,b}} \leq a\, C_g,$$

which gives the result.

Remark.

Using the relation (7.II.9) the function $g_{a,b}$ can also be written for almost all $x \geq 0$, in the form

(7.II.17) $$g_{a,b}(x) = \frac{1}{a^{2\alpha+\frac{3}{2}}} T_{b/a}(g)(\frac{x}{a}).$$

Proposition 7.II.4. Let g be a wavelet associated with the Bessel operator. Then the mapping $(a, b) \rightarrow g_{a,b}$ is continuous from $]0,+\infty[\times [0,+\infty[$ into $L^2([0,+\infty[, d\mu_\alpha)$.

Proof.

From Proposition 4.II.2 and Theorem 7.II.1.i) the mapping $g \to g_{a,b}$ is linear and continuous from $L^2([0, +\infty[, d\mu_\alpha)$ into itself. Then it suffices to consider the case where g is in $\mathcal{D}_*(\mathbb{R})$. For all $(a_0, b_0) \in]0, +\infty[\times [0, +\infty[$, there exists $M > 0$ such that

$$|a - a_0| < 1, |b - b_0| < 1 \implies \text{supp } g_{a,b} \subset [-M, M].$$

Then

$$\|g_{a,b} - g_{a_0,b_0}\|_{2,\alpha}^2 \leq \left(\int_0^M d\mu_\alpha(x)\right) \sup_{x \in [0,M]} |g_{a,b}(x) - g_{a_0,b_0}(x)|^2. \tag{7.II.18}$$

But from Theorem 5.II.1 and the relation (7.II.16) we have

$$\forall\, x \geq 0 \quad g_{a,b}(x) = \sqrt{a} \int_0^\infty j_\alpha(\lambda x)\, j_\alpha(\lambda b)\, \mathcal{F}_B(g)(a\lambda) d\mu_\alpha(\lambda).$$

This relation shows that the mapping $(a, b, x) \to g_{a,b}(x)$ is continuous on $]0, +\infty[\times [0, +\infty[\times [0, +\infty[$. Thus the second member of (7.II.18) tends to zero when (a, b) tends to (a_0, b_0).

7.II.2. Continuous wavelet transform associated with the Bessel operator.

In this subsection we consider a wavelet g associated with the Bessel operator in $L^2([0, +\infty[, d\mu_\alpha)$.

Definition 7.II.2. A continuous wavelet transform Ψ_g associated with the Bessel operator is defined for regular functions f on $[0, +\infty[$, by

$$\Psi_g(f)(a,b) = \int_0^\infty f(x)\overline{g_{a,b}(x)}\, d\mu_\alpha(x), \quad a > 0, \quad b \geq 0. \tag{7.II.19}$$

Remark.

The relation (7.II.19) can also be written in the form

$$\Psi_g(f)(a,b) = \sqrt{a}\, f *_B \bar{g}_a(b), \tag{78.II.20}$$

where $*_B$ is the generalized convolution product given by the relation (4.III.1).

Proposition 7.II.5.

i) For all $a \in]0, +\infty[$ and f in $L^q([0, +\infty[, d\mu_\alpha)$, $q \in [1, 2]$, the function $b \to \Psi_g(f)(a,b)$ is defined almost everywhere on $[0,+\infty[$, belongs to $L^r([0, +\infty[, d\mu_\alpha)$, $r \in [1, +\infty]$ such that $\frac{1}{r} = \frac{1}{q} - \frac{1}{2}$, and we have

(7.II.21) $$\|\Psi_g(f)(a,.)\|_{r,\alpha} \le a^{-\alpha-\frac{1}{2}} \|f\|_{q,\alpha} \|g\|_{2,\alpha}.$$

ii) For all $a \in]0, +\infty[$ and f in $L^2([0, +\infty[, d\mu_\alpha)$, the function $b \to S_g(f)(a,b)$ is continuous on $[0, +\infty[$, tends to zero at the infinity and we have

(7.II.22) $$\sup_{b\ge 0} |\Psi_g(f)(a,b)| \le a^{-\alpha-\frac{1}{2}} \|f\|_{2,\alpha} \|g\|_{2,\alpha}.$$

iii) If g is in $\mathcal{S}_*(\mathbb{R})$(resp. $\mathcal{D}_*(\mathbb{R})$), then for all $a \in]0, +\infty[$ and f in $\mathcal{S}_*(\mathbb{R})$(resp. $\mathcal{D}_*(\mathbb{R})$), the function $b \to \Psi_g(f)(a,b)$ belongs to $\mathcal{S}_*(\mathbb{R})$(resp. $\mathcal{D}_*(\mathbb{R})$).

Proof.

The i) and ii) derive from Propositions 4.III.4, 4.III.5 and Theorem 7.II.1.i).

The iii) can be deduced from Corollaries 4.III.1, 5.III.1.

Proposition 7.II.6. Let f be in $L^2([0, +\infty[, d\mu_\alpha)$. Then the mapping $(a, b) \to \Psi_g(f)(a,b)$ is continuous on $]0, +\infty[\times [0, +\infty[$.

Proof.

From Definition 7.II.2 we have

$$|\Psi_g(f)(a,b) - \Psi_g(f)(a',b')| \le \int_0^\infty |f(x)| \, |\overline{g_{a,b}(x)} - \overline{g_{a',b'}(x)}| \, d\mu_\alpha(x).$$

Using Hölder's inequality we obtain

$$|\Psi_g(f)(a,b) - \Psi_g(f)(a',b')| \le \|f\|_{2,\alpha} \|g_{a,b} - g_{a',b'}\|_{2,\alpha}.$$

We deduce the result from Proposition 7.II.4.

Proposition 7.II.7. When the wavelet g associated with the Bessel operator belongs to $\mathcal{S}_*(\mathbb{R})$, we have for all f in $\mathcal{S}_*(\mathbb{R})$ the following relation

(7.II.23) $$\forall a > 0, \Psi_g(f)(a,.) = W_\alpha^{-1}[S_{W_\alpha(g)}(W_\alpha(f))(a,.)].$$

Proof.

Let $a \in]0, +\infty[$. Using the relation (7.II.20) and Proposition 5.III.7 we obtain

$$\Psi_g(f)(a,.) = \sqrt{a}\, f *_B \bar{g}_a,$$
$$= \sqrt{a}\, W_\alpha^{-1}[W_\alpha(f) *_0 W_\alpha(\bar{g}_a)].$$

But from the relation (7.II.13) we have

$$W_\alpha(\bar{g}_a) = \frac{1}{\sqrt{a}} \overline{H_a \circ W_\alpha(g)}.$$

Then, using the relation (7.I.15), we deduce

$$\Psi_g(f)(a,.) = W_\alpha^{-1}[W_\alpha(f) *_0 \overline{H_a \circ W_\alpha(g)}],$$
$$= W_\alpha^{-1}[S_{W_\alpha(g)}(W_\alpha(f))(a,.)].$$

Theorem 7.II.3.

i) Plancherel formula for Ψ_g.

For all f in $L^2([0, +\infty[, d\mu_\alpha)$ we have

$$\int_0^\infty |f(x)|^2 d\mu_\alpha(x) = \frac{1}{C_g}\int_0^\infty \int_0^\infty |\Psi_g(f)(a,b)|^2 d\mu_\alpha(b)\frac{da}{a^2}. \tag{7.II.24}$$

ii) Parseval formula for Ψ_g.

For all f_1, f_2 in $L^2([0, +\infty[, d\mu_\alpha)$ we have

$$\int_0^\infty f_1(x)\, \overline{f_2(x)}\, d\mu_\alpha(x) = \frac{1}{C_g}\int_0^\infty \int_0^\infty \Psi_g(f_1)(a,b)\overline{\Psi_g(f_2)(a,b)}\, d\mu_\alpha(b)\frac{da}{a^2}. \tag{7.II.25}$$

Proof.

i) Using Fubini-Tonelli's Theorem, the relation (7.II.20) and Corollary 5.IV.2 we obtain

$$\frac{1}{C_g}\int_0^\infty \int_0^\infty |\Psi_g(f)(a,b)|^2 d\mu_\alpha(b)\frac{da}{a^2} = \frac{1}{C_g}\int_0^\infty \left(\int_0^\infty |f *_B \bar{g}_a|^2 d\mu_\alpha(b)\right)\frac{da}{a},$$
$$= \frac{1}{C_g}\int_0^\infty \left(\int_0^\infty |\mathfrak{F}_B(f)(\lambda)|^2 |\mathfrak{F}_B(\bar{g}_a)(\lambda)|^2 d\mu_\alpha(\lambda)\right)\frac{da}{a},$$
$$= \int_0^\infty |\mathfrak{F}_B(f)(\lambda)|^2 \left(\frac{1}{C_g}\int_0^\infty |\mathfrak{F}_B(g)(a\lambda)|^2 \frac{da}{a}\right) d\mu_\alpha(\lambda).$$

But from Definition 7.II.1 we have for almost all $\lambda \geq 0$:

$$\frac{1}{C_g}\int_0^\infty |\mathfrak{F}_B(g)(a\lambda)| \frac{da}{a} = 1. \tag{7.II.26}$$

Thus

$$\frac{1}{C_g}\int_0^\infty\int_0^\infty |\Psi_g(f)(a,b)|^2 d\mu_\alpha(b)\,\frac{da}{a^2} = \int_0^\infty |\mathscr{F}_B(f)(a\lambda)|^2\, d\mu_\alpha(\lambda).$$

We obtain the relation (7.II.24) from Theorem 5.IV.3.

ii) We deduce the result from the i).

The following result is the so-called Calderon's reproducing formula.

Theorem 7.II.4. We suppose that the wavelet g associated with the Bessel operator is such that the function $\mathscr{F}_B(g)$ belongs to $L^\infty([0, +\infty[, d\mu_\alpha)$. Then for f in $L^2([0, +\infty[, d\mu_\alpha)$ and $0 < \varepsilon < \delta < +\infty$, the function

(7.II.27) $$f_{\varepsilon,\delta}(x) = \frac{1}{C_g}\int_\varepsilon^\delta \int_0^\infty \Psi_g(f)(a,b)\, g_{a,b}(x)\, d\mu_\alpha(b)\,\frac{da}{a^2},$$

belongs to $L^2([0, +\infty[, d\mu_\alpha)$ and satisfies

(7.II.28) $$\lim_{\substack{\varepsilon\to 0^+\\ \delta\to +\infty}} \|f_{\varepsilon,\delta} - f\|_{2,\alpha} = 0.$$

To prove this theorem we need the following Lemmas.

Lemma 7.II.1. We suppose that the wavelet g associated with the Bessel operator is such that $\mathscr{F}_B(g)$ belongs to $L^\infty([0, +\infty[, d\mu_\alpha)$. For $0 < \varepsilon < \delta < +\infty$, we put

(7.II.29) $$\forall\, x \geq 0,\ G_{\varepsilon,\delta}(x) = \frac{1}{C_g}\int_\varepsilon^\delta g_a *_B \bar{g}_a(x)\,\frac{da}{a},$$

and for almost all $\lambda \geq 0$:

(7.II.30) $$K_{\varepsilon,\delta}(\lambda) = \frac{1}{C_g}\int_\varepsilon^\delta |\mathscr{F}_B(g)\,(a\lambda)|^2\,\frac{da}{a}.$$

Then

i) $G_{\varepsilon,\delta} \in L^p([0, +\infty[, d\mu_\alpha)$, $p \in [2, +\infty]$.

ii) $K_{\varepsilon,\delta} \in L^p([0, +\infty[, d\mu_\alpha)$, $p \in [1, +\infty]$.

iii) We have

(7.II.31) $$\mathscr{F}_B(G_{\varepsilon,\delta}) = K_{\varepsilon,\delta}\,.$$

Proof.

i) - For $p = +\infty$, the result follows from Proposition 4.III.5 and Theorem 6.II.1.i).

- For $p \in [2, +\infty[$, we use Hölder's inequality for the measure $\frac{da}{a}$ and we obtain

$$\forall\, x \geq 0,\ |G_{\varepsilon,\delta}(x)|^p \leq \frac{1}{(C_g)^p}\Big(\int_\varepsilon^\delta \frac{da}{a}\Big)^{p-1}\int_\varepsilon^\delta |g_a *_B \bar{g}_a(x)|^p \frac{da}{a}.$$

So, by Fubini-Tonelli's theorem, Proposition 4.III.5 and the relation (7.II.10) we have

$$\int_0^\infty |G_{\varepsilon,\delta}(x)|^p\, d\mu_\alpha(x) \leq \frac{1}{(C_g)^p}\Big(\int_\varepsilon^\delta \frac{da}{a}\Big)^{p-1}\int_\varepsilon^\delta \int_0^\infty |g_a *_B \bar{g}_a(x)|^p\, d\mu_\alpha(x)\frac{da}{a},$$

$$\leq \frac{1}{(C_g)^p}\Big(\int_\varepsilon^\delta \frac{da}{a}\Big)^{p-1}\int_\varepsilon^\delta \|g_a *_B \bar{g}_a(x)\|_{\infty,\alpha}^{p-2}\Big(\int_0^\infty |g_a *_B \bar{g}_a(x)|^2\, d\mu_\alpha(x)\Big)\frac{da}{a},$$

$$\leq \frac{1}{(C_g)^p}\Big(\int_\varepsilon^\delta \frac{da}{a}\Big)^{p-1}\|g\|_{2,\alpha}^{2(p-2)}\int_\varepsilon^\delta \Big(\int_0^\infty |g_a *_B \bar{g}_a(x)|^2\, d\mu_\alpha(x)\Big)\frac{da}{a^{2(p-2)(\alpha+1)+1}}.$$

But from Corollary 5.IV.2 and Theorems 5.IV.3, 7.II.1, we have

$$\int_0^\infty |g_a *_B \bar{g}_a(x)|^2\, d\mu_\alpha(x) = \int_0^\infty |\mathfrak{F}_B(g_a)(\lambda)|^4 d\mu_\alpha(\lambda),$$

$$\leq \|\mathfrak{F}_B(g_a)\|_{\infty,\alpha}^2 \int_0^\infty |\mathfrak{F}_B(g_a)(\lambda)|^2 d\mu_\alpha(\lambda),$$

$$\leq \|\mathfrak{F}_B(g_a)\|_{\infty,\alpha}^2 \|g_a\|_{2,\alpha}^2,$$

$$\leq \frac{1}{a^{2(\alpha+1)}}\|\mathfrak{F}_B(g)\|_{\infty,\alpha}^2 \|g\|_{2,\alpha}^2.$$

Thus

$$\int_0^\infty |G_{\varepsilon,\delta}(x)|^p\, d\mu_\alpha(x) \leq \frac{1}{(C_g)^p}\|\mathfrak{F}_B(g)\|_{\infty,\alpha}^2 \|g\|_{2,\alpha}^{2(p-1)}\Big(\int_\varepsilon^\delta \frac{da}{a}\Big)\int_\varepsilon^\delta \frac{da}{a^{2(p-1)(\alpha+1)+1}} < +\infty.$$

ii) - For $p = +\infty$, we have from the relation (7.II.26) :

$$\forall\, \lambda > 0,\ |K_{\varepsilon,\delta}(\lambda)| \leq \frac{1}{C_g}\int_0^\infty |\mathfrak{F}_B(g)(a\lambda)|^2 \frac{da}{a} = 1.$$

- For $p \in [1, +\infty[$, we use Hölder's inequality for the measure $\frac{da}{a}$ and we obtain

$$\forall\, \lambda > 0,\ |K_{\varepsilon,\delta}(\lambda)|^p \leq \frac{1}{(C_g)^p}\Big(\int_\varepsilon^\delta \frac{da}{a}\Big)^{p-1}\int_\varepsilon^\delta |\mathfrak{F}_B(g)(a\lambda)|^{2p}\frac{da}{a}.$$

Using Fubini-Tonelli's theorem and Theorems 5.IV.3, 7.II.1, we find

$$\int_0^\infty |K_{\varepsilon,\delta}(\lambda)|^p\, d\mu_\alpha(\lambda) \leq \frac{1}{(C_g)^p}\left(\int_\varepsilon^\delta \frac{da}{a}\right)^{p-1}\int_\varepsilon^\delta \int_0^\infty |\mathfrak{F}_B(g)(a\lambda)|^{2p}\, d\mu_\alpha(\lambda)\, \frac{da}{a},$$

$$\leq \frac{1}{(C_g)^p}\left(\int_\varepsilon^\delta \frac{da}{a}\right)^{p-1}\|\mathfrak{F}_B(g)\|_{\infty,\alpha}^{2(p-1)} \times$$

$$\int_\varepsilon^\delta \left(\int_0^\infty |\mathfrak{F}_B(g)(a\lambda)|^2\, d\mu_\alpha(\lambda)\right) \frac{da}{a},$$

$$\leq \frac{1}{(C_g)^p}\left(\int_\varepsilon^\delta \frac{da}{a}\right)^{p-1}\|\mathfrak{F}_B(g)\|_{\infty,\alpha}^{2(p-1)}\int_\varepsilon^\delta \|g_a\|_{2,\alpha}^2 \frac{da}{a},$$

$$\leq \frac{1}{(C_g)^p}\left(\int_\varepsilon^\delta \frac{da}{a}\right)^{p-1}\|\mathfrak{F}_B(g)\|_{\infty,\alpha}^{2(p-1)}\|g\|_{2,\alpha}^2\int_\varepsilon^\delta \frac{da}{a^{2\alpha+3}} < +\infty.$$

iii) From Proposition 5.IV.4 we get

$$\forall\, x \geq 0,\ g_a *_B \bar{g}_a(x) = \int_0^\infty |\mathfrak{F}_B(g_a)(\lambda)|^2 j_\alpha(\lambda x)d\mu_\alpha(\lambda).$$

So

$$\forall\, x \geq 0,\ G_{\varepsilon,\delta}(x) = \frac{1}{C_g}\int_\varepsilon^\delta \left(\int_0^\infty |\mathfrak{F}_B(g_a)|^2 j_\alpha(\lambda x)d\mu_\alpha(\lambda)\right) \frac{da}{a}.$$

Using the fact that

$$\forall\, x \geq 0,\ \forall\, \lambda \geq 0,\ |j_\alpha(\lambda x)| \leq 1,$$

Fubini-Tonelli's theorem and Theorems 5.IV.3, 7.II.1, we see that

$$\int_\varepsilon^\delta \int_0^\infty |\mathfrak{F}_B(g_a)(\lambda)|^2\, |j_\alpha(\lambda x)|d\mu_\alpha(x)\, \frac{da}{a} \leq \int_\varepsilon^\delta \left(\int_0^\infty |\mathfrak{F}_B(g_a)(\lambda)|^2\, d\mu_\alpha(\lambda)\right) \frac{da}{a},$$

$$\leq \int_\varepsilon^\delta \|g_a\|_{2,\alpha}^2 \frac{da}{a},$$

$$\leq \|g\|_{2,\alpha}\int_\varepsilon^\delta \frac{da}{a^{2\alpha+3}} < +\infty.$$

Then from Fubini's theorem we find

$$\forall\, x \geq 0,\ G_{\varepsilon,\delta}(x) = \int_0^\infty \left(\frac{1}{C_g}\int_\varepsilon^\delta |\mathfrak{F}_B(g)(a\lambda)|^2 \frac{da}{a}\right) j_\alpha(\lambda x)d\mu_\alpha(\lambda),$$

$$= \int_0^\infty K_{\varepsilon,\delta}(\lambda)j_\alpha(\lambda x)d\mu_\alpha(\lambda),$$

and iii) is proved.

Lemma 7.II.2. We suppose that the wavelet g associated with the Bessel operator is such that $\mathfrak{F}_B(g)$ belongs to $L^\infty([0, +\infty[, d\mu_\alpha)$. Then the function $f_{\varepsilon,\delta}$ defined in Theorem 7.II.4 belongs to $(L^\infty \cap L^2)([0, +\infty[, d\mu_\alpha)$ and satisfies the relations

(7.II.32) $$f_{\varepsilon,\delta} = f *_B G_{\varepsilon,\delta}.$$

(7.II.33) $$\mathfrak{F}_B(f_{\varepsilon,\delta}) = \mathfrak{F}_B(f) \,.\, K_{\varepsilon,\delta}.$$

Proof.

For $x \in [0, +\infty[$, we have from Hölder's inequality

$$\int_\varepsilon^\delta \int_0^\infty |\Psi_g(f)(a,b)|\, |g_{a,b}(x)| d\mu_\alpha(b) \frac{da}{a^2} \leq \left(\int_\varepsilon^\delta \int_0^\infty |\Psi_g(f)(a,b)|^2 d\mu_\alpha(b) \frac{da}{a^2}\right)^{1/2} \times \left(\int_\varepsilon^\delta \int_0^\infty |g_{a,b}(x)|^2\, d\mu_\alpha(b) \frac{da}{a^2}\right)^{1/2}.$$

But from Theorem 7.II.3 i), Fubini-Tonelli's theorem and the relations (7.II.14), (4.II.14), (7.II.10) we have

$$\int_\varepsilon^\delta \int_0^\infty |\Psi_g(f)(a,b)|^2 d\mu_\alpha(b) \frac{da}{a^2} \leq \int_0^\infty \int_0^\infty |\Psi_g(f)(a,b)|^2 d\mu_\alpha(b) \frac{da}{a^2},$$
$$\leq C_g \, \|f\|_{2,\alpha}^2 ,$$

and

$$\int_\varepsilon^\delta \int_0^\infty |g_{a,b}(x)|^2\, d\mu_\alpha(b) \frac{da}{a^2} \leq \int_\varepsilon^\delta \left(\int_0^\infty |g_a(b)|^2\, d\mu_\alpha(b)\right) \frac{da}{a},$$
$$\leq \|g\|_{2,\alpha}^2 \int_\varepsilon^\delta \frac{da}{a^{2\alpha+3}} .$$

Thus

$$\int_\varepsilon^\delta \int_0^\infty |\Psi_g(f)(a,b)|\, |g_{a,b}(x)|\, d\mu_\alpha(b) \frac{da}{a^2} \leq C_g^{1/2} \left(\int_\varepsilon^\delta \frac{da}{a^{2\alpha+3}}\right)^{1/2} \|f\|_{2,\alpha} \|g\|_{2,\alpha} < +\infty.$$

From this relation we deduce that the function $f_{\varepsilon,\delta}$ belongs to $L^\infty([0, +\infty[, d\mu_\alpha)$, and using Fubini's theorem we obtain

$$f_{\varepsilon,\delta}(x) = \frac{1}{C_g} \int_\varepsilon^\delta \left(\int_0^\infty \Psi_g(f)(a,b)\, g_{a,b}(x)\, d\mu_\alpha(b)\right) \frac{da}{a^2}.$$

The relation (7.II.20) and Theorem 5.IV.4 show that the function $b \to \Psi_g(f)(a,b)$ belongs to $L^2([0, +\infty[, d\mu_\alpha)$ and for almost all $\lambda \in [0, +\infty[$ we have

$$\mathfrak{F}_B(\Psi_g(f)(a,.))(\lambda) = \sqrt{a}\, \mathfrak{F}_B(f)(\lambda). \overline{\mathfrak{F}_B(g_a)(\lambda)}.$$

On the other hand from the relations (7.II.14), (5.II.8), (4.II.9) and Theorem 5.IV.2, we have for almost all $\lambda \in [0, +\infty[$:

(7.II.34) $$\mathfrak{F}_B(g_{a,b})(\lambda) = \sqrt{a}\, j_\alpha(\lambda x)\, \mathfrak{F}_B(g_a)(\lambda).$$

Thus from Theorem 5.IV.3 we deduce that

$$f_{\varepsilon,\delta}(x) = \frac{1}{C_g}\int_\varepsilon^\delta \left(\int_0^\infty \mathfrak{F}_B(f)(\lambda)\, |\mathfrak{F}_B(g_a)(\lambda)|^2\, j_\alpha(\lambda x)d\mu_\alpha(\lambda)\right)\frac{da}{a}.$$

But using the fact that

$$\forall\ x \geq 0,\ \lambda \geq 0,\ |j_\alpha(\lambda x)| \leq 1,$$

Hölder's inequality, Fubini-Tonelli's theorem and Theorems 5.IV.3, 7.II.1.i), we have

$$\int_\varepsilon^\delta \int_0^\infty |\mathfrak{F}_B(f)(\lambda)|\, |\mathfrak{F}_B(g_a)(\lambda)|^2\, |j_\alpha(\lambda x)|\, d\mu_\alpha(\lambda)\, \frac{da}{a}$$

$$\leq \|\mathfrak{F}_B(g)\|_{\infty,\alpha}\, \|\mathfrak{F}_B(f)\|_{2,\alpha} \int_\varepsilon^\delta \|\mathfrak{F}_B(g_a)\|_{2,\alpha}\frac{da}{a},$$

$$\leq \|\mathfrak{F}_B(g)\|_{\infty,\alpha}\, \|f\|_{2,\alpha}\, \|g\|_{2,\alpha}\int_\varepsilon^\delta \frac{da}{a^{\alpha+2}} < +\infty.$$

Thus from Fubini's Theorem we have

$$f_{\varepsilon,\delta}(x) = \int_0^\infty \left(\frac{1}{C_g}\int_\varepsilon^\delta |\mathfrak{F}_B(g_a)(\lambda)|^2\, \frac{da}{a}\right) \mathfrak{F}_B(f)(\lambda)\, j_\alpha(\lambda x)d\mu_\alpha(\lambda).$$

Using Lemma 7.II.1 we have

$$f_{\varepsilon,\delta}(x) = \int_0^\infty \mathfrak{F}_B(f)(\lambda)\, K_{\varepsilon,\delta}(\lambda)\, j_\alpha(\lambda x)d\mu_\alpha(\lambda).$$

Thus

$$f_{\varepsilon,\delta}(x) = \mathfrak{F}_B(\mathfrak{F}_B(f).K_{\varepsilon,\delta})(x).$$

By applying Lemma 7.II.1, Theorem 5.IV.2 and Proposition 5.IV.3 we deduce that the function $f_{\varepsilon,\delta}$ belongs to $L^2([0, +\infty[, d\mu_\alpha)$ and satisfies the relations (7.II.32), (7.II.33).

Proof of Theorem 7.II.4.

From the relation (7.II.33) and Theorem 5.IV.3 we obtain

$$\|f_{\varepsilon,\delta} - f\|_{2,\alpha} = \|\mathfrak{F}_B(f)\ K_{\varepsilon,\delta} - \mathfrak{F}_B(f)\|_{2,\alpha},$$

$$= \|\mathfrak{F}_B(f)(1 - K_{\varepsilon,\delta})\|_{2,\alpha}.$$

But by (7.II.1) we have for almost all $\lambda > 0$:

$$\lim_{\substack{\varepsilon\to 0^+\\ \delta\to+\infty}} K_{\varepsilon,\delta}(\lambda) = 1.$$

So, the relation (7.II.28) follows from the dominated convergence theorem.

Next we formulate a pointwise reconstruction formula for the transform Ψ_g.

Theorem 7.II.5. For f in $L^1([0, +\infty[, d\mu_\alpha)$ such that $\mathcal{F}_B(f)$ belongs to $L^1([0, +\infty[, d\mu_\alpha)$ we have

$$f(x) = \frac{1}{C_g}\int_0^\infty \left(\int_0^\infty \Psi_g(f)(a,b)\, g_{a,b}(x)d\mu_\alpha(b)\right)\frac{da}{a^2}, \text{ a.e. on } [0, +\infty[. \tag{7.II.35}$$

Where, for each $x \geq 0$, both the inner integral and the outer integral are absolutely convergent, but possible not the double integral.

Proof.

We put for $a > 0$ and $x \geq 0$:

$$I(a,x) = \int_0^\infty \Psi_g(f)(a,b)\, g_{a,b}(x)\, d\mu_\alpha(b),$$

and

$$J(x) = \frac{1}{C_g}\int_0^\infty I(a,x)\,\frac{da}{a^2}.$$

We shall prove that the integrals $I(a,x)$ and $J(x)$ are absolutely convergent and we have

$$J(x) = \int_0^\infty \mathcal{F}_B(f)(\lambda)\, j_\alpha(\lambda x)d\mu_\alpha(\lambda). \tag{7.II.36}$$

From the relations (7.II.14) and (7.II.20) we have

$$\Psi_g(f)(a,b)\, g_{a,b}(x) = a\, f *_B \bar{g}_a(b)\, T_x(g_a)(b).$$

Propositions 4.II.2, 4.III.4 and Theorem 7.II.1.i) show that the functions $b \to f *_B \bar{g}_a(b)$ and $b \to T_x(g_a)(b)$ belong to $L^2([0, +\infty[, d\mu_\alpha)$. We deduce from Hölder's inequality that the integral $I(a,x)$ converges absolutely. On the other hand from Proposition 5.IV.3 we have for almost all $\lambda \geq 0$:

$$\mathcal{F}_B(f *_B \bar{g}_a)(\lambda) = \mathcal{F}_B(f)(\lambda).\ \overline{\mathcal{F}_B(g_a)(\lambda)},$$

and from the relation (7.II.34)

$$\mathcal{F}_B(T_x(g_a))(\lambda) = j_\alpha(\lambda x)\, \mathcal{F}_B(g_a)(\lambda).$$

Then using Theorem 5.IV.3 we obtain

$$I(a,x) = a\int_0^\infty \mathcal{F}_B(f)(\lambda)|\, \mathcal{F}_B(g_a)(\lambda)|^2\, j_\alpha(\lambda x)d\mu_\alpha(\lambda). \tag{7.II.37}$$

From this equality, Fubini-Tonelli's theorem, the fact that

$$\forall\, x \geq 0,\ \lambda \geq 0,\ |j_\alpha(\lambda x)| \leq 1,$$

and the relation (7.II.26) we deduce that

(7.II.38) $$\frac{1}{C_g}\int_0^\infty |I(a,x)|\frac{da}{a^2} \leq \int_0^\infty |\mathcal{F}_B(f)(\lambda)| \left(\frac{1}{C_g}\int_0^\infty |\mathcal{F}_B(g)(a\lambda)|^2 \frac{da}{a}\right)d\mu_\alpha(\lambda),$$

$$\leq \|\mathcal{F}_B(f)\|_{1,\alpha} < +\infty.$$

This shows that the integral $J(x)$ converges absolutely.

We prove now the relation (7.II.36). From the relation (7.II.37) we have

(7.II.39) $$J(x) = \frac{1}{C_g}\int_0^\infty \left(\int_0^\infty \mathcal{F}_B(f)(\lambda)|\mathcal{F}_B(g_a)(\lambda)|^2 j_\alpha(\lambda x)d\mu_\alpha(\lambda)\right) \frac{da}{a}.$$

The inequality (7.II.38) shows that the double integral (7.II.39) is absolutely convergent. Thus using Fubini's theorem and the relation (7.II.26) we obtain

$$J(x) = \int_0^\infty \mathcal{F}_B(f)(\lambda)\left(\frac{1}{C_g}\int_0^\infty |\mathcal{F}_B(g)(a\lambda)|^2 \frac{da}{a}\right) j_\alpha(\lambda x)d\mu_\alpha(\lambda),$$

$$= \int_0^\infty \mathcal{F}_B(f)(\lambda)\ j_\alpha(\lambda x)d\mu_\alpha(\lambda).$$

We deduce the relation (7.II.35) from Theorem 5.II.2.

Theorem 7.II.3 shows that the continuous wavelet transform Ψ_g associated with the Bessel operator, is an isometry from the Hilbert space $L^2([0, +\infty[, d\mu_\alpha)$ into the Hilbert space $L^2(]0, +\infty[\times [0, +\infty[, \frac{da}{a^2C_g}d\mu_\alpha(b))$ (the space of square integrable functions on $]0, +\infty[\times [0, +\infty[$ with respect to the measure $\frac{da}{a^2C_g}d\mu_\alpha(b)$). For the characterization of the image of Ψ_g, we interpret the vectors $g_{a,b}$, $(a, b) \in]0, +\infty[\times [0, +\infty[$, as a set of coherent states in the Hilbert space $L^2([0, +\infty[, d\mu_\alpha)$. (See [51] p. 37-38) .

Definition 7.II.3. A set of coherent states in a Hilbert space $\mathcal{H}$ is a subset $\{g_\ell\}_{\ell\in\mathcal{L}}$ of $\mathcal{H}$ such that

i) $\mathcal{L}$ is a locally compact topological space and the mapping $\ell \to g_\ell : \mathcal{L} \to \mathcal{H}$ is continuous.

ii) There is a positive Borel measure $d\ell$ on $\mathcal{L}$ such that, for f in $\mathcal{H}$:

(7.II.40) $$\|f\|^2 = \int_{\mathfrak{L}} |(f, g_\ell)|^2 \, d\ell ,$$

where $(.,.)$ and $\|.\|$ are respectively the scalar product and the norm of $\mathcal{H}$.

Using the relation (7.II.40), we define an isometry Ψ from $\mathcal{H}$ into $L^2(\mathfrak{L}, d\ell)$ by

(7.II.41) $$\Psi(f)(\ell) = (f, g_\ell), \quad f \in \mathcal{H}.$$

Lemma 7.II.3. Let $\mathcal{H}_1$ and $\mathcal{H}_2$ be two Hilbert spaces and U an isometry from $\mathcal{H}_1$ into $\mathcal{H}_2$. Then

i) Im U is a closed subspace of $\mathcal{H}_2$.

ii) $U.U^*$ is an orthogonal projection of $\mathcal{H}_2$ onto ImU. Where U^* is the adjoint operator of U.

Theorem 7.II.6. Let $\{g_\ell\}_{\ell \in \mathfrak{L}}$ be a set of coherent states in a Hilbert space $\mathcal{H}$, and Ψ the isometry defined by the relation (7.II.41). Let F be in $L^2(\mathfrak{L}, d\ell)$. Then F belongs to Im Ψ if and only if

(7.II.42) $$F(\ell) = \int_{\mathfrak{L}} F(\ell') \, (g_{\ell'}, g_\ell) \, d\ell'.$$

Proof.

From Lemma 7.II.3, the function F belongs to ImΨ if and only if we have

$$\Psi\Psi^*(F) = F,$$

where Ψ^* is the adjoint operator of Ψ.

Using the definition of Ψ and Ψ^*, we obtain

$$\begin{aligned}\Psi\Psi^*(F)(\ell) &= (\Psi^*(F), g_\ell) , \\ &= \int_{\mathfrak{L}} F(\ell') \overline{\Psi(g_\ell)(\ell')} \, d\ell' , \\ &= \int_{\mathfrak{L}} F(\ell') \, (g_{\ell'}, g_\ell) \, d\ell'.\end{aligned}$$

Which gives the result.

Let now $\mathcal{H} = L^2([0, +\infty[, d\mu_\alpha)$, $\mathfrak{L} =]0, +\infty[\times [0, +\infty[$. Choose a nonzero function g in $L^2([0, +\infty[, d\mu_\alpha)$ and let $g_\ell = g_{a,b}$ be given by the relation

(7.II.14), if $\ell \in \mathfrak{L}$. Then we have a set of coherent states. Indeed, i) of Definition 7.II.3 is satisfied because of Proposition 7.II.4, and ii) of Definition 7.II.3 is satisfied for the measure $\frac{da}{a^2 C_g} d\mu_\alpha(b)$ on $]0, +\infty[\times [0, +\infty[$ (see Theorem 7.II.3.i)).

Theorem 7.II.7. Let F be in $L^2[0, +\infty[\times [0, +\infty[, \frac{da}{a^2 C_g} d\mu_\alpha)$. Then F belongs to Im Ψ_g if and only if

$$F(a, b) = \frac{1}{C_g}\int_0^\infty \int_0^\infty F(a',b')\left(\int_0^\infty g_{a',b'}(x)\, \overline{g_{a,b}(x)}\, d\mu_\alpha(x)\right) \frac{da'}{(a')^2}\, d\mu_\alpha(b') . \tag{4.II.43}$$

Proof.

Apply Theorem 7.II.6 with $\mathcal{H} = L^2([0, +\infty[, d\mu_\alpha)$, $\mathfrak{L} =]0, +\infty[\times [0, +\infty[$, coherent states $g_{a,b}$ and measure $d\ell$ given by $\frac{da}{a^2 C_g} d\mu_\alpha(b)$.

7.III. Inversion formulas for the operators $\mathfrak{R}_\alpha$ and W_α.

Notations. We denote by

- $\mathcal{S}^0_*(\mathbb{R})$ the subspace of $\mathcal{S}_*(\mathbb{R})$ consisting of functions f such that

$$\forall\, n \in \mathbb{N}, \frac{d^{2n}}{dx^{2n}} f(0) = 0. \tag{7.III.1}$$

- $\mathcal{S}_{*,-\frac{1}{2}}(\mathbb{R})$ the subspace of $\mathcal{S}_*(\mathbb{R})$ consisting of functions f such that

$$\forall\, n \in \mathbb{N}, \int_0^\infty f(x)\, x^{2n}\, dx = 0. \tag{7.III.2}$$

- $\mathcal{S}_{*,\alpha}(\mathbb{R})$ the subspace of $\mathcal{S}_*(\mathbb{R})$ consisting of functions f such that

$$\forall\, n \in \mathbb{N}, \int_0^\infty f(x)\, x^{2n}\, d\mu_\alpha(x) = 0. \tag{7.III.3}$$

Theorem 7.III.1. The Fourier-cosine transform $\mathcal{F}_0$ is a topological isomorphism from $\mathcal{S}_{*,-\frac{1}{2}}(\mathbb{R})$ onto $\mathcal{S}^0_*(\mathbb{R})$.

Proof.

From Theorem 3.II.2 the transform $\mathcal{F}_0$ is a topological isomorphism from $\mathcal{S}_*(\mathbb{R})$ onto itself. Moreover for all f in $\mathcal{S}_*(\mathbb{R})$ and $n \in \mathbb{N}$, we have

$$\frac{d^{2n}}{d\lambda^{2n}} \mathcal{F}_0(f)(0) = (-1)^n \int_0^\infty f(x)\, x^{2n} dx.$$

This gives the result.

Theorem 7.III.2. The Fourier-Bessel transform $\mathfrak{F}_B$ is a topological isomorphism from $\mathcal{S}_{*,\alpha}(\mathbb{R})$ onto $\mathcal{S}^0_*(\mathbb{R})$.

Proof.

From Theorem 5.III.1 the transform $\mathfrak{F}_B$ is a topological isomorphism from $\mathcal{S}_*(\mathbb{R})$ onto itself. Moreover for all f in $\mathcal{S}_*(\mathbb{R})$ and $n \in \mathbb{N}$, we deduce from the relation (1.II.1) :

$$\frac{d^{2n}}{d\lambda^{2n}} \mathfrak{F}_B(f)(0) = \frac{(-1)^n \Gamma(\alpha+1)}{2^{2n}\ n!\Gamma(\alpha+n+1)} \int_0^\infty f(x)\ x^{2n} d\mu_\alpha(x).$$

This gives the result.

Corollary 7.III.1. The transform W_α is a topological isomorphism from $\mathcal{S}_{*,\alpha}(\mathbb{R})$ onto $\mathcal{S}_{*,-\frac{1}{2}}(\mathbb{R})$.

Proof.

From the relation (5.III.1) we have

(7.III.4) $$W_\alpha(f) = 2^\alpha\,\Gamma(\alpha+1)\ \mathfrak{F}_0^{-1} \circ \mathfrak{F}_B(f).$$

We deduce the result from this relation and Theorems 7.III.1 and 7.III.2.

Proposition 7.III.1.

i) For all f in $\mathcal{S}_{*,-\frac{1}{2}}(\mathbb{R})$ and g in $\mathcal{S}_*(\mathbb{R})$ the function $f *_0 g$ belongs to $\mathcal{S}_{*,-\frac{1}{2}}(\mathbb{R})$.

ii) For all f in $\mathcal{S}_{*,\alpha}(\mathbb{R})$ and g in $\mathcal{S}_*(\mathbb{R})$ the function $f *_B g$ belongs to $\mathcal{S}_{*,\alpha}(\mathbb{R})$.

Proof.

i) From the relation (3.II.4) we have the relation

(7.III.5) $$f *_0 g = \mathfrak{F}_0^{-1}\,(\mathfrak{F}_0(f).\ \mathfrak{F}_0(g)).$$

To obtain the result it suffices to use this relation and Theorem 7.III.1.

ii) Using the relation (5.II.9) we deduce that

(7.III.6) $$f *_B g = \mathfrak{F}_B(\mathfrak{F}_B(f).\ \mathfrak{F}_B(g)).$$

This relation and Theorem 7.III.2 give the result.

Proposition 7.III.2. The operator K_0 given by

$$K_0(f) = \mathfrak{F}_0^{-1}\left[\frac{\pi\lambda^{2\alpha+1}}{2^{\alpha+1}\Gamma(\alpha+1)}\mathfrak{F}_0(f)\right], \tag{7.III.7}$$

is a topological isomorphism from $\mathcal{S}_{*,-\frac{1}{2}}(\mathbb{R})$ onto itself.

Proof.

The multiplication operator

$$f \to \frac{\pi\lambda^{2\alpha+1}}{2^{\alpha+1}\Gamma(\alpha+1)} f,$$

is a topological isomorphism from $\mathcal{S}_*^0(\mathbb{R})$ onto itself. The inverse operator is given by

$$f \to \frac{2^{\alpha+1}\Gamma(\alpha+1)}{\pi\lambda^{2\alpha+1}} f.$$

The result derives from Theorem 7.III.1.

Proposition 7.III.3. The operator K_B given by

$$K_B(f) = \mathfrak{F}_B\left[\frac{\pi\lambda^{2\alpha+1}}{2^{\alpha+1}\Gamma(\alpha+1)}\mathfrak{F}_B(f)\right], \tag{7.III.8}$$

is a topological isomorphism from $\mathcal{S}_{*,\alpha}(\mathbb{R})$ onto itself.

Proof.

From the relations (5.III.1) and (7.III.7) we have for all f in $\mathcal{S}_{*,\alpha}(\mathbb{R})$:

$$K_B(f) = W_\alpha^{-1} \circ K_0 \circ W_\alpha(f). \tag{7.III.9}$$

We deduce the result from Proposition 7.III.2 and Corollary 7.III.1.

Remark.

For $\alpha = k + \frac{1}{2}$, $k \in \mathbb{N}$, the operators K_0 and K_B are given respectively by

$$\forall\, x \geq 0,\ K_0(f)(x) = \frac{(-1)^{k+1}\pi}{2^{\alpha+1}\Gamma(\alpha+1)}\left(\frac{d^2}{dx^2}\right)^{k+1} f(x),\ f \in \mathcal{S}_{*,-\frac{1}{2}}(\mathbb{R}). \tag{7.III.10}$$

$$\forall\, x \geq 0,\ K_B(f)(x) = \frac{(-1)^{k+1}\pi}{2^{\alpha+1}\Gamma(\alpha+1)}\ \ell_\alpha^{k+1}(f)(x),\ \ f \in \mathcal{S}_{*,\alpha}(\mathbb{R}). \tag{7.III.11}$$

Where ℓ_α is the Bessel operator given by

$$\ell_\alpha = \frac{d^2}{dx^2} + \frac{2\alpha+1}{x}\frac{d}{dx}. \tag{7.III.12}$$

Proposition 7.III.4.

i) For all f in $\mathcal{S}_{*,-\frac{1}{2}}(\mathbb{R})$ and g in $\mathcal{S}_*(\mathbb{R})$ we have

(7.III.13) $$K_0(f *_0 g) = K_0(f) *_0 g.$$

ii) For all f in $\mathcal{S}_{*,\alpha}(\mathbb{R})$ and g in $\mathcal{S}_*(\mathbb{R})$ we have

(7.III.14) $$K_B(f *_B g) = K_B(f) *_B g.$$

Proof.

i) From the relations (7.III.7) and (3.II.4) we have

$$\begin{aligned} K_0(f *_0 g) &= \mathcal{F}_0^{-1}\Big[\frac{\pi\,\lambda^{2\alpha+1}}{2^{\alpha+1}\,\Gamma(\alpha+1)}\mathcal{F}_0(f *_0 g)\Big], \\ &= \mathcal{F}_0^{-1}\Big[\frac{\pi\,\lambda^{2\alpha+1}}{2^{\alpha+1}\,\Gamma(\alpha+1)}\mathcal{F}_0(f)\mathcal{F}_0(g)\Big], \\ &= \Big\{\mathcal{F}_0^{-1}\Big[\frac{\pi\,\lambda^{2\alpha+1}}{2^{\alpha+1}\,\Gamma(\alpha+1)}\mathcal{F}_0(f)\Big]\Big\} *_0 g, \\ &= K_0(f) *_0 g. \end{aligned}$$

ii) Using the relation (7.III.8) and (5.II.9) we obtain

$$\begin{aligned} K_B(f *_B g) &= \mathcal{F}_B\Big[\frac{\pi\,\lambda^{2\alpha+1}}{2^{\alpha+1}\,\Gamma(\alpha+1)}\mathcal{F}_B(f *_B g)\Big], \\ &= \mathcal{F}_B\Big[\frac{\pi\,\lambda^{2\alpha+1}}{2^{\alpha+1}\,\Gamma(\alpha+1)}\mathcal{F}_B(f)\,.\,\mathcal{F}_B(g)\Big], \\ &= \Big\{\mathcal{F}_B\Big[\frac{\pi\,\lambda^{2\alpha+1}}{2^{\alpha+1}\,\Gamma(\alpha+1)}\mathcal{F}_B(f)\Big]\Big\} *_B g. \\ &= K_B(f) *_B g. \end{aligned}$$

Theorem 7.III.3. For all f in $\mathcal{S}_{*,\alpha}(\mathbb{R})$ we have the following inversion formulas for the operator $\mathfrak{R}_\alpha$:

(7.III.15) $$\mathfrak{R}_\alpha K_0 W_\alpha(f) = 2^\alpha\Gamma(\alpha+1)f\,.$$

(7.III.16) $$\mathfrak{R}_\alpha W_\alpha K_0(f) = 2^\alpha\Gamma(\alpha+1)f.$$

Proof.

Using Theorem 5.II.2, the relations (2.I.15), (5.III.1), Propositions 2.I.1 and (7.III.2) we obtain

$$\begin{aligned} \forall\, x \geq 0,\ f(x) &= \int_0^\infty \mathcal{F}_B(f)(\lambda)\, j_\alpha(\lambda x)\, d\mu_\alpha(\lambda), \\ &= \mathfrak{R}_\alpha\Big[\int_0^\infty \mathcal{F}_B(f)(\lambda)\cos(\lambda.)\, d\mu_\alpha(\lambda)\Big](x)\,, \end{aligned}$$

$$= 2^{-\alpha}(\Gamma(\alpha+1))^{-1}\mathcal{R}_\alpha[\frac{2}{\pi}\int_0^\infty \frac{\pi\,\lambda^{2\alpha+1}}{2^{\alpha+1}\,\Gamma(\alpha+1)}\mathfrak{F}_0 \circ W_\alpha(f)(\lambda)\cos(\lambda.)d\lambda](x)\,,$$

$$= 2^{-\alpha}(\Gamma(\alpha+1))^{-1}\mathcal{R}_\alpha\{\mathfrak{F}_0^{-1}\ [\frac{\pi\,\lambda^{2\alpha+1}}{2^{\alpha+1}\,\Gamma(\alpha+1)}\mathfrak{F}_0\,(W_\alpha(f))]\}(x).$$

Thus

$$\forall\ x \geq 0,\ f(x) = 2^{-\alpha}(\Gamma(\alpha+1))^{-1}\mathcal{R}_\alpha\ K_0\,(W_\alpha(f))(x).$$

Which is the relation (7.III.15).

We deduce the relation (7.III.16) from the relations (7.III.9) and (7.III.15).

Corollary 7.III.2. The transform $\mathcal{R}_\alpha$ is a topological isomorphism from $\mathcal{S}_{*,-\frac{1}{2}}(\mathbb{R})$ onto $\mathcal{S}_{*,\alpha}(\mathbb{R})$.

Proof.

We deduce the result from Proposition 7.III.2, Corollary 7.III.1 and the relation (7.III.15).

Theorem 7.III.4. For all f in $\mathcal{S}_{*,-\frac{1}{2}}(\mathbb{R})$, we have the following inversion formulas for the operator W_α:

(7.III.17) $$W_\alpha\,\mathcal{R}_\alpha\,K_0\,(f) = 2^\alpha\Gamma(\alpha+1)f\,.$$

(7.III.18) $$W_\alpha\ K_B\,\mathcal{R}_\alpha(f) = 2^\alpha\Gamma(\alpha+1)f\,.$$

Proof.

From Corollary 7.III.1 (resp. 7.III.2) the function $W_\alpha^{-1}(f)$ (resp. $\mathcal{R}_\alpha(f)$) belongs to $\mathcal{S}_{*,\alpha}(\mathbb{R})$. Then by writing the fomula (7.III.15) (resp.7.III.16) for the function $W_\alpha^{-1}(f)$(resp. $\mathcal{R}_\alpha(f)$), we obtain the relation (7.III.17) (resp. 7.III.18).

Corollary 7.III.3. For all f in $\mathcal{S}_{*,-\frac{1}{2}}(\mathbb{R})$ and g in $\mathcal{S}_*(\mathbb{R})$ we have

(7.III.19) $$\mathcal{R}_\alpha(f *_0 g) = \mathcal{R}_\alpha(f\,) *_B W_\alpha^{-1}(g).$$

Proof.

Using the relation (7.III.16) and Propositions 7.III.4 ii), 5.III.7 we obtain

$$\mathcal{R}_\alpha^{-1}[\mathcal{R}_\alpha(f) *_B W_\alpha^{-1}(f)] = 2^{-\alpha}(\Gamma(\alpha+1))^{-1}\,W_\alpha\,K_B[\mathcal{R}_\alpha(f\,) *_B W_\alpha^{-1}(g)],$$

$$= 2^{-\alpha}(\Gamma(\alpha+1))^{-1}W_\alpha[K_\alpha\mathcal{R}_\alpha(f\,) *_B\,W_\alpha^{-1}(g)],$$

$$= 2^{-\alpha}(\Gamma(\alpha+1))^{-1}W_\alpha\,K_B\mathcal{R}_\alpha(f\,) *_0 g.$$

But from the relation (7.III.18) we have

$$2^{-\alpha}(\Gamma(\alpha+1))^{-1}W_\alpha K_B \mathcal{R}_\alpha(f) = f.$$

Thus

$$\mathcal{R}_\alpha^{-1}[\mathcal{R}_\alpha(f) *_B W_\alpha^{-1}(f)] = f *_0 g.$$

Which gives the result.

7.IV. Inversion formulas for the operator $\mathcal{R}_\alpha$ and W_α using wavelets associated with the Bessel operator.

Theorem 7.IV.1. Let g be a wavelet associated with the Bessel operator in $\mathcal{S}_{*,\alpha}(\mathbb{R})$. Then

i) For all f in $\mathcal{S}_{*,\alpha}(\mathbb{R})$ we have

$$\forall\, a>0,\ \forall\, b\geq 0,\ \Psi_g(f)(a,b) = \mathcal{R}_\alpha[\,S_{W_\alpha(g)}(\mathcal{R}_\alpha^{-1}(f))(a,.)](b).$$

ii) For all f in $\mathcal{S}_{*,-\frac{1}{2}}(\mathbb{R})$ we have

$$\forall\, a>0,\ \forall\, b\geq 0,\ S_{W_\alpha(g)}(f)(a,b) = 2^\alpha\,\Gamma(\alpha+1)\,W_\alpha\,[\Psi_g(W_\alpha^{-1}(f))(a,.)](b).$$

Proof.

i) using the relation (7.II.20) and Corollary 7.III.3, we obtain

$$\forall\, a>0,\ \forall\, b\geq 0,\ \Psi_g(f)(a,b) = \sqrt{a}\, f *_B \bar{g}_a(b),$$
$$= \sqrt{a}\ \mathcal{R}_\alpha[\mathcal{R}_\alpha^{-1}(f) *_0 W_\alpha(\bar{g}_a)](b).$$

But from Proposition 7.II.3, we have

$$W_\alpha(\bar{g}_a) = \frac{1}{\sqrt{a}}\ H_a \circ \overline{W_\alpha(g_a)}\ .$$

Then

$$\forall\, a>0,\ \forall\, b\geq 0,\ \Psi_g(f)(a,b) = \mathcal{R}_\alpha[\,\mathcal{R}_\alpha^{-1}(f) *_0 \overline{H_a \circ W_\alpha(g)}\,](b),$$
$$= \mathcal{R}_\alpha[S_{W_\alpha(g)}(\mathcal{R}_\alpha^{-1}(f))\ (a,.)](b).$$

Where S is the classical continuous wavelet transform on $[0,+\infty[$.

ii) We have

$$\forall\, a>0,\ \forall\, b\geq 0,\ S_{W_\alpha(g)}(f)(a,b) = [f *_0 H_a \circ \overline{W_\alpha(g)}\,](b).$$

By Proposition 5.III.7 we obtain

$$\forall\, a > 0,\ \forall\, b \geq 0,\ S_{W_\alpha(g)}(f)(a,b) = 2^{\alpha}\Gamma(\alpha+1) \times \\ W_\alpha[W_\alpha^{-1}(f) *_B \overline{W_\alpha^{-1} \circ H_a \circ W_\alpha(g)}]\ (b).$$

But from Proposition 7.II.3 we have

$$W_\alpha^{-1} \circ H_a \circ W_\alpha(g) = \sqrt{a}\, g_a.$$

Thus

$$\forall\, a > 0,\ \forall\, b \geq 0,\ S_{W_\alpha(g)}(f)(a,b) = 2^{\alpha}\Gamma(\alpha+1) W_\alpha[\, W_\alpha^{-1}(f) *_B \bar{g}_a](b), \\ = 2^{\alpha}\Gamma(\alpha+1) W_\alpha[\Psi_g(\, W_\alpha^{-1}(f))(a,.)](b).$$

Proposition 7.IV.1.

i) If g is a classical wavelet on $[0, +\infty[$ in $\mathcal{S}_{*,-\frac{1}{2}}(\mathbb{R})$. Then the function $K_0(g)$ is a classical wavelet on $[0, +\infty[$ in $\mathcal{S}_{*,-\frac{1}{2}}(\mathbb{R})$ and we have for all $a > 0$:

(7.VI.1) $$K_0 H_a(g) = \frac{1}{a^{2\alpha+1}} H_a K_0(g).$$

ii) If g is a wavelet associated with the Bessel operator in $\mathcal{S}_{*,\alpha}(\mathbb{R})$. Then the function $K_B(g)$ is a wavelet associated with the Bessel operator in $\mathcal{S}_{*,\alpha}(\mathbb{R})$ and we have for all $a > 0$:

(7.IV.2) $$K_B(g_a) = \frac{1}{a^{2\alpha+1}} (K_B(g))_a\,.$$

Proof.

i) Let g be a classical wavelet on $[0, +\infty[$ in $\mathcal{S}_{*,-\frac{1}{2}}(\mathbb{R})$. From the relation (7.III.7) we have

(7.IV.3) $$\forall\, \lambda \geq 0,\ \mathcal{F}_0(K_0(g))(\lambda) = \frac{\pi\, \lambda^{2\alpha+1}}{2^{\alpha+1}\, \Gamma(\alpha+1)} \mathcal{F}_0(g)(\lambda).$$

Proposition 7.I.1 implies that $K_0(g)$ is a classical wavelet on $[0, +\infty[$.

We deduce the relation (7.IV.1) from (7.IV.3) and Proposition 7.I.3.

ii) Let g be a wavelet associated with the Bessel operator in $\mathcal{S}_{*,\alpha}(\mathbb{R})$. From the relation (7.III.8) we have

(7.IV.4) $$\forall\, \lambda \geq 0,\ \mathcal{F}_B(K_B(g))(\lambda) = \frac{\pi\, \lambda^{2\alpha+1}}{2^{\alpha+1}\, \Gamma(\alpha+1)} \mathcal{F}_B(g)(\lambda).$$

Using Proposition 7.II.1, we deduce that $K_B(g)$ is a wavelet associated with the Bessel operator.

On the other hand from the relation (7.IV.4) we have

$$\forall \lambda \geq 0,\ \mathfrak{F}_B(K_B(g_a))(\lambda) = \frac{\pi\,\lambda^{2\alpha+1}}{2^{\alpha+1}\,\Gamma(\alpha+1)}\mathfrak{F}_B(g_a)(\lambda),$$

$$= \frac{\pi\,\lambda^{2\alpha+1}}{2^{\alpha+1}\,\Gamma(\alpha+1)}\mathfrak{F}_B(g)(a\lambda),$$

$$\forall \lambda \geq 0,\ \mathfrak{F}_B(K_B(g_a))(\lambda) = \frac{1}{a^{2\alpha+1}}\mathfrak{F}_B(K_B(g)_a)(\lambda).$$

From Proposition 7.III.3 and Theorem 5.III.1 we deduce that

$$K_B(g_a) = \frac{1}{a^{2\alpha+1}}\,(K_B(g))_a.$$

Theorem 7.IV.2. Let g be a wavelet associated with the Bessel operator in $\mathcal{S}_{*,\alpha}(\mathbb{R})$.

Then

i) For all f in $\mathcal{S}_{*,\alpha}(\mathbb{R})$, we have for all $a > 0$ and $b \geq 0$:

$$\text{(7.IV.5)}\qquad \Psi_g(f)(a,b) = \frac{2^{-\alpha}(\Gamma(\alpha+1))^{-1}}{a^{2\alpha+1}}\,\mathfrak{R}_\alpha[\,S_{K_0W_\alpha(g)}(W_\alpha(f))\ (a,.)](b).$$

ii) For all f in $\mathcal{S}_{*,-\frac{1}{2}}(\mathbb{R})$, we have for all $a > 0$ and $b \geq 0$:

$$\text{(7.IV.6)}\qquad S_{W_\alpha(g)}(f)(a,b) = \frac{1}{a^{2\alpha+1}}W_\alpha[\,\Psi_{K_B(g)}(\mathfrak{R}_\alpha(f))(a,.)](b).$$

Proof.

i) From the relation (7.II.20) we have

$$\forall a > 0,\ \ \forall b \geq 0,\ \Psi_g(f)(a,b) = \sqrt{a}\, f *_B \bar{g}_a(b).$$

Using Corollary 7.III.3, we obtain

$$\forall a > 0,\ \ \forall b \geq 0,\ \Psi_g(f)(a,b) = \sqrt{a}\,\mathfrak{R}_\alpha[W_\alpha(f) *_0 \mathfrak{R}_\alpha^{-1}(\bar{g}_a)](b).$$

But from the relations (7.III.15) and (7.II.13) we have

$$\mathfrak{R}_\alpha^{-1}(\bar{g}_a) = 2^{-\alpha}(\Gamma(\alpha+1))^{-1}K_0\,W_\alpha(\bar{g}_a),$$

$$= \frac{2^{-\alpha}(\Gamma(\alpha+1))^{-1}}{\sqrt{a}}K_0\,H_aW_\alpha(\bar{g}),$$

$$= \frac{2^{-\alpha}(\Gamma(\alpha+1))^{-1}}{a^{2\alpha+3/2}}\overline{H_aK_0W_\alpha(g)}\ .$$

Thus

$$\forall\, a>0,\ \forall\, b\geq 0,\ \Psi_g(f)(a,b) = \frac{2^{-\alpha}(\Gamma(\alpha+1))^{-1}}{a^{2\alpha+1}} \mathcal{R}_\alpha[W_\alpha(f) *_0 \overline{H_a K_0 W_\alpha(g)}](b),$$
$$= \frac{2^{-\alpha}(\Gamma(\alpha+1))^{-1}}{a^{2\alpha+1}} \mathcal{R}_\alpha[S_{K_0 W_\alpha(g)}(W_\alpha(f))(a,.)](b).$$

ii) From Theorem 7.IV.1. ii), we have

$$\forall\, a>0,\ \forall\, b\geq 0,\ S_{W_\alpha(g)}(f)(a,b) = 2^\alpha\Gamma(\alpha+1)\ \sqrt{a}\, W_\alpha\,[W_\alpha^{-1}(f) *_B \bar{g}_a](b).$$

But from the relation (7.III.18) we have

$$W_\alpha^{-1}(f) = 2^{-\alpha}(\Gamma(\alpha+1))^{-1} K_B\, \mathcal{R}_\alpha(f).$$

Thus

$$\forall\, a>0,\ \forall\, b\geq 0,\ S_{W_\alpha(g)}(f)(a,b) = \sqrt{a}\, W_\alpha\,[K_B\, \mathcal{R}_\alpha(f) *_B \bar{g}_a](b).$$

Using Propositions 7.III.4 and 7.IV.1.ii) we obtain

$$\forall\, a>0,\ \forall\, b\geq 0,\ S_{W_\alpha(g)}(f)(a,b) = \sqrt{a}\, W_\alpha\,[\mathcal{R}_\alpha(f) *_B \overline{K_B(g_a)}](b),$$
$$= \frac{1}{a^{2\alpha+\frac{1}{2}}} W_\alpha\,[\mathcal{R}_\alpha(f) *_B \overline{(K_B(g))_a}](b),$$
$$= \frac{1}{a^{2\alpha+1}} W_\alpha\,[\Psi_{K_B(g)}(\mathcal{R}_\alpha(f))(a,.)](b).$$

Theorem 7.IV.3. Let g be a wavelet associated with the Bessel operator in $\mathcal{S}_{*,\alpha}(\mathbb{R})$. Then

i) For all f in $\mathcal{S}_{*,-\frac{1}{2}}(\mathbb{R})$, we have

(7.IV.7) $$\forall\, x\geq 0,\ W_\alpha^{-1}(f)(x) = \frac{2^{-\alpha}(\Gamma(\alpha+1))^{-1}}{C_g}\int_0^\infty\left(\int_0^\infty \mathcal{R}_\alpha[S_{K_0 W_\alpha(g)}(f)(a,.)](b)\times g_{a,b}(x)\,d\mu_\alpha(b)\right)\frac{da}{a^{2\alpha+3}}.$$

ii) For all f in $\mathcal{S}_{*,\alpha}(\mathbb{R})$, we have

(7.IV.8) $$\forall\, x\geq 0,\ \mathcal{R}_\alpha^{-1}(f)(x) = \frac{1}{C_g^0}\int_0^\infty\left(\int_0^\infty W_\alpha[\Psi_{K_B(g)}(f)(a,.)](b)(W_\alpha(g))^0_{a,b}(x)\right)\frac{da}{a^{2\alpha+3}}.$$

Proof.

i) We deduce the result from Theorems 7.IV.2 and 7.II.5.

ii) The result derives from Theorems 7.IV.2 and 7.I.3.

In the following we shall give another inversion formula for the operator W_α using the Calder'on reproducing formula given in Theorem 7.II.4.

Proposition 7.IV.2. Let f be in $L^1([0, +\infty[, d\mu_\alpha)$ and g in $L^\infty([0, +\infty[, dx)$. Then

(7.IV.9) $$\mathcal{R}_\alpha(g *_0 W_\alpha(f)) = 2^\alpha \Gamma(\alpha+1)\, (\mathcal{R}_\alpha(g)) *_B f.$$

Proof.

From Theorem 2.I.4, the function $\mathcal{R}_\alpha(g)$ belongs to $L^\infty([0, +\infty[, dx)$. Then we have

$$\forall\, x \geq 0,\ \mathcal{R}_\alpha(g) *_B f(x) = \int_0^\infty T_x(f)\,(y)\, \mathcal{R}_\alpha(g)(y)d\mu_\alpha(y).$$

Using the relation (2.II.2) we obtain

$$\forall\, x \geq 0,\ \mathcal{R}_\alpha(g) *_B f(x) = \frac{1}{2^\alpha\ \Gamma(\alpha+1)} \int_0^\infty W_\alpha(T_x(f))(y)g(y)dy.$$

But from Proposition 5.III.6, the density of $\mathcal{D}_*(\mathbb{R})$ in $L^1([0, +\infty[, d\mu_\alpha)$, the continuity of the operator W_α from $L^1([0, +\infty[, d\mu_\alpha)$ into $L^1([0, +\infty[, dx)$, and the continuity of the operators σ_y and $\mathcal{R}_\alpha$ from $L^1([0, +\infty[\ dx)$ into itself, we deduce that

$$W_\alpha(T_x(f))(y) = \mathcal{R}_\alpha\, (\sigma_y\, W_\alpha(f))(y).$$

Hence

$$\forall\, x \geq 0,\ \mathcal{R}_\alpha(g) *_B f(x) = 2^{-\alpha}(\Gamma(\alpha+1))^{-1} \int_0^\infty \mathcal{R}_\alpha(\sigma_y\, W_\alpha(f))(y)\ g(y)dy,$$

$$\forall\, x \geq 0,\ \mathcal{R}_\alpha(g) *_B f(x) = 2^{-\alpha}(\Gamma(\alpha+1))^{-1}\mathcal{R}_\alpha[\int_0^\infty \sigma_y\, W_\alpha(f)(.)g(y)\ dy](x),$$

$$= 2^{-\alpha}(\Gamma(\alpha+1))^{-1}\mathcal{R}_\alpha(g\ *_0 W_\alpha(f))(x).$$

Proposition 7.IV.3. Let g be in $(L^1 \cap L^2)$ $([0, \infty[, dx)$ such that $\mathcal{F}_0(g)$ belongs to $L^1([0, +\infty[, \frac{2}{\pi}d\lambda)$ and satisfies

(7.IV.10) $$\exists\, \gamma > \alpha + \frac{1}{2},\ \text{such that } \mathcal{F}_0(g)(\lambda) = O(\lambda^\gamma),\ \lambda \to 0^+.$$

Then the function $\mathcal{R}_\alpha(g)$ belongs to $L^1([0, +\infty[, d\mu_\alpha)$ and we have for almost all $\lambda > 0$:

(7.IV.11) $$\mathcal{F}_B(\mathcal{R}_\alpha(g))(\lambda) = \frac{2^{\alpha+1}\ \Gamma(\alpha+1)}{\pi}\, \frac{\mathcal{F}_0(g)(\lambda)}{\lambda^{2\alpha+1}}.$$

Proof.

By the formula (3.II.7) we have

$$g(x) = \frac{2}{\pi}\int_0^\infty \mathfrak{F}_0(g)(\lambda)\cos\lambda x\, d\lambda, \text{ a.e.}$$

So, using (2.I.15) and Theorem 2.I.4, we have

$$\mathfrak{R}_\alpha(g)(x) = \frac{2}{\pi}\int_0^\infty \mathfrak{F}_0(g)(\lambda)\, j_\alpha(\lambda x) d\lambda, \text{ a.e.}$$

$$\mathfrak{R}_\alpha(g)(x) = \int_0^\infty h(\lambda)\, j_\alpha(\lambda x) d\mu_\alpha(\lambda), \text{ a.e.}$$

With

$$\forall\, \lambda > 0,\ h(\lambda) = \frac{2^{\alpha+1}\Gamma(\alpha+1)}{\pi\, \lambda^{2\alpha+1}}\mathfrak{F}_0(g)(\lambda).$$

In view of Theorem 5.IV.2 it suffices to prove that the function h belongs to $L^2([0, +\infty[, d\mu_\alpha)$. We have

$$\int_0^\infty |h(\lambda)|^2 d\mu_\alpha(\lambda) = \frac{2^{\alpha+2}\Gamma(\alpha+1)}{\pi^2}\int_0^\infty |\mathfrak{F}_0(g)(\lambda)|^2 \frac{d\lambda}{\lambda^{2\alpha+1}},$$

$$= \frac{2^{\alpha+2}\Gamma(\alpha+1)}{\pi^2}\int_0^1 |\mathfrak{F}_0(g)(\lambda)|^2 \frac{d\lambda}{\lambda^{2\alpha+1}} + \frac{2^{\alpha+2}\Gamma(\alpha+1)}{\pi^2}\int_1^\infty |\mathfrak{F}_0(g)(\lambda)|^2 \frac{d\lambda}{\lambda^{2\alpha+1}} = I_1 + I_2.$$

But

$$I_1 \leq \frac{2^{\alpha+2}\Gamma(\alpha+1)}{\pi^2}\int_0^1 \frac{d\lambda}{\lambda^{2\alpha-2\gamma+1}} < +\infty,$$

and from Theorem 3.II.4 :

$$I_2 \leq \frac{2^{\alpha+2}\Gamma(\alpha+1)}{\pi^2}\int_0^\infty |\mathfrak{F}_0(g)(\lambda)|^2 d\lambda,$$

$$\leq \frac{2^{\alpha+1}\Gamma(\alpha+1)}{\pi}\|g\|_2^2 < +\infty.$$

This completes the proof.

Proposition 7.IV.4. Let g be a nonzero function in $(L^1 \cap L^2)([0, +\infty[, dx)$ such that $\mathfrak{F}_0(g)$ belongs to $L^1([0, +\infty[, \frac{2}{\pi}d\lambda)$ and satisfies

(7.IV.12) $\exists\, \gamma > 2\alpha + 1$, such that $\mathfrak{F}_0(g) = O(\lambda^\gamma),\ \lambda \to 0^+$.

Then the function $\mathfrak{R}_\alpha(g)$ is a wavelet associated with the Bessel operator in

$L^2([0, +\infty[, d\mu_\alpha)$ and the function $\mathfrak{F}_B(\mathcal{R}_\alpha(g))$ belongs to $L^\infty([0, +\infty[, dx)$.

Proof.

By Proposition 7.IV.3 the function $\mathcal{R}_\alpha(g)$ belongs to $L^2([0, +\infty[, d\mu_\alpha)$ and we have for almost all $\lambda > 0$:

$$\mathfrak{F}_B(\mathcal{R}_\alpha(g))(\lambda) = \frac{2^{\alpha+1}\Gamma(\alpha+1)}{\pi} \cdot \frac{\mathfrak{F}_0(g)(\lambda)}{\lambda^{2\alpha+1}}.$$

Using Proposition 5.II.1 and the relation (7.IV.12) we see that the function $\mathfrak{F}_B(\mathcal{R}_\alpha(g))$ is bounded and that

$$\mathfrak{F}_B(\mathcal{R}_\alpha(g))(\lambda) = O(\lambda^{\gamma-2\alpha-1}),\ \lambda \to 0^+.$$

Thus, in view of Proposition 7.II.1, the function $\mathfrak{F}_B(\mathcal{R}_\alpha(g))$ is a wavelet associated with the Bessel operator.

Proposition 7.IV.5. Let g be in $(L^1 \cap L^2)([0, +\infty[, dx)$ such that the function $\mathfrak{F}_0(g)$ belongs to $L^1([0, +\infty[, \frac{2}{\pi}d\lambda)$ and satisfies (7.IV.10). Then for all $a > 0$ we have

(7.IV.13) $$(\mathcal{R}_\alpha(g))_a = \frac{1}{a^{2\alpha+3/2}}\mathcal{R}_\alpha(g_a^0),$$

where g_a^0 is given by

$$\forall\, x \geq 0,\ g_a^0(x) = H_a(g)(x) = \frac{1}{\sqrt{a}}\, g(\frac{x}{a}).$$

Proof.

Using Proposition 7.IV.3 and the relation (7.II.11) we have for almost all $\lambda > 0$:

$$\mathfrak{F}_B((\mathcal{R}_\alpha(g))_a)(\lambda) = \mathfrak{F}_B(\mathcal{R}_\alpha(g))(a\lambda),$$

$$= \frac{2^{\alpha+1}\Gamma(\alpha+1)}{\pi a^{2\alpha+1}} \cdot \frac{\mathfrak{F}_0(g)(a\lambda)}{\lambda^{2\alpha+1}}.$$

Since

$$\mathfrak{F}_0(g_a^0)(\lambda) = \sqrt{a}\,\mathfrak{F}_0(g)(a\lambda).$$

Thus

$$\mathfrak{F}_B((\mathcal{R}_\alpha(g))_a)(\lambda) = \frac{2^{\alpha+1}\Gamma(\alpha+1)}{\pi a^{2\alpha+3/2}} \cdot \frac{\mathfrak{F}_0(g_a^0)(\lambda)}{\lambda^{2\alpha+1}},$$

$$= \frac{1}{a^{2\alpha+3/2}}\mathfrak{F}_B(\mathcal{R}_\alpha(g_a^0))(\lambda).$$

We deduce the result from Theorem 5.IV.2.

Theorem 7.IV.4. Let g be a classical wavelet on $[0, +\infty[$ in $(L^1 \cap L^2)([0, +\infty[, dx)$ such that the function $\mathfrak{F}_0(g)$ belongs to $L^1([0, +\infty[, \frac{2}{\pi}d\lambda)$ and satisfies (7.IV.12). Then for all f in $L^1([0, +\infty[, d\mu_\alpha)$ we have

(7.IV.14) $$\forall a > 0, \forall b \geq 0, \Psi_{\mathcal{R}_\alpha(g)}(f)(a,b) = \frac{2^{-\alpha}(\Gamma(\alpha+1))^{-1}}{a^{2\alpha+1}} \mathcal{R}_\alpha[S_g(W_\alpha(f))(a,.)](b).$$

Proof.

By the relation (7.II.20) and Proposition 7.IV.5 we have

$$\forall a > 0, \forall b \geq 0, \Psi_{\mathcal{R}_\alpha(g)}(f)(a,b) = \sqrt{a}\, f *_B \overline{(\mathcal{R}_\alpha(g))_a}(b),$$
$$= \frac{1}{a^{2\alpha+1}} f *_B \overline{\mathcal{R}_\alpha(g_a^0)}(b).$$

Using Propostion 7.IV.2

$$\forall a > 0, \forall b \geq 0, \Psi_{\mathcal{R}_\alpha(g)}(f)(a,b) = \frac{2^{-\alpha}(\Gamma(\alpha+1))^{-1}}{a^{2\alpha+1}} \mathcal{R}_\alpha[W_\alpha(f) *_0 \overline{g_a^0}](b),$$
$$= \frac{2^{-\alpha}(\Gamma(\alpha+1))^{-1}}{a^{2\alpha+1}} \mathcal{R}_\alpha[S_g(W_\alpha(f))(a,.)](b).$$

Theorem 7.IV.5. Let g be a classical wavelet on $[0, +\infty[$ in $(L^1 \cap L^2)([0, +\infty[, dx)$ such that the function $\mathfrak{F}_0(g)$ belongs to $L^1([0, +\infty[, \frac{2}{\pi}d\lambda)$ and satisfies (7.IV.12). Then for f in $(L^1 \cap L^2)([0, +\infty[, d\mu_\alpha)$ and $0 < \varepsilon < \delta < +\infty$, the function

$$f_{\varepsilon,\delta}(x) = \frac{2^{-\alpha}(\Gamma(\alpha+1))^{-1}}{C_{\mathcal{R}_\alpha(g)}} \int_\varepsilon^\delta \int_0^\infty \mathcal{R}_\alpha[S_g(W_\alpha(f))(a,.)](b)\, (\mathcal{R}_\alpha(g))_{a,b}(x) d\mu_\alpha(b) \frac{da}{a^{2\alpha+3}},$$

possesses the following property

$$\lim_{\substack{\varepsilon \to 0^+ \\ \delta \to +\infty}} \|f_{\varepsilon,\delta} - f\|_{2,\alpha} = 0.$$

Proof.

We deduce the result from Theorems 7.IV.4 and 7.II.4.

7.V. Problems.

Problem 7.V.1.

A classical wavelet on $\mathbb{R}^n$, $n \geq 1$, is a function g in $L^2(\mathbb{R}^n)$ (the space of square integrable functions on $\mathbb{R}^n$ with respect to the Lebesgue measure) satisfying the following admissibility condition : There exists a constant C_g^0 such that

$$0 < C_g^0 = \int_0^{\infty} |\mathcal{F}(g)(a\lambda)|^2 \frac{da}{a} < +\infty,$$

for almost all $\lambda \in \mathbb{R}^n$. Here $\mathcal{F}$ is the usual Fourier transform on $\mathbb{R}^n$ given by

$$\mathcal{F}(f)(\xi) = \int_{\mathbb{R}^n} f(x)\, e^{-i(x,\xi)}\, dx.$$

Where $(.,.)$ is the scalar product of $\mathbb{R}^n$.

1°) Let $t \in]0, +\infty[$ and $k \in \mathbb{N}^*$.

i) Show that the function

$$\forall\, x \geq 0,\ g(x) = (-1)^k \frac{d^k}{dt^k}(e^{-t\|x\|^2}),$$

is a classical wavelet on $\mathbb{R}^n$.

ii) Give the value of C_g^0.

2°) Let g be an even function in $(L^1 \cap L^2)(\mathbb{R})$ such that $\mathcal{F}(g) \in L^1(\mathbb{R})$ (the space of integrable functions on $\mathbb{R}$ with respect to the Lebesgue measure) and satisfying

$$\exists\, \gamma > n-1 \text{ such that } \mathcal{F}(g)(\lambda) = O(\lambda^{\gamma}), \text{ as } \lambda \to 0^+.$$

Prove that the function ${}^t R(g)$ is a radial classical wavelet on $\mathbb{R}^n$ and $\mathcal{F}({}^t R(g)) \in L^{\infty}(\mathbb{R}^n)$ (the space of essentially bounded functions on $\mathbb{R}^n$ with respect to the Lebesgue measure). Where ${}^t R$ is the dual Radon transform defined in Problem 2.VI.1.

3°) Let g be a classical wavelet on $\mathbb{R}^n$ in $L^2(\mathbb{R}^n)$ such that $\mathcal{F}(g)$ is in $L^{\infty}(\mathbb{R}^n)$. For $a \in]0, +\infty[$ and $0 < \varepsilon < \delta < +\infty$, define the following functions

$$g_a(x) = a^{-n/2} g\left(\frac{x}{a}\right),\quad x \in \mathbb{R}^n.$$

$$G_{\varepsilon,\delta}(x) = \frac{1}{C_g^0} \int_{\varepsilon}^{\delta} g_a * \bar{g}_a(x) \frac{da}{a^{n+1}},\ x \in \mathbb{R}^n.$$

$$K_{\varepsilon,\delta}(\lambda) = \frac{1}{C_g^0} \int_{\varepsilon}^{\delta} |\mathcal{F}(g)(a\lambda)|^2 \frac{da}{a},\ \lambda \in \mathbb{R}^n.$$

Where $*$ is the usual convolution product on $\mathbb{R}^n$.
Prove that $G_{\varepsilon,\delta} \in L^2(\mathbb{R}^n)$, $K_{\varepsilon,\delta} \in L^1(\mathbb{R}^n)$, and $\mathcal{F}(G_{\varepsilon,\delta}) = K_{\varepsilon,\delta}$.

(See [52] [66]).

Problem 7.V.2.

Let g be a classical wavelet on $\mathbb{R}^n$ in $L^2(\mathbb{R}^n)$ given in Problem 7.V.1. For $(a,b) \in]0,+\infty[\times \mathbb{R}^n$, define the function $g^0_{a,b}$ on $\mathbb{R}^n$ by

$$g^0_{a,b}(x) = a^{-n/2} g\left(\frac{x-b}{a}\right).$$

Denote $g^0_{a,0}$ by g^0_a.

The classical n-dimensional continuous wavelet transform is defined for regular functions on $\mathbb{R}^n$ by

$$S^{(n)}_g(f)(a,b) = \int_{\mathbb{R}^n} f(x)\overline{g^0_{a,b}(x)}\,dx.$$

1°) Let g be a classical wavelet on $\mathbb{R}^n$ in $(L^1 \cap L^2)(\mathbb{R}^n)$. Show that for f in $L^2(\mathbb{R}^n)$ we have

$$f(.) = \frac{1}{C^0_g}\int_0^\infty \int_{\mathbb{R}^n} S^{(n)}_g(f)(a,b)\, g^0_{a,b}(.)db\, \frac{da}{a^{n+1}},$$

strongly in $L^2(\mathbb{R}^n)$.

2°) Let g be a classical wavelet on $\mathbb{R}^n$ in $L^2(\mathbb{R}^n)$ such that $\mathcal{F}(g) \in L^\infty(\mathbb{R}^n)$. Prove that for $f \in L^2(\mathbb{R}^n)$ and $0 < \varepsilon < \delta < +\infty$, the function

$$f^{\varepsilon,\delta}(x) = \frac{1}{C^0_g}\int_0^\infty \int_{\mathbb{R}^n} S^{(n)}_g(f)(a,b) g_{a,b}(x)db\, \frac{da}{a^{n+1}},$$

belongs to $L^2(\mathbb{R}^n)$ and we have

$$\lim_{\substack{\varepsilon\to 0^+\\ \delta\to+\infty}} \|f^{\varepsilon,\delta} - f\|_2 = 0.$$

3°) Let g be a classical wavelet in $L^2(\mathbb{R}^n)$.Prove that for f in $L^1(\mathbb{R}^n)$, such that $\mathcal{F}(f)$ is in $L^1(\mathbb{R}^n)$ we have

$$f(x) = \frac{1}{C^0_g}\int_0^\infty \left(\int_{\mathbb{R}^n} S^{(n)}_g(f)(a,b) g_{a,b}(x)db\right) \frac{da}{a^{n+1}}.$$

Where, for each $x \in \mathbb{R}^n$, both the inner and the outer integral are absolutely convergent, but possibly not the double integral.

(See [52] [66]).

Problem 7.V.3.

Let g be an even classical wavelet on $\mathbb{R}$ (see Problem 7.V.1) in $(L^1 \cap L^2)(\mathbb{R})$ such that $\mathfrak{F}(g) \in L^1(\mathbb{R})$ and satisfying

$$\exists\, \gamma > n - 1 \text{ such that } \mathfrak{F}(g)(\lambda) = O(\lambda^\gamma), \text{ as } \lambda \to 0^+.$$

1°) Prove that for $f \in L^1(\mathbb{R}^n)$ we have

$$S^{(n)}_{{}^tR(g)}(f)(a,b) = a^{(1-n)/2} \int_{S^{n-1}} S^{(1)}_g(R_\theta(f))(a, (b, \theta))d\theta.$$

Where $R_\theta(f)$ is the Radon transform of f defined in Problem 2.VI.1, $(.,.)$ the scalar product of $\mathbb{R}^n$, and $S^{(n)}$ the n-dimensional continuous wavelet transform given in Problem 7.V.2.

2°) Let f be in $L^1(\mathbb{R}^n)$ such that $\mathfrak{F}(f)$ is in $L^1(\mathbb{R}^n)$. Show that in any point x where f is continuous we have

$$f(x) = \frac{1}{C^0_{{}^tR(g)}} \int_0^\infty \Big[\int_{\mathbb{R}^n} \Big(\int_{S^{n-1}} S^{(1)}_g(R_\theta(f))(a, (b, \theta))d\theta\Big) \times ({}^tR(g))^0_{a,b}(x)db\Big] \frac{da}{a^{(3n+1)/2}}.$$

3°) Prove that for f in $(L^1 \cap L^2)(\mathbb{R}^n)$ and $0 < \varepsilon < \delta < +\infty$, the function

$$f^{\varepsilon,\delta}(x) = \frac{1}{C^0_{{}^tR(g)}} \int_\varepsilon^\delta \int_{\mathbb{R}^n} \Big(\int_{S^{n-1}} S^{(1)}_g(R_\theta(f))(a, (b, \theta))d\theta\Big) \times ({}^tR(g))^0_{a,b}(x)db \frac{da}{a^{(3n+1)/2}}$$

satisfies

$$\lim_{\substack{\varepsilon \to 0^+ \\ \delta \to +\infty}} \|f^{\varepsilon,\delta} - f\|_2 = 0.$$

(See [52] [66]).

Problem 7.V.4.

Let $\nu \in M^b([0, +\infty[)$ and $a \in]0, +\infty[$. We define the dilated measure ν_a of ν by

$$\int_0^\infty f(x)\, d\nu_a(x) = \int_0^\infty f(ax) d\nu(x),\ f \in C_c([0, +\infty[).$$

1°) Show that when $\nu = g\mu_\alpha$, with $g \in L^1([0, +\infty[, d\mu_\alpha)$ we have

$$\nu_a = g_a \mu_\alpha,\ a > 0,$$

with

$$g_a(x) = \frac{1}{a^{2\alpha+2}}\, g(\frac{x}{a}) \quad x \geq 0.$$

2°) Let $\nu \in M^b([0, +\infty[)$ and $a \in]0, +\infty[$. Prove that

$$\forall \lambda \geq 0,\ \mathcal{F}_B(\nu_a)(\lambda) = \mathcal{F}_B(\nu)(a\lambda).$$

3°) We consider $\nu \in M^b([0, +\infty[)$ and $f \in L^p([0, +\infty[, d\mu_\alpha)$, $1 \leq p < +\infty$.

Prove that

$$\lim_{\substack{a\to 0\\ a>0}} \nu_a *_B f = \nu([0, +\infty[).f.$$

Where the limit is in $L^p([0, +\infty[, d\mu_\alpha)$, $1 \leq p < +\infty$.

4°) Let $g \in L^1([0, +\infty[, d\mu_\alpha)$ and $f \in L^p([0, +\infty[, d\mu_\alpha)$, $1 < p < +\infty$.

Show that

$$\lim_{a\to +\infty} f *_B g_a = 0,$$

where the limit is in $L^p([0, +\infty[, d\mu_\alpha)$, $1 < p < +\infty$.

(See [65]).

Problem 7.V.5.

Let $\nu \in M^b([0, +\infty[)$ such that

$$\int_0^\infty \mathcal{F}_B(\nu)(\lambda) \frac{d\lambda}{\lambda} < +\infty.$$

We denote by C_ν this integral.

1°) Let $\nu \in M^b([0, +\infty[)$ such that

$$\int_0^\infty |\nu([0, y])| \frac{dy}{y} < +\infty.$$

Prove that the integral C_ν is finite and admits the representation

$$C_\nu = \int_0^\infty \nu([0, y])\frac{dy}{y}.$$

2°) Let $\nu \in M^b([0, +\infty[)$. Show that the following two assumptions on ν are equivalent

i) $\displaystyle\int_0^1 |\nu|([0, y])\frac{dy}{y} < +\infty$ and $\displaystyle\int_1^\infty |\nu|((y, +\infty))\frac{dy}{y} < +\infty.$

ii) $\displaystyle\int_0^\infty |\mathrm{Log}x|\, d|\nu|(x) < +\infty.$

3°) Let $\nu \in M^b([0, +\infty[)$ satisfying $\nu([0, +\infty)) = 0$ and

$$\int_0^\infty |\mathrm{Log}x|\, d|\nu|(x) < +\infty.$$

Prove that the integral C_ν is finite and represented as

$$C_\nu = \int_0^\infty \mathrm{Log}\,\frac{1}{x}\, d\nu(x).$$

(See [65]).

Problem 7.V.6.

1°) Let $\nu \in M^b([0, +\infty[)$. For $0 < \varepsilon < \delta < +\infty$, define

$$G_{\varepsilon,\delta}(x) = \frac{2^\alpha \Gamma(\alpha+1)}{x^{2\alpha+2}}\nu([x/\delta, x/\varepsilon]), \quad x > 0.$$

$$K_{\varepsilon,\delta}(\lambda) = \int_\varepsilon^\delta \mathcal{F}_B(\nu)(a\lambda)\frac{da}{a}, \quad \lambda \geq 0.$$

Prove that $G_{\varepsilon,\delta} \in L^1([0, +\infty[, d\mu_\alpha)$, and we have

$$\mathcal{F}_B(G_{\varepsilon,\delta}) = K_{\varepsilon,\delta} - \nu(\{0\})\,\mathrm{Log}\,\frac{\delta}{\varepsilon}.$$

2°) Let $\nu \in M^b([0, +\infty[)$. We consider for f in $L^1([0, +\infty[, d\mu_\alpha)$, $1 \leq p \leq +\infty$, and $0 < \varepsilon < \delta < +\infty$, the function

$$f^{\varepsilon,\delta}(x) = \int_\varepsilon^\delta f *_B \nu_a(x)\frac{da}{a}, \quad x \geq 0,$$

where ν_a, $a > 0$, is the measure defined in Problem 7.V.4. Show that this function belongs to $L^p([0, +\infty[, d\mu_\alpha)$, $1 \leq p \leq +\infty$, and has the form

$$f^{\varepsilon,\delta}(x) = f *_B G_{\varepsilon,\delta}(x) + \nu(\{0\})\, f(x)\mathrm{Log}\,\frac{\delta}{\varepsilon}.$$

3°) Prove that for f in $L^1([0,+\infty[, d\mu_\alpha)$, we have

$$\mathcal{F}_B(f^{\varepsilon,\delta}) = \mathcal{F}_B(f) . K_{\varepsilon,\delta}$$

4°) Let $\nu \in M^b([0,+\infty[)$ be such that the integeral C_ν defined in Problem 7.V.5 is finite. Show that for all f in $L^1([0,+\infty[, d\mu_\alpha)$ we have

$$\lim_{\substack{\varepsilon\to 0^+\\ \delta\to+\infty}} \|f^{\varepsilon,\delta} - f\|_2 = 0.$$

5°) Let $\nu \in M^b([0,+\infty[)$ satisfying the condition

$$\int_0^1 |\nu([0,y])| \frac{dy}{y} < +\infty.$$

Prove that for all $f \in L^p([0,+\infty[, d\mu_\alpha)$, $1 < p < +\infty$, we have

$$\lim_{\substack{\varepsilon\to 0^+\\ \delta\to+\infty}} f^{\varepsilon,\delta} = C_\nu f,$$

where the limit is in $L^p([0,+\infty[, d\mu_\alpha)$, $1 < p < +\infty$.

(See [65]).

Problem 7.V.7.

1°) Let $\nu \in M^b([0,+\infty[)$ satisfying

$$\int_0^1 |\nu| ([0,y]) \frac{dy}{y} < +\infty.$$

$$\int_1^\infty |\nu ([0,y])| \frac{dy}{y} < +\infty.$$

and

$$\operatorname*{ess\,sup}_{t\geq x} \left| \frac{\nu([0,t])}{t^{2\alpha+2}} \right| \in L^1([0,+\infty[, d\mu_\alpha).$$

We consider $f \in L^1([0,+\infty[, d\mu_\alpha)$, $1 \leq p < +\infty$, and $f^{\varepsilon,\delta}$ the function defined in Problem 7.V.6. Prove that for each Lebesgue point x of f (see Problem 4.IV.9), we have

$$\lim_{\substack{\varepsilon\to 0^+\\ \delta\to+\infty}} f^{\varepsilon,\delta}(x) = C_\nu f(x).$$

Where C_ν is the integral given in Problem 7.V.5.

2°) i) Let f be an essentially bounded function on $[0, +\infty[$ that is weakly oscillating around zero, i.e.

$$\lim_{t \to +\infty} \frac{1}{t^{2\alpha+2}} \int_0^t T_x(f)(y)\, d\mu_\alpha(y) = 0, \quad x \geq 0.$$

Where T_x, $x \geq 0$, are the generalized translation operators given by the relation (4.II.3). Let $g \in L^1([0, +\infty[, d\mu_\alpha)$. Prove that

$$\lim_{a \to +\infty} f *_B g_a(x) = 0,$$

uniformly for $x \in [0, +\infty[$. Where g_a, $a > 0$, is the function given by

$$g_a(x) = \frac{1}{a^{2\alpha+2}} g\left(\frac{x}{a}\right), \quad x \geq 0.$$

ii) Let $\nu \in M^b([0, +\infty[)$ satisfying

$$\int_0^\infty |\nu([0,y])| \frac{dy}{y} < +\infty,$$

and

$$\operatorname*{ess\,sup}_{t \geq x} \left| \frac{\nu([0,t])}{t^{2\alpha+2}} \right| \in L^1([0, +\infty[, d\mu_\alpha).$$

We consider $f \in L^\infty([0, +\infty[, d\mu_\alpha)$, weakly oscillating around zero. Show that in any Lebesgue point x of f we have

$$\lim_{\substack{\varepsilon \to 0^+ \\ \delta \to +\infty}} f^{\varepsilon,\delta}(x) = C_\nu f(x).$$

8 Wavelet Packets Associated With The Bessel Operator

Introduction.

Wavelet theory involves breaking up a complicated function (phenomenon) into many simple pieces at different scales and positions. It allows greater flexibility with more desirable features such as "discretization by wavelet packets", and readiness for better implementation (see [17] [20]).

We present in this chapter a general construction, allowing the development of wavelet packets associated with the Bessel operator from the continuous wavelet analysis associated with the Bessel operator, given in Chapter 7. More precisely we study three types of wavelet packets associated with the Bessel operator, the P-wavelet packet, the M-wavelet packet and the S-wavelet packet, and the corresponding wavelet packet transforms and we prove for these transforms Plancherel formulas, Calderon's reproducing formulas and pointwise reconstruction formulas. Also we define two types of scale discrete scaling functions associated with the Bessel operator, and we establish Plancherel and pointwise reconstruction formulas associated with these functions.

Finally we discuss a multiresolution analysis by means of wavelet packets associated with the Bessel operator.

In Problems 8.VI.1, 8.VI.2 and 8.VI.3 we develop the theory of D-wavelets (See [68]).

The theory of wavelet packets associated with a singular differential operator more general than the Bessel operator is given in [90] with applications.

8.I. The P-wavelet packet transform associated with the Bessel operator.

We consider in this section a wavelet g associated with the Bessel operator which has been studied in the subsection 7.II.1, the function g_a, $a > 0$, given by the relation

(7.II.9), and $\{r_j\}_{j\in\mathbb{Z}}$ a scale sequence in $]0, +\infty[$ which is strictly decreasing and satisfying

$$\lim_{j\to-\infty} r_j = +\infty, \quad \lim_{j\to+\infty} r_j = 0.$$

8.I.1. Plancherel and reconstruction formulas.

Proposition 8.I.1.

i) The function $\lambda \to \left(\frac{1}{C_g}\int_{r_{j+1}}^{r_j} |\mathfrak{F}_B(g_a)(\lambda)|^2 \frac{da}{a}\right)^{1/2}$ belongs to $L^2([0, +\infty[, d\mu_\alpha)$.

ii) There exists a function g_j^P in $L^2([0,+\infty[, d\mu_\alpha)$ such that for almost all $\lambda \geq 0$:

$$\mathfrak{F}_B(g_j^P)(\lambda) = \left(\frac{1}{C_g}\int_{r_{j+1}}^{r_j} |\mathfrak{F}_B(g_a)(\lambda)|^2 \frac{da}{a}\right)^{1/2}. \tag{8.I.1}$$

Proof.

i) From Fubini-Tonelli's theorem, the relation (7.II.11) and Theorem 5.IV.3, we have

$$\int_0^\infty \left(\frac{1}{C_g}\int_{r_{j+1}}^{r_j} |\mathfrak{F}_B(g_a)(\lambda)|^2 \frac{da}{a}\right)d\mu_\alpha(\lambda) = \frac{1}{C_g}\int_{r_{j+1}}^{r_j} \left(\int_0^\infty |\mathfrak{F}_B(g)(a\lambda)|^2 d\mu_\alpha(\lambda)\right) \frac{da}{a}.$$

By the change of variables $u = \lambda a$ we obtain

$$\begin{aligned}\int_0^\infty \left(\frac{1}{C_g}\int_{r_{j+1}}^{r_j} |\mathfrak{F}_B(g_a)(\lambda)|^2 \frac{da}{a}\right)d\mu_\alpha(\lambda),\\
&= \frac{1}{C_g}\int_{r_{j+1}}^{r_j} \left(\int_0^\infty |\mathfrak{F}_B(g)(u)|^2 d\mu_\alpha(u)\right) \frac{da}{a^{2\alpha+3}},\\
&= \frac{\|g\|_{2,\alpha}}{C_g}\int_{r_{j+1}}^{r_j} \frac{da}{a^{2\alpha+3}} < +\infty.\end{aligned}$$

This completes the proof of i).

ii) We deduce the result from Theorem 5.VI.2.

Definition 8.I.1.

i) The sequence $\{g_j^P\}_{j\in\mathbb{Z}}$ is called wavelet packet associated with the Bessel operator (also called P-wavelet packet associated with the Bessel operator).

ii) The function g_j^P, $j \in \mathbb{Z}$, is called P-wavelet packet member of step j, associated with the Bessel operator.

From the relation (8.I.1) we deduce that the P-wavelet packet associated with the Bessel operator, satisfies the following properties.

i) For each $j \in \mathbb{Z}$, the function $\lambda \to \mathfrak{F}_B(g_j^P)(\lambda)$ verifies for almost all $\lambda \geq 0$:

(8.I.2) $$0 \leq \mathfrak{F}_B(g_j^P)(\lambda) \leq 1.$$

ii) We have for almost all $\lambda \geq 0$:

(8.I.3) $$\sum_{j=-\infty}^{\infty} (\mathfrak{F}_B(g_j^P)(\lambda))^2 = 1.$$

Example.

Let g be the wavelet associated with the Bessel operator defined by the relation (7.II.6). From the relation (8.I.1) we deduce that the P-wavelet packet $\{g_j^P\}_{j\in\mathbb{Z}}$ associated with the Bessel operator is given by

(8.I.4) $$\forall\, \lambda \geq 0,\ \mathfrak{F}_B(g_j^P)(\lambda) = [(1 + 2t\,\lambda^2\, r_{j+1}^2)\, e^{-2t\lambda^2 r_{j+1}^2} - (1 + 2t\lambda^2\, r_j^2)e^{-2t\lambda^2 r_j^2}]^{1/2}$$

Let $\{g_j^P\}_{j\in\mathbb{Z}}$ be a P-wavelet packet associated with the Bessel operator. We consider for all $j \in \mathbb{Z}$ and $s \geq 0$, the function $g_{j,s}^P$ given for almost all $x \geq 0$ by

(8.I.5) $$g_{j,s}^P(x) = T_x(g_j^P)(s),$$

where T_x, $x \geq 0$, are the generalized translation operators given in the section 4.II. This function satisfies the following properties.

i) The function $g_{j,s}^P$ belongs to $L^2([0, +\infty[, d\mu_\alpha)$ and we have

(8.I.6) $$\| g_{j,s}^P \|_{2,\alpha} \leq \| g_j^P \|_{2,\alpha}\, .$$

ii) We have for almost all $\lambda \geq 0$:

(8.I.7) $$\mathfrak{F}_B(g_{j,s}^P)(\lambda) = j_\alpha(\lambda s)\, \mathfrak{F}_B(g_j^P)(\lambda).$$

(8.I.8) $$0 \leq \mathfrak{F}_B(g_{j,s}^P)(\lambda) \leq 1.$$

(8.I.9) $$\sum_{j=-\infty}^{\infty} (\mathfrak{F}_B(g^P_{j,s})(\lambda))^2 \leq 1.$$

Definition 8.I.2. Let $\{g^P_j\}_{j\in\mathbb{Z}}$ be a P-wavelet packet associated with the Bessel operator. The P-wavelet packet transform Ψ^P_g associated with the Bessel operator, is defined for regular functions f on $[0, +\infty[$ by

(8.I.10) $$\Psi^P_g(f)(j,s) = \int_0^\infty f(x)\overline{g^P_{j,s}}(x)\, d\mu_\alpha(x),\ j\in\mathbb{Z}, s\geq 0.$$

This transform can also be written in the form

(8.I.11) $$\Psi^P_g(f)(j,s) = f *_B \overline{g^P_j}(s).$$

where $*_B$ is the generalized convolution product studied in the subsection 4.III.1.

Proposition 8.I.2. Let $\{g^P_j\}_{j\in\mathbb{Z}}$ be a P-wavelet packet associacted with the Bessel operator. Then for all $j\in\mathbb{Z}$ and f in $L^q([0, +\infty[, d\mu_\alpha)$, $q\in[1, 2]$, the function $s\to\Psi^P_g(f)(j,s)$ belongs to $L^r([0, +\infty[, d\mu_\alpha)$ with $r\in[1, +\infty]$ such that $\frac{1}{r} = \frac{1}{q}-\frac{1}{2}$, and we have

(8.I.12) $$\|\Psi^P_g(f)(j,.)\|_{r,\alpha} \leq \|f\|_{q,\alpha}\, \|g^P_j\|_{2,\alpha}.$$

Theorem 8.I.1. Let $\{g^P_j\}_{j\in\mathbb{Z}}$ be a P-wavelet packet associated with the Bessel operator.

i) Plancherel formula for Ψ^P_g.

For all f in $L^2([0, +\infty[, d\mu_\alpha)$we have

(8.I.13) $$\int_0^\infty |f(x)|^2 d\mu_\alpha(x) = \sum_{j=-\infty}^{\infty}\int_0^\infty |\Psi^P_g(f)(j,s)|^2\, d\mu_\alpha(s).$$

ii) Parseval formula for Ψ^P_g.

For all f_1 and f_2 in $L^2[(0, +\infty[, d\mu_\alpha)$ we have

(8.I.14) $$\int_0^\infty f_1(x)\overline{f_2(x)}d\mu_\alpha(x) = \sum_{j=-\infty}^{\infty}\int_0^\infty \Psi^P_g(f_1)(j,s)\,\overline{\Psi^P_g(f_2)(j,s)}\, d\mu_\alpha(s).$$

Proof.

i) From the relation (8.I.11) and Corollary 5.IV.2 we obtain for all $j\in\mathbb{Z}$:

(8.I.15) $$\int_0^\infty |\Psi^P_g(f)(j,s)|^2 d\mu_\alpha(s) = \int_0^\infty |\mathfrak{F}_B(f)(\lambda)|^2\,(\mathfrak{F}_B(g^P_j)(\lambda))^2 d\mu_\alpha(\lambda).$$

Using Fubini-Tonelli's theorem we deduce

$$\sum_{j=-\infty}^{\infty} \int_0^{\infty} |\Psi_g^P(f)(j,s)|^2 \, d\mu_\alpha(s).= \int_0^{\infty} |\mathfrak{F}(f)(\lambda)|^2 \Big(\sum_{j=-\infty}^{\infty} (\mathfrak{F}_B(g_j^P)(\lambda))^2 \Big) \, d\mu_\alpha(\lambda).$$

But from the relation (8.I.3) we have

$$\sum_{j=-\infty}^{\infty} (\mathfrak{F}_B(g_j^P)\,(\lambda))^2 = 1.$$

Then

$$\sum_{j=-\infty}^{\infty} \int_0^{\infty} |\Psi_g^P(f)(j,s)|^2 \, d\mu_\alpha(s) = \int_0^{\infty} |\mathfrak{F}_B(f)(\lambda)|^2 d\mu_\alpha(\lambda).$$

Thus Theorem 5.IV.3 gives the relation (8.I.13).

ii) The result is easily deduced from the i).

Theorem 8.I.2. Let $\{g_j^P\}_{j\in\mathbb{Z}}$ be a P-wavelet packet associated with the Bessel operator. For f in $L^p([0, +\infty[, d\mu_\alpha)$, $p = 1, 2$, such that $\mathfrak{F}_B(f)$ belongs to $L^1([0, +\infty[, d\mu_\alpha)$, we have the following reconstruction formula for Ψ_g^P :

$$f(x) = \sum_{j=-\infty}^{\infty} \int_0^{\infty} \Psi_g^P(f)(j,s) \; g_{j,s}^P(x) d\mu_\alpha(s), \quad \text{a.e,} \tag{8.I.16}$$

where, for each $x \geq 0$, both the integral and the series are absolutely convergent, but possibly not the series of integrals.

Proof.

We put for $j \in \mathbb{Z}$ and $x \geq 0$:

$$I(j, x) = \int_0^{\infty} \Psi_g^P(f)(j,s) \; g_{j,s}^P(x) d\mu_\alpha(s).$$

a) We suppose that $f \in L^1([0, +\infty[, d\mu_\alpha)$ is such that $\mathfrak{F}_B(f) \in L^1([0, +\infty[, d\mu_\alpha)$.

From the relations (8.I.5) and (8.I.11) we have

$$\Psi_g^P(f)(j,s) \; g_{j,s}^P(x) = f *_B \overline{g_j^P}(s) \; T_x(g_j^P)(s).$$

Propositions 4.II.2, 4.III.4, and 8.I.1. ii) imply that the functions $s \to f *_B \overline{g_j^P}(s)$ and $s \to T_x(g_j^P)(s)$ belong to $L^2([0, +\infty[, d\mu_\alpha)$, then from Hölder's inequality the integral

I(j,x) is absolutely convergent. On the other hand from Proposition 5.IV.3 we have for almost all $\lambda \geq 0$:

$$\mathfrak{F}_B(f *_B \overline{g_j^P})(\lambda) = \mathfrak{F}_B(f)(\lambda).\mathfrak{F}_B(g_j^P)(\lambda).$$

Using this equality, the relation (8.I.7) and Theorem 5.IV.3 we obtain

(8.I.17) $$I(j,x) = \int_0^\infty \mathfrak{F}_B(f)(\lambda)\,(\mathfrak{F}_B(g_j^P)(\lambda))^2 j_\alpha(\lambda x)d\mu_\alpha(\lambda).$$

By applying the relation (8.I.3) we have

(8.I.18) $$\sum_{j=-\infty}^{\infty} |I(j,x)| \leq \int_0^\infty |\mathfrak{F}_B(f)(\lambda)|\,(\sum_{j=-\infty}^{\infty}(\mathfrak{F}_B(g_j^P)(\lambda))^2)d\mu_\alpha(\lambda),$$

$$\leq \int_0^\infty |\mathfrak{F}_B(f)(\lambda)|\,d\mu_\alpha(\lambda) = \|\mathfrak{F}_B(f)\|_{1,\alpha} < +\infty.$$

Thus the series $\sum_{j=-\infty}^{\infty} |I(j,x)|$ is convergent.

From the relation (8.I.17) we have

(8.I.19) $$\sum_{j=-\infty}^{\infty} I(j,x) = \sum_{j=-\infty}^{\infty} \int_0^\infty \mathfrak{F}_B(f)(\lambda)\,(\mathfrak{F}_B(g_j^P)(\lambda))^2\, j_\alpha(\lambda x)\, d\mu_\alpha(\lambda).$$

But the inequality (8.I.18) shows that we can apply Fubini's theorem to the second member of (8.I.19). Thus

$$\sum_{j=-\infty}^{\infty} I(j,x) = \int_0^\infty \mathfrak{F}_B(f)(\lambda)\,(\sum_{j=-\infty}^{\infty}(\mathfrak{F}_B(g_j^P)(\lambda))^2) j_\alpha(\lambda x)\, d\mu_\alpha(\lambda).$$

The relation (8.I.3) and Theorem 5.II.2 give the relation (8.I.16).

b) We suppose that $f \in L^2([0,+\infty[, d\mu_\alpha)$ is such that $\mathfrak{F}_B(f) \in L^1([0,+\infty[, d\mu_\alpha)$.

From the relation (8.I.2) we have for all $j \in \mathbb{Z}$:

(8.I.20) $$\|\mathfrak{F}_B(f)\mathfrak{F}_B(g_j^P)\|_{2,\alpha} \leq \|\mathfrak{F}_B(f)\|_{2,\alpha} < +\infty.$$

Then Theorem 5.IV.4 implies that the function $f *_B \overline{g_j^P}$ belongs to $L^2([0,+\infty[, d\mu_\alpha)$ and we have

(8.I.21) $$\mathfrak{F}_B(f *_B \overline{g_j^P}) = \mathfrak{F}_B(f).\mathfrak{F}_B(g_j^P).$$

The same proof as for the a) and Theorem 5.IV.2 and Proposition 5.IV.2 give the relation (8.I.16).

8.I.2. Calderon's reproducing formula.

Proposition 8.I.3. Let $\{g_j^P\}_{j\in\mathbb{Z}}$ be a P-wavelet packet associated with the Bessel operator. For $m, n \in \mathbb{Z}$ with $m < n$, we define the functions $G_{m,n}$ and $K_{m,n}$ by

(8.I.22) $$\forall\, x \geq 0,\ G_{m,n}(x) = \sum_{j=m}^{n-1} g_j^P *_B \overline{g_j^P}(x),$$

and for almost all $\lambda \geq 0$:

(8.I.23) $$K_{m,n}(\lambda) = \frac{1}{C_g}\int_{r_n}^{r_m} |\mathfrak{F}_B(g_a)(\lambda)|^2 \frac{da}{a}.$$

Then

(8.I.24) i) $G_{m,n} \in L^2([0, +\infty[, d\mu_\alpha)$.

(8.I.25) ii) $K_{m,n} \in (L^2 \cap L^\infty)([0, +\infty[, d\mu_\alpha)$.

(8.I.26) iii) $\mathfrak{F}_B(G_{m,n}) = K_{m,n}$.

(8.I.27) iv) For almost all $\lambda > 0$, we have

$$\lim_{\substack{m\to-\infty\\ n\to+\infty}} K_{m,n}(\lambda) = 1.$$

Proof.

Using Proposition 8.I.1 ii) and the relation (8.I.2) we see that for all $j \in \mathbb{Z}$, the function g_j^P belongs to $L^2([0, +\infty[, d\mu_\alpha)$ and $\mathfrak{F}_B(g_j^P)$ is in $L^\infty([0, +\infty[, d\mu_\alpha)$. Then from Theorem 5.IV.4, the function $g_j^P *_B \overline{g_j^P}$ belongs to $L^2([0, +\infty[, d\mu_\alpha)$ and we have for almost all $\lambda \geq 0$:

(8.I.28) $$\mathfrak{F}_B(g_j^P *_B \overline{g_j^P})(\lambda) = (\mathfrak{F}_B(g_j^P)(\lambda))^2.$$

Thus the function $G_{m,n}$ is in $L^2([0, +\infty[, d\mu_\alpha)$ and using the relation (8.I.1) we deduce that for almost all $\lambda \geq 0$:

(8.I.29) $$\mathfrak{F}_B(G_{m,n})(\lambda) = \sum_{j=m}^{n-1} (\mathfrak{F}_B(g_j^P)(\lambda))^2,$$
$$= \frac{1}{C_g}\int_{r_n}^{r_m} |\mathfrak{F}_B(g_j^P)(\lambda)|^2 \frac{da}{a} = K_{m,n}(\lambda).$$

This relation shows that $K_{m,n}$ belongs to $L^\infty([0, +\infty[, d\mu_\alpha)$ and applying Definition 7.II.1 we deduce that for almost all $\lambda > 0$:

$$\lim_{\substack{m\to-\infty \\ n\to+\infty}} K_{m,n}(\lambda) = 1.$$

Proposition 8.I.4. Let $\{g_j^P\}_{j\in\mathbb{Z}}$ be a P-wavelet packet associated with the Bessel operator. For f in $L^2([0,+\infty[, d\mu_\alpha)$ and $m, n \in \mathbb{Z}$, with $m < n$, we define the function $f_{m,n}^P$ by

$$\forall\, x \geq 0,\ f_{m,n}^P(x) = \sum_{j=m}^{n-1} \int_0^\infty \Psi_g^P(f)(j,s)\ g_{j,s}^P(x) d\mu_\alpha(s). \tag{8.I.30}$$

Then the function $f_{m,n}^P$ belongs to $L^2([0,+\infty[, d\mu_\alpha)$ and we have

$$f_{m,n}^P = f *_B G_{m,n}. \tag{8.I.31}$$

$$\mathcal{F}_B(f_{m,n}^P) = \mathcal{F}_B(f).K_{m,n}. \tag{8.I.32}$$

Where $G_{m,n}$ and $K_{m,n}$ are the functions defined in Proposition 8.I.3.

Proof.

From the relations (8.I.30) and (8.I.31) we deduce that for all $j \in \mathbb{Z}$, the function $s \to \Psi_g^P(f)(j,s)$ belongs to $L^2([0,+\infty[, d\mu_\alpha)$ and we have

$$\mathcal{F}_B(\Psi_g^P(f)(j,.)) = \mathcal{F}_B(f)\mathcal{F}_B(g_j^P). \tag{8.I.33}$$

Using this equality, the relation (8.I.7) and Theorem 5.IV.3, we obtain for all $j \in \mathbb{Z}$ and $x \geq 0$:

$$\int_0^\infty \Psi_g^P(f)(j,s)\ g_{j,s}^P(x) d\mu_\alpha(s) \tag{8.I.34}$$
$$= \int_0^\infty \mathcal{F}_B(f)(\lambda)\ (\mathcal{F}_B(g_j^P)(\lambda))^2\ j_\alpha(\lambda x) d\mu_\alpha(\lambda).$$

Thus

$$\forall\, x \geq 0,\ f_{m,n}^P(x) = \int_0^\infty \mathcal{F}_B(f)(\lambda)\Big(\sum_{j=m}^{n-1} (\mathcal{F}_B(g_j^P)(\lambda))^2\Big)\ j_\alpha(\lambda x) d\mu_\alpha(\lambda).$$

Using the relation (8.I.29) and Proposition 5.IV.4, we obtain

$$\forall\, x \geq 0,\ f_{m,n}^P(x) = \int_0^\infty \mathcal{F}_B(f)(\lambda).\mathcal{F}_B(G_{m,n})(\lambda) j_\alpha(\lambda x) d\mu_\alpha(\lambda),$$
$$= \mathcal{F}_B(\mathcal{F}_B(f).\ \mathcal{F}_B(G_{m,n}))(x).$$

Then

$$\forall\, x \geq 0,\ f_{m,n}^P(x) = f *_B G_{m,n}(x).$$

By applying Theorem 5.IV.4 and Proposition 8.I.3 we deduce that $f^P_{m,n}$ belongs to $L^2([0, +\infty[, d\mu_\alpha)$ and we have

$$\mathfrak{F}_B(f^P_{m,n}) = \mathfrak{F}_B(f).K_{m,n}\,.$$

The following result is the so-called Calderon's reproducing formula.

Theorem 8.I.3. The function $f^P_{m,n}$ defined by Proposition 8.I.4 satisfies

$$\lim_{\substack{m\to-\infty\\ n\to+\infty}} \| f^P_{m,n} - f\|_{2,\alpha} = 0.$$

Proof.

From Proposition 8.I.4 and Theorem 5.IV.3 we have

$$\begin{aligned}\|f^P_{m,n} - f\|_{2,\alpha} &= \|\mathfrak{F}_B(f^P_{m,n}) - \mathfrak{F}_B(f)\|_{2,\alpha},\\ &= \|\mathfrak{F}_B(f)\,(1 - K_{m,n})\|_{2,\alpha}\,.\end{aligned}$$

We deduce the result from the relation (8.I.27) and the dominated convergence theorem.

8.II. Scale discrete scaling function associated with the Bessel operator.

We consider a P-wavelet packet $\{g^P_j\}_{j\in\mathbb{Z}}$ associated with the Bessel operator which has been studied in section 8.I.

Proposition 8.II.1

i) For all $J\in\mathbb{Z}$ and for almost all $\lambda\geq 0$, we have

$$\sum_{j=-\infty}^{J-1} (\mathfrak{F}_B(g^P_j)(\lambda))^2 = \frac{1}{C_g}\int_{r_J}^{\infty} |\mathfrak{F}_B(g_a)(\lambda)|^2\,\frac{da}{a}\,. \tag{8.II.1}$$

ii) For all $J\in\mathbb{Z}$, the function $\lambda\to\left(\sum_{j=-\infty}^{J-1} (\mathfrak{F}_B(g^P_j)(\lambda))^2\right)^{1/2}$ belongs to $L^2([0, +\infty[, d\mu_\alpha)$.

iii) For all $J\in\mathbb{Z}$, there exists a function G^P_J in $L^2([0, +\infty[, d\mu_\alpha)$ such that for almost all $\lambda\geq 0$:

$$\mathfrak{F}_B(G^P_J)(\lambda) = \left(\sum_{j=-\infty}^{J-1} (\mathfrak{F}_B(g^P_j)(\lambda))^2\right)^{1/2}. \tag{8.II.2}$$

Proof.

i) We obtain the result from the relation (8.I.1).

ii) The relation (8.II.1) and the same proof as for Proposition 8.I.1, give the result.

iii) We deduce the result from Theorem 5.IV.2.

Definition 8.II.1. The sequence $\{G_J^P\}_{J\in\mathbb{Z}}$ is called scale discrete scaling function associated with the Bessel operator.

From the relations (8.II.2) and (8.I.3) we show that the scale discrete scaling function $\{G_J^P\}_{J\in\mathbb{Z}}$ possesses the following properties.

i) For all $J\in\mathbb{Z}$ and for almost all $\lambda\geq 0$, we have

$$0\leq \mathfrak{F}_B(G_J^P)(\lambda)\leq 1. \tag{8.II.3}$$

ii) For almost all $\lambda\geq 0$, we have

$$\lim_{J\to+\infty}\mathfrak{F}_B(G_J^P)(\lambda)=1. \tag{8.II.4}$$

Example.

We consider the P-wavelet packet $\{g_j^P\}_{j\in\mathbb{Z}}$ given by the relation (8.I.4). From the relation (8.II.2) we deduce that the scale discrete scaling function $\{G_J^P\}_{J\in\mathbb{Z}}$ corresponding to $\{g_j^P\}_{j\in\mathbb{Z}}$ is defined by

$$\forall\,\lambda\geq 0,\ \mathfrak{F}_B(G_J^P)(\lambda)=(1+2t\,\lambda^2 r_J^2)^{1/2}\,e^{-t\lambda^2 r_J^2}. \tag{8.II.5}$$

Let $\{G_J^P\}_{J\in\mathbb{Z}}$ be a scale discrete scaling function associated with the Bessel operator. We consider for all $J\in\mathbb{Z}$ and $s\geq 0$, the function $G_{J,s}^P$ defined for almost all $x\geq 0$ by

$$G_{J,s}^P(x)=T_x(G_J^P)(s). \tag{8.II.6}$$

Where T_x, $x\geq 0$, are the generalized translation operators given in the section 4.II.

The following propositions give properties of the function $G_{J,s}^P$.

Proposition 8.II.2.

i) The function $G^P_{J,s}$ belongs to $L^2([0,+\infty[, d\mu_\alpha)$ and we have

$$\|G^P_{J,s}\|_{2,\alpha} \leq \|G^P_J\|_{2,\alpha}. \tag{8.II.7}$$

ii) For almost all $\lambda \geq 0$, we have the following relations

$$\mathfrak{F}_B(G^P_{J,s})(\lambda) = j_\alpha(\lambda s)\, \mathfrak{F}_B(G^P_J)(\lambda). \tag{8.II.8}$$

$$0 \leq \mathfrak{F}_B(G^P_{J,s})(\lambda) \leq 1. \tag{8.II.9}$$

Proposition 8.II.3.

i) For all $J \in \mathbb{Z}$ and for almost all $\lambda \geq 0$, we have

$$(\mathfrak{F}_B(G^P_J)(\lambda))^2 + \sum_{j=J}^{\infty} (\mathfrak{F}_B(g^P_j)(\lambda))^2 = 1. \tag{8.II.10}$$

ii) For all $J \in \mathbb{Z}$ and for almost all $\lambda \geq 0$, we have

$$(\mathfrak{F}_B(g^P_j)(\lambda))^2 = (\mathfrak{F}_B(G^P_{j+1})(\lambda))^2 - (\mathfrak{F}_B(G^P_j)(\lambda))^2. \tag{8.II.11}$$

iii) For almost all $\lambda \geq 0$, we have

$$\sum_{j=-\infty}^{\infty} [(\mathfrak{F}_B(G^P_{j+1})(\lambda))^2 - (\mathfrak{F}_B(G^P_j)(\lambda))^2] = 1. \tag{8.II.12}$$

Proof.

i) From the relations (8.I.1) and (8.I.3) we have for all $J \in \mathbb{Z}$ and for almost all $\lambda \geq 0$:

$$\frac{1}{C_g}\int_0^\infty |\mathfrak{F}_B(g_a)|^2 \frac{da}{a} = \sum_{j=-\infty}^{\infty} (\mathfrak{F}_B(g^P_j)(\lambda))^2,$$
$$= \Big(\sum_{j=-\infty}^{J-1} + \sum_{j=J}^{\infty}\Big)(\mathfrak{F}_B(g^P_j)(\lambda))^2 = 1.$$

The relation (8.II.2) gives the result.

ii) We deduce the result from the relation (8.II.2).

iii) We obtain the relation (8.II.12) from the ii) and (8.I.3).

Remark.

The relation (8.II.11) shows that the P-wavelet packet members associated with the Bessel operator may be regarded as difference of two law-pass filters.

Notations.

We denote by $(.,.)_\alpha$ the scalar product of the Hilbert space $L^2([0, +\infty[, d\mu_\alpha)$. Then for f and g in $L^2([0, +\infty[, d\mu_\alpha)$ we have

(8.II.13) $$(f, g)_\alpha = \int_0^\infty f(x)\overline{g(x)}\, d\mu_\alpha(x).$$

Theorem 8.II.1. Let $\{G_J^P\}_{J\in\mathbb{Z}}$ be a scale discrete scaling function associated with the Bessel operator.

i) Plancherel formula associated with $\{G_J^P\}_{J\in\mathbb{Z}}$.

For all f in $L^2([0, +\infty[, d\mu_\alpha)$ we have

(8.II.14) $$\int_0^\infty |f(x)|^2 d\mu_\alpha(x) = \lim_{J\to+\infty} \int_0^\infty |(f, G_{J,s}^P)_\alpha|^2 d\mu_\alpha(s).$$

ii) Parseval formula associated with $\{G_J^P\}_{J\in\mathbb{Z}}$.

For all f_1 and f_2 in $L^2([0, +\infty[, d\mu_\alpha)$ we have

(8.II.15) $$\int_0^\infty f_1(x)\overline{f_2(x)}\, d\mu_\alpha(x) = \lim_{J\to+\infty} \int_0^\infty (f_1, G_{J,s}^P)_\alpha \overline{(f_2, G_{J,s}^P)_\alpha} d\mu_\alpha(s).$$

Proof.

i) From the relation (4.III.1) we have for all $J \in \mathbb{Z}$. and $s \geq 0$:

(8.II.16) $$(f, G_{J,s}^P)_\alpha = f *_B \overline{G_J^P}(s).$$

Then

$$\int_0^\infty |(f, G_{J,s}^P)_\alpha|^2 d\mu_\alpha(s) = \int_0^\infty |f *_B \overline{G_J^P}(s)|^2 d\mu_\alpha(s).$$

Using Corollary 5.IV.2 we obtain

(8.II.17) $$\int_0^\infty |(f, G_{J,s}^P)_\alpha|^2 d\mu_\alpha(s) = \int_0^\infty |\mathcal{F}_B(f)(\lambda)|^2 (\mathcal{F}_B(G_J^P)(\lambda))^2 d\mu_\alpha(\lambda).$$

Applying relations (8.II.3), (8.II.4) and dominated convergence theorem to the integral of second member, we obtain

$$\lim_{J\to+\infty} \int_0^\infty |(f, G_{J,s}^P)_\alpha|^2 d\mu_\alpha(s) = \int_0^\infty |\mathcal{F}_B(f)(\lambda)|^2\, d\mu_\alpha(\lambda).$$

Then Theorem 5.IV.3 gives the relation (8.II.14).

ii) We deduce the relation (8.II.15) from the i).

The following theorem shows that we can obtain another Plancherel formula by using the scale discrete scaling function $\{G_J^P\}_{J\in\mathbb{Z}}$ associated with the Bessel operator, and the P-wavelet transform Ψ_g^P associated with the Bessel operator, studied in the section 8.I.

Theorem 8.II.2.

i) Plancherel formula associated with $\{G_J^P\}_{J\in\mathbb{Z}}$ and Ψ_g^P.

For all f in $L^2([0, +\infty[, d\mu_\alpha)$, we have

$$\int_0^\infty |f(x)|^2 d\mu_\alpha(x) = \int_0^\infty |(f, G_{J,s}^P)_\alpha|^2 d\mu_\alpha(s) + \sum_{j=J}^\infty \int_0^\infty |\Psi_g^P(f)(j,s)|^2 d\mu_\alpha(s). \tag{8.II.18}$$

ii) Parseval formula associated with $\{G_J^P\}_{J\in\mathbb{Z}}$ and Ψ_g^P.

For all f_1 and f_2 in $L^2([0, +\infty[, d\mu_\alpha)$, we have

$$\int_0^\infty f_1(x)\overline{f_2(x)}d\mu_\alpha(x) = \int_0^\infty (f_1, G_{J,s}^P)_\alpha (f_2, G_{J,s}^P)_\alpha d\mu_\alpha(s) + \sum_{j=J}^\infty \int_0^\infty \Psi_g^P(f_1)(j,s)\overline{\Psi_g^P(f_2)(j,s)}d\mu_\alpha(s). \tag{8.II.19}$$

Proof.

i) From the relations (8.II.17), (8.II.2) and Fubini-Tonelli's theorem we have for all $J\in\mathbb{Z}$:

$$\int_0^\infty |(f, G_{J,s}^P)_\alpha|^2 d\mu_\alpha(s) = \int_0^\infty |\mathfrak{F}_B(f)(\lambda)|^2 \Big(\sum_{j=-\infty}^{J-1} (\mathfrak{F}_B(g_j^P)(\lambda))^2\Big) d\mu_\alpha(\lambda).$$

On the other hand using the relation (8.I.15) and Fubini-Tonelli's theorem we obtain

$$\sum_{j=J}^\infty \int_0^\infty |\Psi_g^P(f)(j,s)|^2 d\mu_\alpha(s) = \int_0^\infty |\mathfrak{F}_B(f)(\lambda)|^2 \Big(\sum_{j=J}^\infty (\mathfrak{F}_B(g_j^P)(\lambda))^2\Big) d\mu_\alpha(\lambda).$$

Thus

$$\int_0^\infty |(f, G_{J,s}^P)_\alpha|^2) d\mu_\alpha(s) + \sum_{j=J}^\infty |\Psi_g^P(f)(j,s)|^2 d\mu_\alpha(s) = \int_0^\infty |\mathfrak{F}_B(f)(\lambda)|^2 \Big(\sum_{j=-\infty}^\infty (\mathfrak{F}_B(g_j^P)(\lambda))^2\Big) d\mu_\alpha(\lambda).$$

We deduce the relation (8.II.18) from (8.I.3) and Theorem 5.IV.3.

ii) We derive (8.II.19) from the i).

Theorem 8.II.3. For f in $L^P([0, +\infty[, d\mu_\alpha)$, p = 1, 2, such that $\mathfrak{F}_B(f)$ belongs to $L^1([0, +\infty[, d\mu_\alpha)$ we have the following reconstruction formulas

$$\text{(8.II.20)} \qquad \text{i) } f(x) = \lim_{J \to +\infty} \int_0^\infty |(f, G^P_{J,s})_\alpha G^P_{J,s}(x) d\mu_\alpha(s), \text{ a.e,}$$

where, for each $x \geq 0$, the integral is absolutely convergent.

$$\text{(8.II.21)} \qquad \text{ii) } f(x) = \int_0^\infty (f, G^P_{J,s})_\alpha G^P_{J,s}(x) d\mu_\alpha(s) + \sum_{j=J}^\infty \int_0^\infty \Psi^P_g(f)(j,s)\, g^P_{j,s}(x)\, d\mu_\alpha(s), \text{ a.e,}$$

where, for each $x \geq 0$, in the second term of the second member, both the integral and the series are absolutely convergent, but possibly not the series of integrals.

Proof.

a) We suppose that $f \in L^1([0, +\infty[, d\mu_\alpha)$ such that $\mathfrak{F}_B(f) \in L^1([0, +\infty[, d\mu_\alpha)$.

Let $x \in [0, +\infty[$. From the relations (8.II.16) and (8.II.6) we have

$$(f, G^P_{J,s})_\alpha G^P_{J,s}(x) = f *_B \overline{G^P_J}(s) T_x(G^P_J)(s).$$

Using Propositions 4.II.2, 4.III.4 and 8.II.2 i) we deduce that the functions $s \to f *_B \overline{G^P_J}(s)$ and $s \to T_x(G^P_J)(s)$ belong to $L^2([0, +\infty[, d\mu_\alpha)$, then from Hölder's inequality the integral

$$\int_0^\infty (f, G^P_{J,s})_\alpha G^P_{J,s}(x) d\mu_\alpha(s)$$

is absolutely convergent.

On the other hand from Proposition 5.IV.3 we have for almost all $\lambda \geq 0$:

$$\mathfrak{F}_B(f *_B \overline{G^P_J})(\lambda) = \mathfrak{F}_B(f)(\lambda).\mathfrak{F}_B(G^P_J)(\lambda).$$

This equality, the relation (8.II.8) and Theorem 5.IV.3 imply that

$$\text{(8.II.22)} \quad \int_0^\infty (f, G^P_{J,s})_\alpha G^P_{J,s}(x) d\mu_\alpha(s) = \int_0^\infty \mathfrak{F}_B(f)(\lambda)\, (\mathfrak{F}_B(G^P_J)(\lambda))^2 j_\alpha(\lambda x) d\mu_\alpha(\lambda).$$

We obtain the relation (8.II.20) by using the relations (8.II.3), (8.II.4), and by applying the dominated convergence theorem.

The relation (8.II.2) implies that the relation (8.II.22) can also be written in the form

$$\int_0^\infty (f, G^P_{J,s})_\alpha G^P_{J,s}(x) d\mu_\alpha(s) = \int_0^\infty \mathfrak{F}_B(f)(\lambda) \left(\sum_{j=-\infty}^{J-1} (\mathfrak{F}_B(g^P_j)(\lambda))^2 \right) j_\alpha(\lambda x) d\mu_\alpha(\lambda).$$

Using the relations (8.I.17), (8.I.18) and (8.I.3) we obtain

$$\int_0^\infty (f, G^P_{J,s})_\alpha G^P_{J,s}(x) d\mu_\alpha(s) + \sum_{j=J}^{\infty} \int_0^\infty \Psi^P_g(f)(j,s)\, g^P_{j,s}(x) d\mu_\alpha(s),$$

$$= \int_0^\infty \mathfrak{F}_B(f)(\lambda)\, j_\alpha(\lambda x) d\mu_\alpha(\lambda).$$

Thus Theorem 5.II.2 gives the relation (8.II.21).

b) We suppose that $f \in L^2([0, +\infty[, d\mu_\alpha)$ such that $\mathfrak{F}_B(f) \in L^1([0, +\infty[, d\mu_\alpha)$.

From the relation (8.II.9) we have for all $J \in \mathbb{Z}$:

(8.II.23) $$\|\mathfrak{F}_B(f)\, \mathfrak{F}_B(G^P_J)\|_{2,\alpha} \le \|\mathfrak{F}_B(f)\|_{2,\alpha} < +\infty.$$

Then Theorem 5.IV.4 implies that the function $f *_B \overline{G^P_J}$ belongs to $L^2([0, +\infty[, d\mu_\alpha)$ and we have

(8.II.24) $$\mathfrak{F}_B(f *_B \overline{G^P_J}) = \mathfrak{F}_B(f).\mathfrak{F}_B(G^P_J).$$

We obtain the relations (8.II.20) and (8.II.21) by using the same proof as for the a) and Theorem 5.IV.2 and Proposition 5.IV.2.

Corollary 8.II.1. For f in $L^p([0, +\infty[, d\mu_\alpha)$, $p = 1, 2$, such that $\mathfrak{F}_B(f)$ belongs to $L^1([0, +\infty[, d\mu_\alpha)$, we have for all $x \ge 0$, the following relation

(8.II.25) $$\int_0^\infty \Psi^P_g(f)(j,s)\, g^P_{j,s}(x) d\mu_\alpha(s) = \int_0^\infty (f, G^P_{j+1,s})_\alpha G^P_{j+1,s}(x) d\mu_\alpha(s) - \int_0^\infty (f, G^P_{j,s})_\alpha G^P_{j,s}(x) d\mu_\alpha(s).$$

Proof.

We deduce the result from the relations (8.I.17), (8.II.21) and (8.II.11).

Remark.

The relation (8.II.25) shows that the jth partial reconstruction of f (i.e. the left hand side of (8.II.25)) may be expressed as difference of two "smoothing" of f at two consecutive scales j+1 and j.

8.III. Modified packets associated with the Bessel operator.

Let $\{G_J^P\}_{J\in\mathbb{Z}}$ be a scale discrete scaling function which has been studied in section 8.II.

We consider the functions g_j^M and $\tilde{g}_j^M$, $j \in \mathbb{Z}$, defined by

(8.III.1) $$g_j^M = G_{j+1}^P - G_j^P.$$

(8.III.2) $$\tilde{g}_j^M = G_{j+1}^P + G_j^P.$$

The following proposition gives properties of these functions.

Proposition 8.III.1.

i) The functions g_j^M and $\tilde{g}_j^M$ belong to $L^2([0, +\infty[, d\mu_\alpha)$.

ii) The functions $\mathfrak{F}_B(g_j^M)$ and $\mathfrak{F}_B(\tilde{g}_j^M)$ are in $(L^2 \cap L^\infty)([0, +\infty[, d\mu_\alpha)$ and we have for almost all $\lambda \geq 0$:

(8.III.3) $$|\mathfrak{F}_B(g_j^M)(\lambda)| \leq 2,$$

and

(8.III.4) $$|\mathfrak{F}_B(\tilde{g}_j^M)(\lambda)| \leq 2.$$

iii) We have for almost all $\lambda \geq 0$:

(8.III.5) $$\sum_{j=-\infty}^{\infty} \mathfrak{F}_B(g_j^M)(\lambda)\mathfrak{F}_B(\tilde{g}_j^M)(\lambda) = 1.$$

Proof.

We deduce these results from Proposition 8.II.2, the relation (8.II.12) and Theorem 5.IV.2.

Definition 8.III.1. The sequences $\{g_j^M\}_{j\in\mathbb{Z}}$ and $\{\tilde{g}_j^M\}_{j\in\mathbb{Z}}$ are called respectively modified wavelet packet associated with the Bessel operator (or M-wavelet packet associated with the Bessel operator) and the corresponding dual modified wavelet packet associated with the Bessel operator (or dual M-wavelet packet associated with the Bessel operator).

Remark

The idea to define a M-wavelet packet associated with the Bessel operator is based on the property (8.III.5) of $\{g_j^M\}_{j\in\mathbb{Z}}$ and $\{\tilde{g}_j^M\}_{j\in\mathbb{Z}}$ which is more general than the property (8.I.3) of the P-wavelet packet $\{g_j^P\}_{j\in\mathbb{Z}}$.

Let $\{g_j^M\}_{j\in\mathbb{Z}}$ be a M-wavelet packet associated with the Bessel operator and its dual $\{\tilde{g}_j^M\}_{j\in\mathbb{Z}}$. We consider for all $j\in\mathbb{Z}$ and $s\geq 0$, the functions $g_{j,s}^M$ and $\tilde{g}_{j,s}^M$ defined for almost all $x\geq 0$, by

(8.III.6) $$g_{j,s}^M(x) = T_x(g_j^M)(s).$$

(8.III.7) $$\tilde{g}_{j,s}^M(x) = T_x(\tilde{g}_j^M)(s).$$

Where T_x, $x\geq 0$, are the generalized translation operators given in the section 4.II.

We give some properties of these functions in the following proposition.

Proposition 8.III.2.

i) The functions $g_{j,s}^M$ and $\tilde{g}_{j,s}^M$ belong to $L^2([0, +\infty[, d\mu_\alpha)$ and we have

(8.III.8) $$\|g_{j,s}^M\|_{2,\alpha} \leq \|g_j^M\|_{2,\alpha}.$$

(8.III.9) $$\|\tilde{g}_{j,s}^M\|_{2,\alpha} \leq \|\tilde{g}_j^M\|_{2,\alpha}.$$

ii) For almost all $\lambda\geq 0$, we have

(8.III.10) $$\mathfrak{F}_B(g_{j,s}^M)(\lambda) = j_\alpha(\lambda s)\mathfrak{F}_B(g_j^M)(\lambda).$$

(8.III.11) $$\mathfrak{F}_B(\tilde{g}_{j,s}^M)(\lambda) = j_\alpha(\lambda s)\mathfrak{F}_B(\tilde{g}_j^M)(\lambda).$$

iii) For almost all $\lambda\geq 0$, we have

(8.III.12) $$|\mathfrak{F}_B(g_j^M)(\lambda)| \leq 2.$$

(8.III.13) $$|\mathfrak{F}_B(\tilde{g}_j^M)(\lambda)| \leq 2.$$

Definition 8.III.2. The M-wavelet packet transform Ψ_g^M associated with the Bessel operator (resp. the dual M-wavelet packet transform $\tilde{\Psi}_g^M$ associated with the Bessel operator) is defined for regular functions f on $[0, +\infty[$ by

(8.III.14) $$\Psi_g^M(f)(j,s) = \int_0^\infty f(x)\overline{g_{j,s}^M}(x)d\mu_\alpha(x), \; j \in \mathbb{Z}, \; s \geq 0.$$

(8.III.15) $$(\text{resp.} \; \tilde{\Psi}_g^M(f)(j,s) = \int_0^\infty f(x)\tilde{g}_{j,s}^M(x)d\mu_\alpha(x), \; j \in \mathbb{Z}, \; s \geq 0).$$

The transform Ψ_g^M (resp. $\tilde{\Psi}_g^M$) can also be written in the form

(8.III.16) $$\Psi_g^M(f)(j,s) = f *_B \overline{g_j^M}(s).$$

(8.III.17) $$(\text{resp.} \; \tilde{\Psi}_g^M(f)(j,s) = f *_B \overline{\tilde{g}_j^M}(s)).$$

Where $*_B$ is the generalized convolution product studied in the subsection 4.III.1.

Proposition 8.III.3. For f in $L^q([0, +\infty[, d\mu_\alpha)$, $q \in [1, 2]$, then for all $j \in \mathbb{Z}$, the functions $s \to \Psi_g^M(f)(j,s)$ and $s \to \tilde{\Psi}_g^M(f)(j,s)$ belong to $L^r([0, +\infty[, d\mu_\alpha)$, with $r \in [1, +\infty]$ such that $\frac{1}{r} = \frac{1}{q} - \frac{1}{2}$, and we have

(8.III.18) $$\|\Psi_g^M(f)(j,.)\|_{r,\alpha} \leq \|f\|_{q,\alpha} \, \| g_j^M\|_{2,\alpha} \, .$$

(8.III.19) $$\|\tilde{\Psi}_g^M(f)(j,.)\|_{r,\alpha} \leq \|f\|_{q,\alpha} \, \| \tilde{g}_j^M\|_{2,\alpha} \, .$$

Proof.

We deduce these results from the relations (8.III.16), (8.III.17) and Propositions 4.III.4, 4.III.5.

Theorem 8.III.1. (Plancherel formula). For all f in $L^2([0, +\infty[, d\mu_\alpha)$ we have

(8.III.20) $$\int_0^\infty |f(x)|^2 d\mu_\alpha(x) = \sum_{j=-\infty}^{\infty} \int_0^\infty \Psi_g^M(f)(j,s) \, \overline{\tilde{\Psi}_g^M(f)(j,s)} d\mu_\alpha(s).$$

Proof.

From the relations (8.III.16), (8.III.17), (8.III.3), (8.III.4), Theorem 5.IV.4 and Corollary 5.IV.3, we obtain for all $j \in \mathbb{Z}$:

(8.III.21) $$\int_0^\infty \Psi_g^M(f)(j,s) \, \overline{\tilde{\Psi}_g^M(f)(j,s)} d\mu_\alpha(s) =$$

$$\int_0^\infty |\mathfrak{F}(f)(\lambda)|^2 \mathfrak{F}_B(g_j^M)(\lambda) \mathfrak{F}_B(\tilde{g}_j^M)(\lambda) d\mu_\alpha(\lambda).$$

By using the relations (8.III.1), (8.III.2) and (8.II.11) we show that for almost all $\lambda \geq 0$:

$$\mathfrak{F}_B(g_j^M)(\lambda)\mathfrak{F}_B(\tilde{g}_j^M)(\lambda) \geq 0.$$

Then by applying Fubini-Tonelli's Theorem we obtain

$$\sum_{j=-\infty}^{\infty} \int_0^{\infty} \Psi_g^M(f)(j,s)\, \overline{\tilde{\Psi}_g^M(f)(j,s)}\, d\mu_\alpha(s)$$

$$= \int_0^{\infty} |\mathfrak{F}_B(f)(\lambda)|^2 \Big(\sum_{j=-\infty}^{\infty} \mathfrak{F}_B(g_j^M)(\lambda)\mathfrak{F}_B(\tilde{g}_j^M)(\lambda)\Big) d\mu_\alpha(\lambda).$$

We deduce the result from the relation (8.III.5) and Theorem 5.IV.3.

Lemma 8.III.1. Let f be in $L^p([0, +\infty[, d\mu_\alpha)$, p = 1, 2. For all $j \in \mathbb{Z}$ and $x \geq 0$, we have the following relations

$$\text{(8.III.22)} \qquad \int_0^{\infty} \Psi_g^M(f)(j,s)\tilde{g}_{j,s}^M(x)d\mu_\alpha(s) = \int_0^{\infty} \tilde{\Psi}_g^M(f)(j,s)\, g_{j,s}^M(x)d\mu_\alpha(s)$$

$$= \int_0^{\infty} \Psi_g^P(f)(j,s)\, g_{j,s}^P(x)d\mu_\alpha(s).$$

Where Ψ_g^P and $g_{j,s}^P$ are given by the relations (8.I.10) and (8.I.5).

Proof.

a) If $f \in L^1([0, +\infty[, d\mu_\alpha)$.

From the relations (8.III.16), (8.III.17), (8.III.6),, (8.III.7), and Propositions 8.III.3, 8.III.2, 5.IV.3, for all $j \in \mathbb{Z}$ and for almost all $x \geq 0$, the functions $s \to f *_B \overline{g_j^M}(s)$, $s \to f *_B \overline{\tilde{g}_j^M}(s)$, $s \to T_x(g_j^M)(s)$ and $s \to T_x(\tilde{g}_j^M)(s)$ belong to $L^2([0, +\infty[, d\mu_\alpha)$ and we have for almost all $\lambda \geq 0$:

$$\text{(8.III.23)} \quad \mathfrak{F}_B(f *_B \overline{g_j^M})(\lambda).\mathfrak{F}_B(T_x(\tilde{g}_j^M))(\lambda) = \mathfrak{F}_B(f *_B \overline{\tilde{g}_j^M})(\lambda)\mathfrak{F}_B(T_x(g_j^M))(\lambda)$$

$$= \mathfrak{F}_B(f)(\lambda)\mathfrak{F}_B(g_j^M)(\lambda).\mathfrak{F}_B(\tilde{g}_j^M)(\lambda)j_\alpha(\lambda x).$$

Thus from Theorem 5.IV.3 we have for almost all $x \geq 0$:

$$\text{(8.III.24)} \quad \int_0^{\infty} \Psi_g^M(f)(j,s)\tilde{g}_{j,s}^M(x)d\mu_\alpha(s) = \int_0^{\infty} \tilde{\Psi}_g^M(f)(j,s)\, g_{j,s}^M(x)d\mu_\alpha(s)$$

$$= \int_0^{\infty} \mathfrak{F}_B(f)(\lambda)\mathfrak{F}_B(g_j^M)(\lambda)\mathfrak{F}_B(\tilde{g}_j^M)(\lambda)j_\alpha(\lambda x)d\mu_\alpha(s).$$

On the other hand from the relations (8.III.1), (8.III.2) and (8.II.11) we have for almost all $\lambda \geq 0$:

(8.III.25) $$\mathfrak{F}_B(g_j^M)(\lambda).\mathfrak{F}_B(\tilde{g}_j^M)(\lambda) = (\mathfrak{F}_B(g_j^P))^2 .$$

But from the relation (8.I.16) we deduce that for all $x \geq 0$:

(8.III.26) $$\int_0^\infty \mathfrak{F}_B(f)(\lambda)\mathfrak{F}_B(g_j^M)(\lambda)\mathfrak{F}_B(\tilde{g}_j^M)(\lambda)j_\alpha(\lambda x)d\mu_\alpha(s) = \int_0^\infty \Psi_g^P(f)(j,s)\, g_{j,s}^P(x)d\mu_\alpha(s).$$

We obtain (8.III.22) from the relations (8.III.24) and (8.III.26).

b) If $f \in L^2([0, +\infty[, d\mu_\alpha)$.

We obtain the relation (8.III.22) by the same proof as for the a), and by using the fact that the functions $\mathfrak{F}_B(g_j^M)$ and $\mathfrak{F}_B(\tilde{g}_j^M)$ satisfy the relations (8.III.12), (8.III.13) and by applying Theorem 5.IV.4.

The following result is the so-called Calderon's reproducing formulas.

Theorem 8.III.2. For f in $L^2([0, +\infty[, d\mu_\alpha)$ and $m, n \in \mathbb{Z}$, with $m < n$, we define the functions $f_{m,n}^M$ and $\tilde{f}_{m,n}^M$ by

(8.III.27) $$f_{m,n}^M(x) = \sum_{j=m}^{n-1} \int_0^\infty \Psi_g^M(f)(j,s)\, \tilde{g}_{j,s}^M(x)d\mu_\alpha(s), \quad x \geq 0.$$

(8.III.28) $$\tilde{f}_{m,n}^M(x) = \sum_{j=m}^{n-1} \int_0^\infty \tilde{\Psi}_g^M(f)(j,s)\, g_{j,s}^M(x)d\mu_\alpha(s), \quad x \geq 0.$$

Then these functions belong to $L^2([0, +\infty[, d\mu_\alpha)$ and we have

(8.III.29) $$\lim_{\substack{m\to-\infty\\ n\to+\infty}} \|f_{m,n}^M - f\|_{2,\alpha} = 0.$$

(8.III.30) $$\lim_{\substack{m\to-\infty\\ n\to+\infty}} \|\tilde{f}_{m,n}^M - f\|_{2,\alpha} = 0.$$

Proof.

We obtain these results from Lemma 8.III.1 and Theorem 8.I.3.

Next we formulate a pointwise reconstruction formulas for the transforms Ψ_g^M and $\tilde{\Psi}_g^M$.

Theorem 8.III.3. For f in $L^p([0, +\infty[, d\mu_\alpha)$, p = 1, 2, such that $\mathfrak{F}_B(f)$ belongs to $L^1([0, +\infty[, d\mu_\alpha)$ we have

$$f(x) = \sum_{j=-\infty}^{\infty} \int_0^\infty \Psi_g^M(f)(j,s)\, \tilde{g}_{j,s}^M(x) d\mu_\alpha(s), \text{ a.e,} \tag{8.III.31}$$

$$f(x) = \sum_{j=-\infty}^{\infty} \int_0^\infty \tilde{\Psi}_g^M(f)(j,s)\, g_{j,s}^M(x) d\mu_\alpha(s), \text{ a.e,} \tag{8.III.32}$$

where for each $x \geq 0$, both the integral and the series are absolutely convergent, but possibly not the series of integrals.

Proof.

Lemma 8.III.1 and Theorem 8.I.2 give the relations (8.III.31) and (8.III.32).

Proposition 8.III.4. The scale discrete scaling function $\{G_J^P\}_{J\in\mathbb{Z}}$ given by Definition 8.II.1 possesses the following property

$$\mathfrak{F}_B(G_J^P)(\lambda) = \Big(\sum_{j=-\infty}^{J-1} \mathfrak{F}_B(g_j^M)(\lambda).\mathfrak{F}_B(\tilde{g}_j^M)(\lambda)\Big)^{1/2}, \text{ a.e.} \tag{8.III.33}$$

Proof.

We deduce (8.III.33) from the relations (8.II.2) and (8.III.25).

Theorem 8.III.4. For f in $L^p([0, +\infty[, d\mu_\alpha)$, p = 1, 2, such that $\mathfrak{F}_B(f)$ belongs to $L^1([0, +\infty[, d\mu_\alpha)$, we have the following reconstruction formulas

$$f(x) = \int_0^\infty (f, G_{J,s}^P)_\alpha G_{J,s}^P(x) d\mu_\alpha(s) + \sum_{j=J}^{\infty} \int_0^\infty \Psi_g^M(f)(j,s)\, \tilde{g}_{j,s}(x) d\mu_\alpha(s), \text{ a.e,} \tag{8.III.34}$$

$$f(x) = \int_0^\infty (f, G_{J,s}^P)_\alpha G_{J,s}^P(x) d\mu_\alpha(s) + \sum_{j=J}^{\infty} \int_0^\infty \tilde{\Psi}_g^M(f)(j,s) g_{j,s}(x) d\mu_\alpha(s), \text{ a.e,} \tag{8.III. 35}$$

where, for each $x \geq 0$, in the first term of the second member, the integral is absolutely convergent and in the second term, both the integral and the series are absolutely convergent, but possibly not the series of integrals.

Proof.

We obtain the relations (8.III.34) and (8.III.35) by applying Lemma 8.III.1 and Theorem 8.II.3. ii).

8.IV S-wavelet packet associated with the Bessel operator.

In this section we construct a pair $(\{g_j^s\}_{j\in\mathbb{Z}}, \{\tilde{g}_j^s\}_{j\in\mathbb{Z}})$ different from $(\{g_j^M\}_{j\in\mathbb{Z}}, \{\tilde{g}_j^M\}_{j\in\mathbb{Z}})$ satisfying

$$\sum_{j=-\infty}^{\infty} \mathcal{F}_B(g_j^s)(\lambda).\mathcal{F}_B(\tilde{g}_j^s)(\lambda) = 1, \text{ a.e.}$$

Next, we define and study a S-wavelet transform associated with the Bessel operator, and we obtain for this transform Plancherel and recosntruction formulas.

Definition 8.IV.1. Let $\{g_j^s\}_{j\in\mathbb{Z}}$, be a sequence in $L^2([0, +\infty[, d\mu_\alpha)$. This sequence is called S-wavelet packet associated with the Bessel operator if it satisfies the following conditions

i) For all $j \in \mathbb{Z}$ and for almost all $\lambda \geq 0$, the function $\mathcal{F}_B(g_j^s)(\lambda)$ is real.

ii) "Stability conditions"

$$a \leq \sum_{j=-\infty}^{\infty} (\mathcal{F}_B(g_j^s)(\lambda))^2 \leq b, \text{ a.e.} \tag{8.IV.1}$$

where a, b with $0 < a < b < +\infty$, are the stability constants.

Definition 8.IV.2. Let $\{g_j^s\}_{j\in\mathbb{Z}}$ be a S-wavelet packet associated with the Bessel operator. We define the corresponding dual S-wavelet packet $\{\tilde{g}_j^s\}_{j\in\mathbb{Z}}$ associated with the Bessel operator, by

$$\mathcal{F}_B(\tilde{g}_j^s)(\lambda) = \frac{\mathcal{F}_B(g_j^s)(\lambda)}{\sum_{j=-\infty}^{\infty}(\mathcal{F}_B(g_j^s)(\lambda))^2}, \text{ a.e.} \tag{8.IV.2}$$

The following propositions give properties of the S-wavelet packet $\{g_j^s\}_{j\in\mathbb{Z}}$ associated with the Bessel operator and its dual $\{\tilde{g}_j^s\}_{j\in\mathbb{Z}}$.

Proposition 8.IV.1.

i) For almost all $\lambda \geq 0$, we have

(8.IV.3) $$|\mathfrak{F}_B(g_j^s)(\lambda)| \leq b^{1/2}.$$

(8.IV.4) $$|\mathfrak{F}_B(\tilde{g}_j^s)(\lambda)| \leq \frac{b^{1/2}}{a}.$$

ii) For almost all $\lambda \geq 0$, we have

(8.IV.5) $$\sum_{j=-\infty}^{\infty} \mathfrak{F}_B(g_j^s)(\lambda)\, \mathfrak{F}_B(\tilde{g}_j^s)(\lambda) = 1.$$

iii) For almost all $\lambda \geq 0$, we have

(8.IV.6) $$\sum_{j=-\infty}^{\infty} (\mathfrak{F}_B(g_j^s)(\lambda))^2 = \Big(\sum_{j=-\infty}^{\infty} (\mathfrak{F}_B(\tilde{g}_j^s)(\lambda))^2\Big)^{-1}.$$

Proposition 8.IV.2.

i) For all $J \in \mathbb{Z}$ and for almost all $\lambda \geq 0$, we have

(8.IV.7) $$\frac{\sum_{j=-\infty}^{J-1}(\mathfrak{F}_B(\tilde{g}_j^s)(\lambda))^2}{\sum_{j=-\infty}^{\infty}(\mathfrak{F}_B(\tilde{g}_j^s)(\lambda))^2} = \frac{\sum_{j=-\infty}^{J-1}(\mathfrak{F}_B(g_j^s)(\lambda))^2}{\sum_{j=-\infty}^{\infty}(\mathfrak{F}_B(g_j^s)(\lambda))^2}.$$

ii) For all $J \in \mathbb{Z}$ and for almost all $\lambda \geq 0$, we have

(8.IV.8) $$\sum_{j=-\infty}^{J-1} \mathfrak{F}_B(g_j^s)(\lambda)\, \mathfrak{F}_B(\tilde{g}_j^s)(\lambda) = \frac{\sum_{j=-\infty}^{J-1}(\mathfrak{F}_B(g_j^s)(\lambda))^2}{\sum_{j=-\infty}^{\infty}(\mathfrak{F}_B(g_j^s)(\lambda))^2}.$$

Proof.

We deduce these results from the relation (8.IV.2).

Proposition 8.IV.3.

i) The dual S-wavelet packet $\{\tilde{g}_j^s\}_{j\in\mathbb{Z}}$ associated with the Bessel operator satisfies the conditions

(8.IV.9) $$\frac{1}{b} \leq \sum_{j=-\infty}^{\infty} (\mathfrak{F}_B(\tilde{g}_j^s)(\lambda))^2 \leq \frac{1}{a}, \text{ a.e.}$$

where a and b are the stability constants given by the relation (8.IV.1).

ii) If for all $J \in \mathbb{Z}$ the function $\lambda \to (\sum_{j=-\infty}^{J-1} (\mathfrak{F}_B(g_j^s)(\lambda))^2)^{1/2}$ is in $L^2([0, +\infty[, d\mu_\alpha)$. Then the functions

$$\lambda \to (\sum_{j=-\infty}^{J-1} (\mathfrak{F}_B(\tilde{g}_j^s)(\lambda))^2)^{1/2} \text{ and } \lambda \to (\sum_{j=-\infty}^{J-1} \mathfrak{F}_B(g_j^s)(\lambda)\mathfrak{F}_B(\tilde{g}_j^s)(\lambda))^{1/2}$$

belong to $L^2([0, +\infty[, d\mu_\alpha)$.

This last result and Theorem 5.IV.2 permit to state the following definition.

Definition 8.IV.3. Let $\{g_j^s\}_{j\in\mathbb{Z}}$ be a S-wavelet packet associated with the Bessel operator and its dual $\{\tilde{g}_j^s\}_{j\in\mathbb{Z}}$. We suppose that for all $J \in \mathbb{Z}$, the function $\lambda \to (\sum_{j=-\infty}^{J-1} (\mathfrak{F}_B(g_j^s)(\lambda))^2)^{1/2}$ is in $L^2([0, +\infty[, d\mu_\alpha)$. We define the scale discrete scaling function $\{G_J^s\}_{J\in\mathbb{Z}}$ by

$$\mathfrak{F}_B(G_J^s)(\lambda) = (\sum_{j=-\infty}^{J-1} \mathfrak{F}_B(g_j^s)(\lambda).\mathfrak{F}_B(\tilde{g}_j^s)(\lambda))^{1/2}. \tag{8.IV.10}$$

Some properties of the function $\{G_J^s\}_{J\in\mathbb{Z}}$ are given in the following proposition.

Proposition 8.IV.4.

i) For all $J \in \mathbb{Z}$ the function G_J^s belongs to $L^2([0, +\infty[, d\mu_\alpha)$ and for almost all $\lambda \geq 0$, we have

$$0 \leq \mathfrak{F}_B(G_J^s)(\lambda) \leq 1. \tag{8.IV.11}$$

ii) For almost all $\lambda \geq 0$, we have

$$\lim_{J\to+\infty} \mathfrak{F}_B(G_J^s)(\lambda) = 1. \tag{8.IV.12}$$

Proof.

We obtain these results from the relation (8.IV.8).

Let $\{g_j^s\}_{j\in\mathbb{Z}}$ be a S-wavelet packet associated with the Bessel operator, and its dual $\{\tilde{g}_j^s\}_{j\in\mathbb{Z}}$. We consider for all $j \in \mathbb{Z}$ and $y \geq 0$, the functions $g_{j,y}^s$ and $\tilde{g}_{j,y}^s$ defined for almost all $x \geq 0$, by

(8.IV.13) $$g_{j,y}^s(x) = T_x(g_j^s)(y).$$

(8.IV.14) $$\tilde{g}_{j,y}^s(x) = T_x(\tilde{g}_j^s)(y).$$

Where T_x, $x \geq 0$, are the generalized translation operators given in the section 4.II.

These functions have the following properties.

Proposition 8.IV.5.

i) The functions $g_{j,y}^s$ and $\tilde{g}_{j,y}^s$ belong to $L^2([0, +\infty[, d\mu_\alpha)$ and we have

(8.IV.15) $$\|g_{j,y}^s\|_{2,\alpha} \leq \|g_j^s\|_{2,\alpha}.$$

(8.IV.16) $$\|\tilde{g}_{j,y}^s\|_{2,\alpha} \leq \|g_j^s\|_{2,\alpha}.$$

ii) For almost all $\lambda \geq 0$, we have

(8.IV.17) $$\mathfrak{F}_B(g_{j,y}^s)(\lambda) = j_\alpha(\lambda y)\, \mathfrak{F}_B(g_j^s)(\lambda).$$

(8.IV.18) $$\mathfrak{F}_B(g_{j,y}^s)(\lambda) = j_\alpha(\lambda y)\, \mathfrak{F}_B(g_j^s)(\lambda).$$

Definition 8.IV.4. The S-wavelet transform Ψ_g^s associated with the Bessel operator (resp. the dual S-wavelet transform $\tilde{\Psi}_g^s$ associated with the Bessel operator) is defined for regular functions f on $[0, +\infty[$ by

(8.IV.19) $$\Psi_g^s(f)(j,y) = \int_0^\infty f(x)\, \overline{g_{j,y}^s(x)}\, d\mu_\alpha(x),\ j \in \mathbb{Z},\ y \geq 0.$$

(8.IV.20) $$(\text{resp. } \tilde{\Psi}_g^s(f)(j,y) = \int_0^\infty f(x)\, \overline{\tilde{g}_{j,y}^s(x)}\, d\mu_\alpha(x),\ j \subset \mathbb{Z},\ y \geq 0).$$

The transform Ψ_g^s (resp. $\tilde{\Psi}_g^s$) can be written in the form

(8.IV.21) $$\Psi_g^s(f)(j,y) = f *_B \overline{g_j^s}\,(y).$$

(8.IV.22) $$(\text{resp. } \tilde{\Psi}_g^s(f)(j,y) = f *_B \overline{\tilde{g}_j^s}\,(y)).$$

Where $*_B$ is the generalized convolution product studied in the subsection 4.III.1.

Proposition 8.IV.5. Let f be in $L^q([0, +\infty[, d\mu_\alpha)$, $q \in [1, 2]$. Then for all $j \in \mathbb{Z}$, the functions $y \to \Psi_g^s(f)(j,y)$ and $y \to \tilde{\Psi}_g^s(f)(j,y)$ belong to $L^r([0, +\infty[, d\mu_\alpha)$, $r \in [1, +\infty]$ such that $\frac{1}{r} = \frac{1}{q} - \frac{1}{2}$, and we have

(8.IV.23) $$\|\Psi_g^s(f)(j,.)\|_{r,\alpha} \leq \|f\|_{q,\alpha}\, \|g_j^s\|_{2,\alpha}.$$

(8.IV.24) $$\|\tilde{\Psi}^s_g(f)(j,.)\|_{r,\alpha} \leq \|f\|_{q,\alpha} \, \|\tilde{g}^s_j\|_{2,\alpha}.$$

Using the relations (8.IV.3), (8.IV.4) and Theorem 5.IV.4 we deduce the following results.

Proposition 8.IV.6. Let f be in $L^2([0, +\infty[, d\mu_\alpha)$. Then for all $j \in \mathbb{Z}$, the functions $y \to \Psi^s_g(f)(j,y)$ and $\tilde{\Psi}^s_g(f)(j,y)$ belong to $L^2([0, +\infty[, d\mu_\alpha)$ and we have for almost all $\lambda \geq 0$:

(8.IV.25) $$\mathcal{F}_B(\Psi^s_g(f)(j,.))(\lambda) = \mathcal{F}_B(f)(\lambda)\, \mathcal{F}_B(g^s_j)(\lambda).$$

(8.IV.26) $$\mathcal{F}_B(\tilde{\Psi}^s_g(f)(j,.))(\lambda) = \mathcal{F}_B(f)(\lambda)\, \mathcal{F}_B(\tilde{g}^s_j)(\lambda).$$

Using the methods of the precedent sections we obtain the following Plancherel and reconstruction formulas for the transforms Ψ^s_g and $\tilde{\Psi}^s_g$.

Theorem 8.IV.1. (Plancherel formula). For all f in $L^2([0, +\infty[, d\mu_\alpha)$ we have

(8.IV.27) $$\int_0^\infty |f(x)|^2 d\mu_\alpha(x) = \sum_{j=-\infty}^{\infty} \int_0^\infty \Psi^s_g(f)(j,y)\overline{\tilde{\Psi}^s_g(f)(j,y)}\, d\mu_\alpha(y).$$

Theorem 8.IV.2. For f in $L^p([0, +\infty[, d\mu_\alpha)$, $p = 1, 2$, such that $\mathcal{F}_B(f)$ belongs to $L^1([0, +\infty[, d\mu_\alpha)$, we have the following construction formulas

(8.IV.28) $$f(x) = \sum_{j=-\infty}^{\infty} \int_0^\infty \Psi^s_g(f)(j,y)\, \tilde{g}^s_{j,y}(x) d\mu_\alpha(y), \text{ a.e,}$$

(8.IV.29) $$f(x) = \sum_{j=-\infty}^{\infty} \int_0^\infty \tilde{\Psi}^s_g(f)(j,y)\, g^s_{j,y}(x) d\mu_\alpha(y), \text{ a.e,}$$

where for each $x \geq 0$, both the integral and the series are absolutely convergent, but possibly not the series of integrals.

Theorem 8.IV.3. Let $\{G^s_J\}_{J\in\mathbb{Z}}$ the scale discrete scaling function given by Definition 8.IV.3. For f in $L^p([0, +\infty[, d\mu_\alpha)$, $p = 1, 2$, such that $\mathcal{F}_B(f)$ belongs to $L^1([0, +\infty[, d\mu_\alpha)$, we have the following reconstruction formula

(8.IV.30) $$f(x) = \lim_{J\to+\infty} \int_0^\infty (f, G^s_J)_\alpha\, G^s_{J,y}(x) d\mu_\alpha(y), \text{ a.e,}$$

where for each $x \geq 0$, the integral is absolutely convergent.

Remark.

The S-wavelet packet associated with the Bessel operator should be discussed for a = b. In this case we have

$$\tilde{g}_j^s = \frac{1}{a} g_j^s. \tag{8.IV.31}$$

This shows us that decomposition and reconstruction are provided by the same S-wavelet packet associated with the Bessel operator.

8.V. Multiresolution analysis by means of wavelet packets associated with the Bessel operator.

Three types of wavelet packets associated with the Bessel operator have been discussed in the above sections. All wavelet packets associated with the Bessel operator show the fact that efficient algorithms for reconstruction and decomposition can be given by using the notion of multiresolution analysis associated with the Bessel operator. This will be done in this section.

In the following we use notations and results of the precedent sections.

Definition 8.V.1.

i) For all j and J in $\mathbb{Z}$ we define the convolution operators Q_j^E and P_J^E, $E \in \{P,S\}$, on $L^2([0, +\infty[, d\mu_\alpha)$, by

$$Q_j^E(f) = \begin{cases} g_j^P *_B g_j^P *_B f = g_j^M *_B \tilde{g}^M *_B f, & \text{for } E = P. \\ g_j^s *_B \tilde{g}_j^s *_B f & \text{, for } E = S. \end{cases}$$

$$P_J^E(f) = \begin{cases} G_J^P *_B G_J^P *_B f, & \text{for } E = P. \\ G_J^s *_B G_J^s *_B f, & \text{for } E = S. \end{cases}$$

Or, equivalently for almost all $\lambda \geq 0$:

$\mathfrak{F}_B(Q_j^E)(f))(\lambda) =$

$$\begin{cases} \mathfrak{F}_B(f)(\lambda)(\mathfrak{F}_B(g_j^P)(\lambda))^2 = \mathfrak{F}_B(f)(\lambda)\mathfrak{F}_B(g_j^M)(\lambda)\, \mathfrak{F}_B(\tilde{g}_j^M)(\lambda), & \text{for } E = P. \\ \mathfrak{F}_B(f)\mathfrak{F}_B(g_j^s)(\lambda)\mathfrak{F}_B(\tilde{g}_j^s)(\lambda) & \text{, for } E = S. \end{cases}$$

$$\mathfrak{F}_B(P_J^E(f))(\lambda) = \begin{cases} \mathfrak{F}_B(f)(\lambda).(\mathfrak{F}_B(G_J^P)(\lambda))^2 , & \text{for } E = P. \\ \mathfrak{F}_B(f)(\lambda).(\mathfrak{F}_B(G_J^s)(\lambda))^2 , & \text{for } E = S. \end{cases}$$

ii) The "scale spaces " $\mathcal{V}_J^E$ and the "detail spaces" $\mathcal{W}_j^E$, E = {P, S}, are defined respectively by

$$\mathcal{V}_J^E = P_J^E(L^2([0, +\infty[, d\mu_\alpha)).$$

$$\mathcal{W}_j^E = Q_j^E(L^2([0, +\infty[, d\mu_\alpha)).$$

iii) The collection $\{\mathcal{V}_J^E\}_{J\in\mathbb{Z}}$, $E \in \{P, S\}$, of all spaces $\mathcal{V}_J^E$, $J \in \mathbb{Z}$, is called E-multiresolution analysis of $L^2([0, +\infty[, d\mu_\alpha)$, associated with the Bessel operator.

Remarks.

i) According to our definition, the scale spaces $\mathcal{V}_J^E$ are the images of $L^2([0, +\infty[, d\mu_\alpha)$ by P_J^E. Loosly spoken, $\mathcal{V}_J^E$ contains all J-scale smooth functions of $L^2([0, +\infty[, d\mu_\alpha)$.

The notion "detail space" means that $\mathcal{W}_j^E$ contains the "detail" information needed to go from an approximation at resolution j to an approximation at resolution j + 1.

ii) $\mathcal{W}_J^E$ denotes the complementary space of $\mathcal{V}_J^E$ in $\mathcal{V}_{J+1}^E$.

$$\mathcal{V}_{J+1} = \mathcal{V}_J + \mathcal{W}_J^E.$$

This indicates that the definition of detail spaces is independent of the choice of P-wavelet or M-wavelet packets associated with the Bessel operator.

Proposition 8.V.1. The collection $\{\mathcal{V}_J^E\}_{J\in\mathbb{Z}}$, $E \in \{P, S\}$, satisfies the following properties

i) $\mathcal{V}_J^E \subset \mathcal{V}_{J'}^E \subset L^2([0, +\infty[, d\mu_\alpha)$, $-\infty < J \leq J' < +\infty$.

ii) $\lim\limits_{J\to-\infty} \mathcal{V}_J^E = \{0\}$.

iii) $\lim\limits_{J\to+\infty} \mathcal{V}_J^E = L^2([0, +\infty[, d\mu_\alpha)$.

We note that a multiresolution analysis associated with the Bessel operator allows us to decompose $L^2([0, +\infty[, d\mu_\alpha)$ by a nested sequence of subspaces which, in turn, illustrates how to approximate a given function f in $L^2([0, +\infty[, d\mu_\alpha)$ by an approximation in each of the space $\mathcal{V}_J^E$.

Explicitely written out in terms of the functions

$$\tilde{h}_j^E = \begin{cases} \Psi_g^P(f)(j,.) , & \text{for } E = P. \\ \Psi_g^M(f)(j,.) , & \text{for } E = M. \\ \Psi_g^s(f)(j,.) , & \text{for } E = S. \end{cases}$$

$$h_j^E = \begin{cases} (\tilde{h}_j^P, g_j^P)_\alpha , & \text{for } E = P. \\ (\tilde{h}_j^M, \tilde{g}_j^M)_\alpha , & \text{for } E = M. \\ (\tilde{h}_j^s, \tilde{g}_j^s)_\alpha , & \text{for } E = S. \end{cases}$$

$$\tilde{f}_J^E = \begin{cases} (f, G_J^P)_\alpha , & \text{for } E = P. \\ (f, G_J^P)_\alpha , & \text{for } E = M. \\ (f, G_J^s)_\alpha , & \text{for } E = S. \end{cases}$$

$$f_J^E = \begin{cases} (\tilde{f}_J^P, G_J^P)_\alpha , & \text{for } E = P. \\ (\tilde{f}_J^M, G_J^P)_\alpha , & \text{for , } E = M. \\ (\tilde{f}_J^s, G_J^s)_\alpha , & \text{for , } E = S. \end{cases}$$

with $j, J \in \mathbb{Z}$ and f in $L^2([0, +\infty[, d\mu_\alpha)$.

We have

$$f_J^E \in \mathcal{V}_J^P ,\ E \in \{P, M\} ;\ f_J^E \in \mathcal{V}_J^S ,\ E = S.$$

and

$$h_j^E \in \mathcal{W}_j^P ,\ E \in \{P, M\} ;\ h_j^E \in \mathcal{W}_j^S , E = S.$$

Theorem 8.V.2. A function f in $L^2([0, +\infty[, d\mu_\alpha)$ can be approximated in twofold sense

$$f = \lim_{J\to+\infty} f_J^E , E \in \{P, M, S\}.$$

and

$$f = \lim_{J\to+\infty} \sum_{j=-\infty}^{J-1} h_j^E ,\ E \in \{P, M, S\}.$$

8.VI. Problems.

Problem 8.VI.1.

Let Ψ be a real valued function belonging to $L^2(\mathbb{R}^n)$ (the space of square integrable functions on $\mathbb{R}^n$, $n \geq 2$, with respect to the Lebesgue measure), and $\{r_j\}_{j\in\mathbb{Z}}$ a scale sequence, i.e. a strictly decreasing sequence in $]0, +\infty[$ such that

$$\lim_{j\to-\infty} r_j = +\infty \quad \text{and} \quad \lim_{j\to+\infty} r_j = 0.$$

For $j \in \mathbb{Z}$, we define the function Ψ_j^D on $\mathbb{R}^n$ by

$$\Psi_j^D(x) = \frac{1}{(r_j)^n}\Psi\left(\frac{x}{r_j}\right).$$

The sequence $\{\Psi_j^D\}_{j\in\mathbb{Z}}$ is called Daubechies wavelet (or briefly : D-wavelet) if it satisfies the D-stability conditions

$$A \leq \sum_{j=-\infty}^{\infty} (\mathfrak{F}(\Psi_j^D)(\xi))^2 \leq B, \text{ for almost all } \xi \in \mathbb{R}^n.$$

Where $0 < A \leq B < +\infty$, are the "D-stability constants", and $\mathfrak{F}$ denotes the usual Fourier transform on $\mathbb{R}^n$, given by

$$\mathfrak{F}(f)(\xi) = \int_{\mathbb{R}^n} f(x)\, e^{-i(x,\xi)}\, dx,$$

with $(.,.)$ the scalar product of $\mathbb{R}^n$.

1°) Let $p \in \mathbb{N}^*$ and ϕ be a differentiable and decreasing function on $[0, +\infty[$ satisfying

$$\exists\, \gamma > \frac{1}{2p} \text{ such that } \phi(\lambda) = O(\lambda^{-\gamma}), \text{ as } \lambda \to +\infty.$$

We define the function Ψ on $\mathbb{R}$ by

$$\mathfrak{F}(\Psi)(\lambda) = \lambda^p\left(-\frac{d}{d\lambda}\phi(\lambda^{2p})\right)^{1/2}.$$

and

$$r_j = r_0\, \rho^{-j}, \quad \rho \in]1, +\infty[.$$

Show that the sequence $\{\Psi_j^D\}_{j\in\mathbb{Z}}$ is a D-wavelet on $\mathbb{R}$.

2°) Let $\{\Psi_j^D\}_{j\in\mathbb{Z}}$ be a D-wavelet on $\mathbb{R}^n$. The corresponding dual D-wavelet $\{\tilde{\Psi}_j^D\}_{j\in\mathbb{Z}}$ is defined by

$$\mathfrak{F}(\tilde{\Psi}_j^D)(\xi) = \frac{\mathfrak{F}(\Psi_j^D)(\xi)}{\sum_{k=-\infty}^{\infty}(\mathfrak{F}(\Psi_k^D)(\xi))^2}, \quad \xi \in \mathbb{R}^n.$$

i) Show that

$$\sum_{j=-\infty}^{\infty} \mathfrak{F}(\Psi_j^D)(\xi).\,\mathfrak{F}(\tilde{\Psi}_j^D)(\xi) = 1, \text{ for almost all } \xi \in \mathbb{R}^n.$$

ii) Prove that

$$\sum_{j=-\infty}^{\infty} (\mathfrak{F}(\tilde{\Psi}_j^D)(\xi))^2 = \Big(\sum_{j=-\infty}^{\infty} (\mathfrak{F}(\Psi_j^D)(\xi))^2\Big)^{-1}, \text{ for almost all } \xi \in \mathbb{R}^n.$$

and deduce that

$$\frac{1}{B} \le \sum_{j=-\infty}^{\infty} (\mathfrak{F}(\tilde{\Psi}_j^D)(\xi))^2 \le \frac{1}{A}, \text{ for almost all } \xi \in \mathbb{R}^n.$$

Where A, B are D-stability constants.

iii) Prove that

$$\frac{1}{B} \|\Psi_j^D\|_2 \le \|\tilde{\Psi}_j^D\|_2 \le \frac{1}{A} \|\Psi_j^D\|_2.$$

Where $\|.\|_2$ is the norm of $L^2(\mathbb{R}^n)$.

(See [68]).

Problem 8.VI.2

In this problem we use notations and results of Problem 8.VI.1

I.1°) Let $\Psi(x) = \Psi_0(\|x\|)$ be a radial function on $\mathbb{R}^n$, $n \ge 2$. Show that the function Ψ belongs to $L^2(\mathbb{R}^n)$ if and only if the function Ψ_0 is in $L^2([0, +\infty[, d\mu_{\frac{n-2}{2}})$ and we have

$$\mathfrak{F}(\Psi)(\xi) = (2\pi)^{n/2}\, \mathfrak{F}_B(\Psi_0)(\|\xi\|),$$

in the L^2-case.

Where $\mathfrak{F}_B$ is the Fourier-Bessel transform corresponding to $\alpha = \frac{n-2}{2}$, defined by the relation (5.II.1).

2°) Prove that the sequence $\{\Psi_j^D\}_{j\in\mathbb{Z}}$ is a D-wavelet on $\mathbb{R}^n$ associated with the function Ψ of the precedent question I.1°), if and only if

$$A \leq \sum_{j=-\infty}^{\infty} \mathcal{F}(\Psi_0)(r_j\lambda))^2 \leq B, \text{ for almost all } \lambda \in [0, +\infty[.$$

II.1°) Let Ψ be an even function in $(L^1 \cap L^2)(\mathbb{R})$ such that $\mathcal{F}(\Psi)$ is in $L^1(\mathbb{R})$ and satisfying

$$\exists\, \gamma > \frac{n-2}{2} \text{ such that } \mathcal{F}(\Psi)(\lambda) = O(\lambda^\gamma), \text{ as } \lambda \to 0^+.$$

Show that $\mathcal{R}_{\frac{n-2}{2}}(\Psi)$ belongs to $L^2([0, +\infty[, d\mu_{\frac{n-2}{2}})$ and we have for almost all $\lambda > 0$:

$$\mathcal{F}_B(\mathcal{R}_{\frac{n-2}{2}}(\Psi))(\lambda) = \frac{2^{n-2}(\Gamma(\frac{n}{2}))^2}{\pi} \frac{\mathcal{F}(\Psi)(\lambda)}{\lambda^{n-1}}.$$

2°) Let $\{\Psi_j^D\}_{j\in\mathbb{Z}}$ be a one dimensional D-wavelet, even if n is odd, odd if n is even. We suppose that Ψ has continuous derivatives up to order n - 1,

$$\frac{d^k}{dx^k}\Psi \in L^1(\mathbb{R}), \text{ for } k = 0, 1, 2, \ldots, n-1,$$

and $\lambda^{n-1}\mathcal{F}(\Psi)(\lambda)$ belongs to $(L^1 \cap L^2)(\mathbb{R})$. Prove that $\{(\mathcal{R}_{\frac{n-2}{2}}(\frac{d^{n-1}}{dx^{n-1}}\Psi))_j^D\}_{j\in\mathbb{Z}}$ is a one dimensional D-wavelet and $\mathcal{F}(\mathcal{R}_{\frac{n-2}{2}}(\frac{d^{n-1}}{dx^{n-1}}\Psi))$ is in $L^\infty(\mathbb{R})$ (the space of essentially bounded functions on $\mathbb{R}$ with respect to the Lebesgue measure).

3°) Let $p \in \mathbb{N}$ and $\phi \in \mathcal{S}(\mathbb{R})$ (the usual Schwartz space) be such that $\frac{d}{d\lambda}\phi(\lambda) < 0$ for all $\lambda \geq 0$. We define the function Ψ on $\mathbb{R}$ by

$$\mathcal{F}(\Psi)(\lambda) = \lambda^{n+2p-1}\left(-\frac{d}{d\lambda}\phi(\lambda^{2n+4p-2})\right)^{1/2},$$

and

$$r_j = r_0\,\rho^{-j},\ \rho \in\,]1, +\infty[.$$

Prove that $\{\Psi_j^D\}_{j\in\mathbb{Z}}$ is a one dimensional D-wavelet satisfying all the conditions of the question II.2°) .

(See [68]).

Problem 8.VI.3.

We use in this problem notations and results of Problems 8.VI.1 and 8.VI.2.

Let $\{\Psi_j^D\}_{j\in\mathbb{Z}}$ be a D-wavelet on $\mathbb{R}^n$, $n \geq 2$, with the corresponding dual D-wavelet $\{\tilde{\Psi}_j^D\}_{j\in\mathbb{Z}}$ defined in Problem 8.VI.1.

The n-dimentional D-wavelet transform is defined by

$$S_\Psi^{(D,n)}(f)(j,b) = \int_{\mathbb{R}^n} f(x)\Psi_{j,b}^D(x)dx, \ \ j \in \mathbb{Z}, \, b \in \mathbb{R}^n.$$

Where, by definition

$$\Psi_{j,b}^D(x) = \Psi_j^D(x-b), \ x \in \mathbb{R}^n.$$

1°) In this question we suppose that the function Ψ is such that $\mathfrak{F}(\Psi)$ belongs to $L^\infty(\mathbb{R}^n)$. Prove that for f in $L^2(\mathbb{R}^n)$ and $I, J \in \mathbb{Z}$, the function

$$f^{I,J}(x) = \sum_{j=I}^{J} \int_{\mathbb{R}^n} S_\Psi^{(D,n)}(f)(j,b)\,\tilde{\Psi}_{j,b}^D(x)db,$$

belongs to $L^2(\mathbb{R}^n)$ and satisfies

$$\lim_{\substack{I\to-\infty \\ J\to+\infty}} \|f^{I,J} - f\|_2 = 0.$$

2) Show that for f in $L^1(\mathbb{R}^n)$ such that $\mathfrak{F}(f)$ belongs to $L^1(\mathbb{R}^n)$, we have

$$f(x) = \sum_{j=-\infty}^{\infty} \left(\int_{\mathbb{R}^n} S_\Psi^{(D,n)}(f)(j,b)\,\tilde{\Psi}_{j,b}^D(x)db\right)$$

whenever f is continuous at x.

(See [68]).

Problem 8.VI.4.

We use in this problem notations and results of the precedent problems.

1°) Let $\{\Psi_j^D\}_{j\in\mathbb{Z}}$ be a one dimensional D-wavelet satisfying the conditions of the question II.1) of Problem 8.VI.2. Let $\phi = \dfrac{d^{n-1}}{dx^{n-1}}\Psi$. Show that for f in $L^1(\mathbb{R}^n)$ we have

$$S_{{}^tR(\phi)}^{(D,n)}(f)(j,b) = \int_{S^{n-1}} S_\phi^{(D,1)}(R_\theta(f))(j, (b, \theta))\, d\theta.$$

Where $R_\theta(f)$ is the Radon transform of f defined in Problem 2.VI.1.

2) Let $\{\Psi_j^D\}_{j\in\mathbb{Z}}$ be a one dimensional D-wavelet satisfying the conditions of the question II.2) of Problem 8.VI.2. Let $\phi = \frac{d^{n-1}}{dx^{n-1}}\Psi$.

i) Prove that $\{({}^tR(\phi))_j^D\}_{j\in\mathbb{Z}}$ is a radial n-dimensional D-wavelet and $\mathfrak{F}({}^tR(\phi))$ is in $L^\infty(\mathbb{R}^n)$.

ii) Prove that for f in $L^1(\mathbb{R}^n)$ such that $\mathfrak{F}(f)$ belongs to $L^1(\mathbb{R}^n)$ we have

$$f(x) = \sum_{j=-\infty}^{\infty} \left[\int_{\mathbb{R}^n} \left(\int_{S^{n-1}} S_\phi^{(D,1)}(R_\theta(f))(j,(b,\theta))\,d\theta\right)({}^tR(\phi))_{j,b}^D(x)db\right],$$

whenever f is continuous at x.

iii) Show that for f in $(L^1\cap L^2)(\mathbb{R}^n)$ and $I, J\in\mathbb{Z}$, the function

$$f^{I,J}(x) = \sum_{j=I}^{J} \int_{\mathbb{R}^n} \left(\int_{S^{n-1}} S_\phi^{(D,1)}(R_\theta(f))(j,(b,\theta))\,d\theta\right) ({}^tR(\phi))_{j,b}(x)db,$$

satisfies

$$\lim_{\substack{I\to-\infty\\ J\to+\infty}} \|f^{I,J} - f\|_2 = 0.$$

(See [68]).

9 Continuous Linear Wavelet Transform Associated With The Bessel Operator And Its Discretization

Introduction

In this chapter we study the continuous linear wavelet analysis associated with the Bessel operator, and we prove in particular a Calderon's reproducing formula and a pointwise reconstruction formula for the continuous linear wavelet transform associated with the Bessel operator.

Next according to [37] (see also [26]) we present a construction of linear wavelet packets and the corresponding linear wavelet packet transform and establish also for this transform a Calderon's formula and a pointwise reconstruction formula.

We conclude this chapter by proving pointwise reconstruction formulas associated with the scale discrete linear scaling function corresponding to a linear wavelet packet associated with the Bessel operator.

The results of this chapter have been generalized in [89] to the case of a singular differential operator more general than the Bessel operator.

9.I.Linear wavelets associated with the Bessel operator.

Definition 9.I.1. A linear wavelet associated with the Bessel operator (also called L-wavelet associated with the Bessel operator) is a measurable function g^L on $[0, +\infty[$ satisfying the conditions

(9.I.1) $$\int_0^\infty |\mathfrak{F}_B(g^L)(\lambda)| \frac{d\lambda}{\lambda} < +\infty.$$

(9.I.2) $$C_{g^L} = \int_0^\infty \mathfrak{F}_B(g^L)(\lambda) \frac{d\lambda}{\lambda} \neq 0.$$

Example.

Let $t \in]0, +\infty[$. From the relation (5.II.2) the function

(9.I.3) $$k_t(x) = \mathfrak{F}_B(e^{-t\lambda^2})(x) = \frac{e^{-\frac{x^2}{4t}}}{(2t)^{\alpha+1}},\ x \geq 0,$$

belongs to $\mathcal{S}_*(\mathbb{R})$. Using Definition 9.I.1 we deduce that the function

(9.I.4) $$g^L(x) = -\frac{d}{dt}k_t(x) = (x^2 - 4(\alpha+1)t)\frac{e^{-\frac{x^2}{4t}}}{(2t)^{\alpha+3}},\ x \geq 0,$$

is a linear wavelet associated with the Bessel operator in $\mathcal{S}_*(\mathbb{R})$ and we have

(9.I.5) $$C_{g^L} = \frac{1}{2t}.$$

Proposition 9.I.1. Let g^L be a L-wavelet associated with the Bessel operator, in $L^2([0, +\infty[, d\mu_\alpha)$. Then the function g_a^L, $a \in]0, +\infty[$, defined for almost all $x \geq 0$, by

(9.I.6) $$g_a^L(x) = \frac{1}{a^{2\alpha+2}}\, g^L\left(\frac{x}{a}\right),$$

is a L-wavelet associated with the Bessel operator in $L^2([0, +\infty[, d\mu_\alpha)$, and we have

(9.I.7) $$C_{g_a^L} = C_{g^L}.$$

Proof.

Using the fact that for almost all $\lambda \geq 0$, we have

(9.I.8) $$\mathfrak{F}_B(g_a^L)(\lambda) = \mathfrak{F}_B(g^L)(a\lambda),$$

and a change of variables we obtain

(9.I.9) $$\int_0^\infty |\mathfrak{F}_B(g_a^L)(\lambda)|\frac{d\lambda}{\lambda} = \int_0^\infty |\mathfrak{F}_B(g^L)(a\lambda)|\frac{d\lambda}{\lambda} = \int_0^\infty |\mathfrak{F}_B(g^L)(u)|\frac{du}{u} < +\infty,$$

and

$$C_{g_a^L} = \int_0^\infty \mathfrak{F}_B(g_a^L)(\lambda)\frac{d\lambda}{\lambda} = \int_0^\infty \mathfrak{F}_B(g^L)(a\lambda)\frac{d\lambda}{\lambda} = \int_0^\infty \mathfrak{F}_B(g^L)(u))\frac{du}{u} \neq 0.$$

Then the function g_a^L, $a \in]0, +\infty[$, is a L-wavelet associated with the Bessel operator, and we have

$$C_{g_a^L} = C_{g^L}.$$

Let g^L be a L-wavelet associated with the Bessel operator in $L^2([0, +\infty[, d\mu_\alpha)$ and g_a^L, $a \in]0, +\infty[$, the function given by the relation (9.I.6). We consider for $y \geq 0$, the function $g_{a,y}^L$ given for almost all $x \geq 0$, by

(9.I.10) $$g^L_{a,y}(x) = T_x(g^L_a)(y).$$

Where T_x, $x \geq 0$, are the generalized translation operators defined by the relation (4.II.3).

Remark.

The function $g^L_{a,y}$ can also be written for almost all $x \geq 0$, in the form

(9.I.11) $$g^L_{a,y}(x) = \frac{1}{a^{2\alpha+2}} T_{y/a}(g^L)(\frac{x}{a}).$$

The function $g^L_{a,y}$ satisfies the following properties.

Proposition 9.I.2.

i) The function $g^L_{a,y}$ belongs to $L^2([0, +\infty[, d\mu_\alpha)$ and we have

(9.I.12) $$\|g^L_{a,y}\|_{2,\alpha} \leq \frac{1}{a^{\alpha+1}} \|g^L\|_{2,\alpha}.$$

ii) For almost all $\lambda \geq 0$, we have

(9.I.13) $$\mathfrak{F}_B(g^L_{a,y})(\lambda) = j_\alpha(\lambda y)\mathfrak{F}_B(g^L_a)(\lambda).$$

Definition 9.I.2. Let g^L be a L-wavelet associated with the Bessel operator in $L^2([0, +\infty[, d\mu_\alpha)$. The continuous L-wavelet transform $\Psi^L_{g^L}$ associated with the Bessel operator is defined for regular functions f on $[0, +\infty[$ by

(9.I.14) $$\Psi^L_{g^L}(f)(a,y) = \int_0^\infty f(x)\, g^L_{a,y}(x)\, d\mu_\alpha(x),\ a > 0,\ y \geq 0.$$

This transform can also be written in the form

(9.I.15) $$\Psi^L_{g^L}(f)(a,y) = f *_B g^L_a(y).$$

Where $*_B$ is the generalized convolution product given by the relation (4.III.1).

Proposition 9.I.3. Let g^L be a L-wavelet associated with the Bessel operator in $L^2([0, +\infty[, d\mu_\alpha)$. Then for all $a \in]0, +\infty[$ and f in $L^q([0, +\infty[, d\mu_\alpha)$, $q \in [1, 2]$, the function $y \to \Psi^L_{g^L}(f)(a,y)$ belongs to $L^r([0, +\infty[, d\mu_\alpha)$ with $r \in [1, +\infty]$, such that $\frac{1}{r} = \frac{1}{q} - \frac{1}{2}$, and we have

(9.I.16) $$\|\Psi^L_{g^L}(f)(a,.)\|_{r,\alpha} \leq \frac{1}{a^{\alpha+1}} \|f\|_{q,\alpha} \|g^L\|_{2,\alpha}.$$

The following result is the so-called Calderon's reproducing formula for the transform $\Psi^L_{g^L}$.

Theorem 9.I.1. Let g^L be a L-wavelet associated with the Bessel operator in $L^2([0,+\infty[, d\mu_\alpha)$. Then for f in $L^2([0,+\infty[, d\mu_\alpha)$ and $0<\varepsilon<\delta<+\infty$, the function

$$f^L_{\varepsilon,\delta}(y) = \frac{1}{C_{g^L}}\int_\varepsilon^\delta \Psi^L_{g^L}(f)(a,y)\frac{da}{a}, \tag{9.I.17}$$

belongs to $L^2([0,+\infty[, d\mu_\alpha)$, and we have

$$\lim_{\substack{\varepsilon\to 0^+\\ \delta\to+\infty}} \|f^L_{\varepsilon,\delta} - f\| = 0. \tag{9.I.18}$$

To prove this theorem we need the following two lemmas.

Lemma 9.I.1. Let g^L be a L-wavelet associated with the Bessel operator in $L^2([0,+\infty[, d\mu_\alpha)$. For $0<\varepsilon<\delta<+\infty$, we define the functions $G^L_{\varepsilon,\delta}$ and $K^L_{\varepsilon,\delta}$ by

$$G^L_{\varepsilon,\delta}(.) = \frac{1}{C_{g^L}}\int_\varepsilon^\delta g^L_a(.)\frac{da}{a}. \tag{9.I.19}$$

$$K^L_{\varepsilon,\delta}(\lambda) = \frac{1}{C_{g^L}}\int_\varepsilon^\delta \mathcal{F}_B(g^L_a)(\lambda)\frac{da}{a},\ \lambda\geq 0. \tag{9.I.20}$$

Then

i) $G^L_{\varepsilon,\delta} \in L^2([0,+\infty[, d\mu_\alpha)$.

ii) $K^L_{\varepsilon,\delta} \in L^\infty([0,+\infty[, d\mu_\alpha)$.

iii) $\mathcal{F}_B(G^L_{\varepsilon,\delta}) = K^L_{\varepsilon,\delta}$.

iv) For almost all $\lambda>0$:

$$\lim_{\substack{\varepsilon\to 0^+\\ \delta\to+\infty}} K^L_{\varepsilon,\delta}(\lambda) = 1.$$

Proof.

i) From the relation (9.I.6) the function $G^L_{\varepsilon,\delta}$ belongs to $L^2([0,+\infty[, d\mu_\alpha)$.

ii) We have for almost all $\lambda\geq 0$:

$$|K^L_{\varepsilon,\delta}(\lambda)| \leq \frac{1}{C_{g^L}}\int_0^\infty |\mathcal{F}_B(g^L_a)(\lambda)|\frac{da}{a}.$$

Using the relations (9.I.8), (9.I.1) and a change of variables we obtain for almost all $\lambda>0$:

$$|K^L_{\varepsilon,\delta}(\lambda)| \leq \frac{1}{C_{g^L}}\int_0^\infty |\mathcal{F}_B(g^L)(u)|\frac{du}{u} < +\infty.$$

Thus the function $K^L_{\varepsilon,\delta}$ belongs to $L^\infty([0,+\infty[, d\mu_\alpha)$.

iii) For all $\Psi \in \mathcal{S}_*(\mathbb{R})$ we have

$$\int_0^\infty G^L_{\varepsilon,\delta}(x)\,\mathfrak{F}_B(\Psi)(x)d\mu_\alpha(x) = \frac{1}{C_{g^L}}\int_0^\infty \Big(\int_\varepsilon^\delta g^L_a(x)\frac{da}{a}\Big)\,\mathfrak{F}_B(\Psi)(x)d\mu_\alpha(x).$$

Using the relation (9.I.6), Fubini-Tonelli's theorem and Hölder's inequality we have

$$\int_0^\infty \int_\varepsilon^\delta |g^L_a(x)|\,|\mathfrak{F}_B(\Psi)(x)|\,d\mu_\alpha(x)\frac{da}{a} \le \|\mathfrak{F}_B(\Psi)\|_{2,\alpha}\|g\|_{2,\alpha}\int_\varepsilon^\delta \frac{da}{a^{\alpha+2}} < +\infty.$$

Then from Fubini-Tonelli's theorem we obtain

$$\int_0^\infty G^L_{\varepsilon,\delta}(x)\,\mathfrak{F}_B(\Psi)(x)d\mu_\alpha(x) = \frac{1}{C_{g^L}}\int_\varepsilon^\delta \Big(\int_0^\infty g^L_a(x)\mathfrak{F}_B(\Psi)(x)d\mu_\alpha(x)\Big)\frac{da}{a}.$$

By applying Theorem 5.IV.2 we have

$$\int_0^\infty G^L_{\varepsilon,\delta}(x)\,\mathfrak{F}_B(\Psi)(x)d\mu_\alpha(x) = \frac{1}{C_{g^L}}\int_\varepsilon^\delta \Big(\int_0^\infty \mathfrak{F}_B(g^L_a)(\lambda)\Psi(\lambda)d\mu_\alpha(\lambda)\Big)\frac{da}{a}.$$

From Fubini-Tonelli's theorem, a change of variables and the relation (9.I.1) we have

$$\int_\varepsilon^\delta \int_0^\infty |\mathfrak{F}_B(g^L_a)(\lambda)|\,|\Psi(\lambda)|\,d\mu_\alpha(\lambda)\frac{da}{a} \le \|\Psi\|_{1,\alpha}\int_0^\infty |\mathfrak{F}_B(g^L)(u)|\frac{du}{u} < +\infty.$$

Thus from Fubini's theorem and the relation (9.I.20) we obtain

$$\begin{aligned}\int_0^\infty G^L_{\varepsilon,\delta}(x)\,\mathfrak{F}_B(\Psi)(x)d\mu_\alpha(x) &= \int_0^\infty \Big(\frac{1}{C_{g^L}}\int_\varepsilon^\delta \mathfrak{F}_B(g^L_a)(\lambda)\frac{da}{a}\Big)\Psi(\lambda)d\mu_\alpha(\lambda),\\ &= \int_0^\infty K^L_{\varepsilon,\delta}(\lambda)\Psi(\lambda)|\,d\mu_\alpha(\lambda).\end{aligned}$$

Then Proposition 5.VII.3 implies that the function $K^L_{\varepsilon,\delta}$ belongs to $L^2([0,+\infty[, d\mu_\alpha)$ and we have

$$\mathfrak{F}_B(G^L_{\varepsilon,\delta}) = K^L_{\varepsilon,\delta}.$$

iv) We obtain the result by a change of variables and the relation (9.I.2).

Lemma 9.I.2. The function $f^L_{\varepsilon,\delta}$ defined in Theorem 9.I.1 belongs to $L^2([0,+\infty[, d\mu_\alpha)$ and satisfies the relations

$$f^L_{\varepsilon,\delta} = f *_B G^L_{\varepsilon,\delta}. \tag{9.I.21}$$

$$\mathfrak{F}_B(f^L_{\varepsilon,\delta}) = \mathfrak{F}(f).\,K^L_{\varepsilon,\delta}. \tag{9.I.22}$$

Proof.

For all $\Psi \in \mathcal{S}_*(\mathbb{R})$ we have from the relation (9.I.15) :

$$\int_0^\infty f^L_{\varepsilon,\delta}(x)\mathfrak{F}_B(\Psi)(x)d\mu_\alpha(x) = \frac{1}{C_{g^L}}\int_0^\infty \Big(\int_\varepsilon^\delta f *_B g^L_a(x)\frac{da}{a}\Big)\, \mathfrak{F}_B(\Psi)(x)d\mu_\alpha(x).$$

Using the relations (9.I.6), (4.III.11) and Fubini-Tonelli's theorem we obtain

$$\int_0^\infty \int_\varepsilon^\delta |f*_B g^L_a(x)|\,|\mathfrak{F}_B(\Psi)(x)|\,d\mu_\alpha(x)\frac{da}{a} \leq \|f\|_{2,\alpha}\,\|g\|_{2,\alpha}\|\mathfrak{F}_B(\Psi)\|_{1,\alpha}\int_\varepsilon^\delta \frac{da}{a^{\alpha+2}} < +\infty.$$

Thus from Fubini's theorem we have

$$\int_0^\infty f^L_{\varepsilon,\delta}(x)\mathfrak{F}_B(\Psi)(x)d\mu_\alpha(x) = \frac{1}{C_{g^L}}\int_\varepsilon^\delta \Big(\int_0^\infty f *_B g^L_a(x)\, \mathfrak{F}_B(\Psi)(x)d\mu_\alpha(x)\Big)\frac{da}{a}.$$

Using Lemma 5.IV.1 we obtain

$$\int_0^\infty f^L_{\varepsilon,\delta}(x)\mathfrak{F}_B(\Psi)(x)d\mu_\alpha(x) = \frac{1}{C_{g^L}}\int_\varepsilon^\delta \Big(\int_0^\infty \mathfrak{F}_B(f)(\lambda)\mathfrak{F}_B(g^L_a)(\lambda)\Psi(\lambda)d\mu_\alpha(\lambda)\Big)\frac{da}{a}.$$

But from Fubini-Tonelli's theorem, Hölder's inequality, Theorem 5.VI.3, a change of variables and the relation (9.I.1), we have

$$\int_\varepsilon^\delta \Big(\int_0^\infty |\mathfrak{F}_B(f)(\lambda)|\,|\mathfrak{F}_B(g^L_a)(\lambda)|\,|\Psi(\lambda)|\,d\mu_\alpha(\lambda)\frac{da}{a}$$

$$\leq \|f\|_{2,\alpha}\,\|\Psi\|_{2,\alpha}\int_0^\infty |\mathfrak{F}_B(g^L)(u)|\frac{du}{u} < +\infty.$$

Then we deduce from Fubini's theorem and the relation (9.I.20) :

$$\int_0^\infty f^L_{\varepsilon,\delta}(x)\mathfrak{F}_B(\Psi)(x)d\mu_\alpha(x) = \int_0^\infty \mathfrak{F}_B(f)(\lambda)\Big(\frac{1}{C_{g^L}}\int_\varepsilon^\delta \mathfrak{F}_B(g^L_a)(\lambda)\frac{da}{a}\Big)\Psi(\lambda)d\mu_\alpha(\lambda),$$

$$= \int_0^\infty \mathfrak{F}_B(f)(\lambda)\, K^L_{\varepsilon,\delta}(\lambda)\Psi(\lambda)\, d\mu_\alpha(\lambda).$$

Thus Proposition 5.VII.3 and Lemma 9.I.1 imply that the function $f^L_{\varepsilon,\delta}$ belongs to $L^2([0,+\infty[, d\mu_\alpha)$ and we have

$$\mathfrak{F}_B(f^L_{\varepsilon,\delta}) = \mathfrak{F}_B(f).K^L_{\varepsilon,\delta}.$$

Using Lemma 9.I.1. iii) we obtain

$$\mathfrak{F}_B(f^L_{\varepsilon,\delta}) = \mathfrak{F}_B(f).\mathfrak{F}_B(G^L_{\varepsilon,\delta}).$$

From Theorems 5.IV.2 and 5.IV.4 we deduce that

$$f^L_{\varepsilon,\delta} = f *_B G^L_{\varepsilon,\delta}.$$

Proof of Theorem 9.I.1.

From Theorem 5.IV.3 and Lemma 9.I.2 we have

$$\|f^L_{\varepsilon,\delta} - f\|_{2,\alpha} = \|\mathfrak{F}_B(f^L_{\varepsilon,\delta}) - \mathfrak{F}_B(f)\|_{2,\alpha},$$

$$= \|\mathcal{F}_B(f)\,(1 - K^L_{\varepsilon,\delta})\|_{2,\alpha}.$$

We deduce the result from Lemma 9.I.1. iv) and the dominated convergence theorem.

Next we formulate a pointwise reconstruction formula for the transform $\Psi^L_{g^L}$.

Theorem 9.I.2. Let g^L be a L-wavelet associated with the Bessel operator, in $L^2([0, +\infty[, d\mu_\alpha)$. For f in $(L^1 \cap L^2)([0, +\infty[, d\mu_\alpha)$ such that $\mathcal{F}_B(f)$ belongs to $L^1([0, +\infty[, d\mu_\alpha)$, we have the following reconstruction formula

$$f(y) = \frac{1}{C_{g^L}} \int_0^\infty \Psi^L_{g^L}(f)(a,y)\, \frac{da}{a},\ y \geq 0, \tag{9.I.23}$$

where for all $y \geq 0$, the integral converges absolutely.

Proof.

Let $y \in [0, +\infty[$. From Definition 9.I.2, Proposition 9.I.2 and Theorem 5.IV.3 we have

$$\Psi^L_{g^L}(f)(a,y) = \int_0^\infty \mathcal{F}_B(f)(\lambda).\mathcal{F}_B(g^L_a)(\lambda) j_\alpha(\lambda y) d\mu_\alpha(\lambda). \tag{9.I.24}$$

Using the fact that

$$\forall\ y \geq 0,\ \lambda \geq 0,\ |j_\alpha(\lambda y)| \leq 1\ ,$$

we obtain

$$\int_0^\infty |\Psi^L_{g^L}(f)(a,y)| \frac{da}{a} \leq \int_0^\infty \int_0^\infty |\mathcal{F}_B(f)(\lambda)|\, |\mathcal{F}_B(g^L_a)(\lambda)|\, d\mu_\alpha(\lambda)\, \frac{da}{a}.$$

From Fubini-Tonelli's theorem we deduce that

$$\int_0^\infty |\Psi^L_{g^L}(f)(a,y)| \frac{da}{a} \leq \int_0^\infty |\mathcal{F}_B(f)(\lambda)|\, \left(\int_0^\infty |\mathcal{F}_B(g^L_a)(\lambda)|\, \frac{da}{a}\right) d\mu_\alpha(\lambda).$$

But from the relation (9.I.8) and a change of variables we have

$$\int_0^\infty |\mathcal{F}_B(g^L_a)(\lambda)|\, \frac{da}{a} = \int_0^\infty |\mathcal{F}_B(g^L)(a\lambda)|\, \frac{da}{a} = \int_0^\infty |\mathcal{F}_B(f)(u)|\, \frac{du}{u}.$$

Thus

$$\int_0^\infty |\Psi^L_{g^L}(f)(a,y)| \frac{da}{a} \leq \left(\int_0^\infty |\mathcal{F}_B(g^L)(u)|\, \frac{du}{u}\right) \|\mathcal{F}_B(f)\|_{1,\alpha} < +\infty.$$

Then the integral given in the relation (9.I.23) converges absolutely. On the other hand from the relation (9.I.24) and Fubini's theorem we deduce that

$$\int_0^\infty \Psi_{g^L}^L(f)(a,y)\frac{da}{a} = \int_0^\infty \mathfrak{F}_B(f)(\lambda) j_\alpha(\lambda y) \left(\int_0^\infty \mathfrak{F}_B(g_a^L)(\lambda)\frac{da}{a}\right)d\mu_\alpha(\lambda).$$

But from the relation (9.I.8) and a change of variables we have

$$\int_0^\infty \mathfrak{F}_B(g_a^L)(\lambda)\frac{da}{a} = \int_0^\infty \mathfrak{F}_B(g^L)(a\lambda)\frac{da}{a} = \int_0^\infty \mathfrak{F}_B(g^L)(u)\frac{du}{u} = C_{g^L}.$$

Then

$$\frac{1}{C_{g^L}}\int_0^\infty \Psi_{g^L}^L(f)(a,y)\frac{da}{a} = \int_0^\infty \mathfrak{F}_B(f)(\lambda)\, j_\alpha(\lambda y)\, d\mu_\alpha(\lambda).$$

We obtain the relation (9.I.23) from Theorem 5.II.2.

9.II. Linear wavelet packets associated with the Bessel operator.

Let g^L be a L-wavelet associated with the Bessel operator which has been studied in the section 9.I. We consider the function g_a^L, $a > 0$, given by the relation (9.I.6), and $\{r_j\}_{j\in\mathbb{Z}}$ a scale sequence in $]0, +\infty[$, which is strictly decreasing and satisfying

$$\lim_{j\to-\infty} r_j = +\infty \ , \quad \lim_{j\to+\infty} r_j = 0.$$

We define the function $g_j^{L,P}$, $j \in \mathbb{Z}$, by

$$g_j^{L,P}(.) = \frac{1}{C_{g^L}}\int_{r_{j+1}}^{r_j} g_a^L(.)\frac{da}{a}. \tag{9.II.1}$$

From Lemma 9.I.1 this function belongs to $L^2([0, +\infty[, d\mu_\alpha)$, and we have for almost all $\lambda \geq 0$:

$$\mathfrak{F}_B(g_j^{L,P})(\lambda) = \frac{1}{C_{g^L}}\int_{r_{j+1}}^{r_j} \mathfrak{F}_B(g_a^L)(\lambda)\frac{da}{a}. \tag{9.II.2}$$

Definition 9.II.1. The sequence $\{g_j^{L,P}\}_{j\in\mathbb{Z}}$ defined by the relation (9.II.1) is called linear wavelet packet associated with the Bessel operator (also called L-wavelet packet associated with the Bessel operator).

Remark.

The function $g_j^{L,P}$, $j \in \mathbb{Z}$, is called L-wavelet packet member of step j, associated with the Bessel operator.

The L-wavelet packet $\{g_j^{L,P}\}_{j\in\mathbb{Z}}$ associated with the Bessel operator, satisfies the following properties.

Proposition 9.II.1.

i) We have for almost all $\lambda \geq 0$:

(9.II.3)
$$\sum_{j=-\infty}^{\infty} |\mathfrak{F}_B(g_j^{L,P})(\lambda)| \leq \frac{1}{|C_{g^L}|}\int_0^{\infty} |\mathfrak{F}_B(g^L)(\lambda)| \frac{d\lambda}{\lambda}.$$

ii) We have for almost all $\lambda \geq 0$:

(9.II.4)
$$\sum_{j=-\infty}^{\infty} \mathfrak{F}_B(g_j^{L,P})(\lambda) = 1.$$

Proof.

We obtain these results from Lemma 9.I.1, the relation (9.I.8) and Definition 9.I.1.

Example.

We consider the L-wavelet packet $\{g_j^{L,P}\}_{j\in\mathbb{Z}}$ associated with the Bessel operator, corresponding to the L-wavelet associated with the Bessel operator, given by the relation (9.I.4). For all $j \in \mathbb{Z}$ we have

$$\int_{r_{j+1}}^{r_j} \mathfrak{F}_B(g_a^L)(\lambda)\frac{da}{a} = \lambda^2\int_{r_{j+1}}^{r_j} ae^{-t\lambda^2a^2}\,da = \frac{1}{2t}[e^{-t\lambda^2 r_{j+1}^2} - e^{-t\lambda^2 r_j^2}].$$

Then from the relations (9.I.5) and (9.II.2) we have

(9.II.5)
$$\forall\,\lambda \geq 0,\ \mathfrak{F}_B(g_j^{L,P})(\lambda) = e^{-t\lambda^2 r_{j+1}^2} - e^{-t\lambda^2 r_j^2}.$$

Let $\{g_j^{L,P}\}_{j\in\mathbb{Z}}$ be a L-wavelet packet associated with the Bessel operator. We consider for all $j \in \mathbb{Z}$ and $y \geq 0$, the function $g_{j,y}^{L,P}$ given for almost all $x \geq 0$, by

(9.II.6)
$$g_{j,y}^{L,P}(x) = T_x(g_j^{L,P})(y),$$

where T_x, $x \geq 0$, are the generalized translation operators given by the relation (4.II.3).

This function satisfies the following properties.

Proposition 9.II.2.

i) The function $g_{j,y}^{L,P}$ belongs to $L^2([0, +\infty[, d\mu_\alpha)$ and we have

(9.II.7) $$\|g^{L,P}_{j,y}\|_{2,\alpha} \leq \|g^{L,P}_{j}\|_{2,\alpha}.$$

ii) For almost all $\lambda \geq 0$, we have

(9.II.8) $$\mathcal{F}_B(g^{L,P}_{j,y})(\lambda) = j_\alpha(\lambda y)\, \mathcal{F}_B(g^{L,P}_{j})(\lambda).$$

(9.II.9) $$\sum_{j=-\infty}^{\infty} |\mathcal{F}_B(g^{L,P}_{j,y})(\lambda)| \leq \frac{1}{|C_{g^L}|}\int_0^\infty |\mathcal{F}_B(g^L)(\lambda)| \frac{d\lambda}{\lambda}.$$

Proof.

We deduce these results from the relations (9.II.6), (9.II.3), the properties of the generalized translation operators T_x, $x \geq 0$, and the fact that

$$\forall\ y \geq 0,\ \lambda \geq 0,\ |j_\alpha(\lambda y)| \leq 1.$$

Definition 9.II.2. Let $\{g^{L,P}_j\}_{j\in\mathbb{Z}}$ be a L-wavelet packet associated with the Bessel operator. The L-wavelet packet transform $\Psi^{L,P}_{g^L}$ associated with the Bessel operator is defined for regular functions f on $[0, +\infty[$ by

(9.II.10) $$\Psi^{L,P}_{g^L}(f)(j,y) = \int_0^\infty f(x)\, g^{L,P}_{j,y}(x)\, d\mu_\alpha(x),\ j\in\mathbb{Z},\ y \geq 0.$$

This transform can also be written in the form

(9.II.11) $$\Psi^{L,P}_{g^L}(f)(j,y) = f *_B g^{L,P}_j(y).$$

Where $*_B$ is the generalized convolution product given by the relation (4.III.1).

Proposition 9.II.3. Let $\{g^{L,P}_j\}_{j\in\mathbb{Z}}$ be a L-wavelet packet associated with the Bessel operator. Then for all f in $L^q([0, +\infty[, d\mu_\alpha)$, $q \in [1, 2]$, and for all $j\in\mathbb{Z}$, the function $y \to \Psi^{L,P}_{g^L}(f)(j,y)$ belongs to $L^r([0, +\infty[, d\mu_\alpha)$, with $r \in [1, +\infty]$, such that $\frac{1}{r} = \frac{1}{q} - \frac{1}{2}$, and we have

(9.II.12) $$\|\Psi^{L,P}_{g^L}\|_{r,\alpha} \leq \|f\|_{q,\alpha}\, \|g^{L,P}_j\|_{2,\alpha}.$$

Proof.

The relations (9.II.11), (4.III.10) and (4.III.11) give the results.

The following result is the so-called Calderon's reproducing formula for the transform $\Psi^{L,P}_{g^L}$.

Theorem 9.II.1. Let $\{g_j^{L,P}\}_{j\in\mathbb{Z}}$ be a L-wavelet packet associated with the Bessel operator. Then for f in $L^2([0,+\infty[, d\mu_\alpha)$ and $m, n \in \mathbb{Z}$, with $m < n$, the function

$$f_{m,n}^{L,P}(y) = \sum_{j=m}^{n-1} \Psi_{g^L}^{L,P}(f)(j,y), \tag{9.II.13}$$

belongs to $L^2([0,+\infty[, d\mu_\alpha)$, and we have

$$\lim_{\substack{m\to-\infty\\ n\to+\infty}} \| f_{m,n}^{L,P} - f\|_{2,\alpha} = 0. \tag{9.II.14}$$

To prove this theorem we need the following Lemma.

Lemma 9.II.1. let $\{g_j^{L,P}\}_{j\in\mathbb{Z}}$ be a L-wavelet packet associated with the Bessel operator. For $m, n \in \mathbb{Z}$, with $m < n$, we have for almost all $\lambda \geq 0$:

$$\mathfrak{F}_B\Big(\sum_{j=m}^{n-1} g_j^{L,P}\Big)(\lambda) = K_{r_n, r_m}^{L}(\lambda). \tag{9.II.15}$$

Where K_{r_n, r_m}^{L} is the function defined in Lemma 9.I.1.

Proof.

From the relation (9.II.2) we have for almost all $\lambda \geq 0$:

$$\begin{aligned}\mathfrak{F}_B\Big(\sum_{j=m}^{n-1} g_j^{L,P}\Big)(\lambda) &= \sum_{j=m}^{n-1} \mathfrak{F}_B(g_j^{L,P})(\lambda),\\ &= \sum_{j=m}^{n-1}\Big(\frac{1}{C_{g^L}}\int_{r_{j+1}}^{r_j} \mathfrak{F}_B(g_a^L)(\lambda)\frac{da}{a}\Big),\\ &= \frac{1}{C_{g^L}}\int_{r_n}^{r_m} \mathfrak{F}_B(g_a^L)(\lambda)\frac{da}{a}.\end{aligned}$$

Thus from the relation (9.I.20), we have for almos all $\lambda \geq 0$:

$$\mathfrak{F}_B\Big(\sum_{j=m}^{n-1} g_j^{L,P}\Big)(\lambda) = K_{r_n, r_m}^{L}(\lambda).$$

Proof of Theorem 9.II.1.

From the relation (9.II.11) we have for all $y \geq 0$:

$$f_{m,n}^{L,P}(y) = \sum_{j=m}^{n-1} \Psi_{g^L}^{L,P}(f)(j,y) = \sum_{j=m}^{n-1} f *_B g_j^{L,P}(y).$$

Thus

(9.II.16) $$f^{L,P}_{m,n}(y) = f *_B (\sum_{j=m}^{n-1} g_j^{L,P})(y).$$

Using the relation (9.II.15) and Lemma 9.I.1 we deduce that the function $\mathfrak{F}_B(f).\mathfrak{F}_B(\sum_{j=m}^{n-1} g_j^{L,P})$ belongs to $L^2([0, +\infty[, d\mu_\alpha)$. Then from Theorem 5.IV.4 and the relation (9.II.16), the function $f^{L,P}_{m,n}$ belongs to $L^2([0, +\infty[, d\mu_\alpha)$ and we have for almost all $\lambda \geq 0$:

$$\mathfrak{F}_B(f^{L,P}_{m,n})(\lambda) = \mathfrak{F}_B(f)(\lambda).\mathfrak{F}_B(\sum_{j=m}^{n-1} g_j^{L,P})(\lambda).$$

Thus from the relation (9.II.15), we have for almost all $\lambda \geq 0$:

(9.II.17) $$\mathfrak{F}_B(f^{L,P}_{m,n})(\lambda) = \mathfrak{F}_B(f)(\lambda) K^L_{r_n, r_m}(\lambda).$$

From Theorem 5.IV.2, we have

$$\|f^{L,P}_{m,n} - f\|_{2,\alpha} = \|\mathfrak{F}_B(f^{L,P}_{m,n}) - \mathfrak{F}_B(f)\|_{2,\alpha}.$$

By applying the relation (9.II.17) we obtain

$$\|f^{L,P}_{m,n} - f\|_{2,\alpha} = \|\mathfrak{F}_B(f)\,(1 - K^L_{r_n, r_m})\|_{2,\alpha}.$$

We deduce the result from Lemma 9.I.1 and the dominated convergence theorem.

Next we give a pointwise reconstruction formula for the transform $\Psi^{L,P}_{g^L}$.

Theorem 9.II.2. Let $\{g_j^{L,P}\}_{j\in\mathbb{Z}}$ be a L-wavelet packet associated with the Bessel operator. For f in $(L^1 \cap L^2)([0, +\infty[, d\mu_\alpha)$, such that $\mathfrak{F}_B(f)$ belongs to $L^1([0, +\infty[, d\mu_\alpha)$, we have

(9.II.18) $$f(y) = \sum_{j=-\infty}^{\infty} \Psi^{L,P}_{g^L}(f)(j,y), \quad y \geq 0,$$

where for all $y \geq 0$, the series converge absolutely.

Proof.

Let $y \in [0, +\infty[$. Using the relation (9.II.10), Proposition 9.II.2 and Theorem 5.IV.3 we deduce

(9.II.19) $$\Psi^{L,P}_{g^L}(f)(j,y) = \int_0^\infty \mathfrak{F}_B(f)(\lambda)\mathfrak{F}_B(g_j^{L,P})(\lambda) j_\alpha(\lambda y) d\mu_\alpha(\lambda).$$

But from the fact that

$$\forall\ y \geq 0,\ \lambda \geq 0,\ |j_\alpha(\lambda y)| \leq 1,$$

and Fubini-Tonelli's theorem we obtain

$$\sum_{j=-\infty}^{\infty} |\Psi_{g^L}^{L,P}(f)(j,y)| \leq \int_0^\infty |\mathfrak{F}_B(f)(\lambda)| \left(\sum_{j=-\infty}^{\infty} |\mathfrak{F}_B(g_j^{L,P})(\lambda)| \right) d\mu_\alpha(\lambda).$$

From Proposition 9.II.1.i) we deduce

$$\sum_{j=-\infty}^{\infty} |\Psi_{g^L}^{L,P}(f)(j,y)| \leq \frac{\|\mathfrak{F}_B(f)\|_{1,\alpha}}{|C_{g^L}|} \int_0^\infty |\mathfrak{F}_B(g^L)(\lambda)| \frac{d\lambda}{\lambda} < +\infty.$$

Thus the series (9.II.18) converges absolutely.

On the other hand the relation (9.II.19) implies for every $M_1, N_1 \in \mathbb{N}$:

$$\sum_{j=-N_1}^{M_1} \Psi_{g^L}^{L,P}(f)(j,y) = \int_0^\infty \mathfrak{F}_B(f)(\lambda) \left(\sum_{j=-N_1}^{M_1} \mathfrak{F}_B(g_j^{L,P})(\lambda) \right) j_\alpha(\lambda y) d\mu_\alpha(\lambda).$$

Using the relations (9.II.9), (9.II.4) and the dominated convergence theorem, we obtain

$$\lim_{M_1,N_1 \to +\infty} \sum_{j=-N_1}^{M_1} \Psi_{g^L}^{L,P}(f)(j,y) = \int_0^\infty \mathfrak{F}_B(f)(\lambda) j_\alpha(\lambda y) d\mu_\alpha(\lambda).$$

We deduce the relation (9.II.18) from Theorem 5.II.2.

9.III. Scale discrete L-scaling function associated with the Bessel operator.

Let g^L be a L-wavelet associated with the Bessel operator in $L^2([0, +\infty[, d\mu_\alpha)$. We consider the function g_a^L, $a > 0$, given by the relation (9.I.6) and the scale sequence $\{r_j\}_{j\in\mathbb{Z}}$ of the section 9.II. For all $J \in \mathbb{Z}$, we define the function $G_J^{L,P}$ by

(9.III.1)
$$G_J^{L,P}(.) = \frac{1}{C_{g^L}} \int_{r_J}^\infty g_a^L(.) \frac{da}{a}.$$

Proposition 9.III.1. The function $G_J^{L,P}$ given by the relation (9.III.1) belongs to $L^2([0, +\infty[, d\mu_\alpha)$, and we have for almost all $\lambda \geq 0$:

(9.III.2)
$$\mathfrak{F}_B(G_J^{L,P})(\lambda) = \frac{1}{C_{g^L}} \int_{r_J}^\infty \mathfrak{F}_B(g_a^L)(\lambda) \frac{da}{a}.$$

Proof.

By applying Hölder's inequality we obtain

$$\left(\int_{r_J}^{\infty} \frac{1}{a}|\mathfrak{F}_B(g_a^L)(\lambda)|da\right)^2 \leq \frac{1}{r_J}\int_{r_J}^{\infty} |\mathfrak{F}_B(g_a^L)(\lambda)|^2 da.$$

Thus Fubini-Tonelli's theorem, Theorem 5.IV.3 and the relation (9.I.6) imply

$$\int_0^{\infty} \left|\int_{r_J}^{\infty} \mathfrak{F}_B(g_a^L)(\lambda)\frac{da}{a}\right|^2 d\mu_\alpha(\lambda) \leq \frac{1}{r_J}\int_{r_J}^{\infty}\left(\int_0^{\infty} |\mathfrak{F}_B(g_a^L)(\lambda)|^2 d\mu_\alpha(\lambda)\right)da,$$

$$\leq \frac{\|g^L\|_{2,\alpha}}{r_J}\int_{r_J}^{\infty}\frac{da}{a^{2\alpha+2}} < +\infty.$$

Thus the function $\lambda \to \int_{r_J}^{\infty} \mathfrak{F}_B(g_a^L)(\lambda)\frac{da}{a}$ belongs to $L^2([0, +\infty[, d\mu_\alpha)$. From this result, Lemma 9.I.1 and Theorem 5.IV.2, the function $G_J^{L,P}$ given by the relation (9.III.1) is well defined, belongs to $L^2([0, +\infty[, d\mu_\alpha)$, and satisfies the relation (9.III.2).

Definition 9.III.1. The sequence $\{G_J^{L,P}\}_{J\in\mathbb{Z}}$ defined by the relation (9.III.1) is called scale discrete L-scaling function associated with the Bessel operator.

Example.

We consider the L-wavelet packet $\{g_j^{L,P}\}_{j\in\mathbb{Z}}$ associated with the Bessel operator given by the relation (9.II.5). From the relation (9.III.2) we deduce that the scale discrete L-scaling function $\{G_J^{L,P}\}_{J\in\mathbb{Z}}$ associated with the Bessel operator corresponding to the previous $\{g_j^{L,P}\}_{j\in\mathbb{Z}}$, satisfies

(9.III.3) $$\forall\, \lambda \geq 0,\ \mathfrak{F}_B(G_J^{L,P})(\lambda) = e^{-t\lambda^2 r_J^2}.$$

The following propositions give properties of the function $G_J^{L,P}$, $J \in \mathbb{Z}$.

Proposition 9.III.2

i) For all $J\in\mathbb{Z}$ and for almost all $\lambda \geq 0$, we have

(9.III.4) $$\mathfrak{F}_B(G_J^{L,P})(\lambda) = \sum_{j=-\infty}^{J-1} \mathfrak{F}_B(g_j^{L,P})(\lambda).$$

ii) For all $J\in\mathbb{Z}$ and for almost all $\lambda \geq 0$, we have

(9.III.5) $$|\mathfrak{F}_B(G_J^{L,P})(\lambda)| \leq \frac{1}{|C_{g^L}|}\int_0^{\infty} |\mathfrak{F}_B(g^L)(\lambda)|\frac{d\lambda}{\lambda}.$$

iii) For almost all $\lambda \geq 0$, we have

(9.III.6) $$\lim_{J \to +\infty} \mathfrak{F}_B(G_J^{L,P})(\lambda) = 1.$$

Proof.

We deduce these results from the relation (9.II.2) and Proposition 9.II.1.

Proposition 9.III.3.

i) For all $J \in \mathbb{Z}$ and for almost all $\lambda \geq 0$, we have

(9.III.7) $$\mathfrak{F}_B(G_J^{L,P})(\lambda) + \sum_{j=J}^{\infty} \mathfrak{F}_B(g_j^{L,P})(\lambda) = 1.$$

ii) For all $J \in \mathbb{Z}$ and for almost all $\lambda \geq 0$, we have

(9.III.8) $$\mathfrak{F}_B(g_j^{L,P})(\lambda) = \mathfrak{F}_B(G_{j+1}^{L,P})(\lambda) - \mathfrak{F}_B(G_j^{L,P})(\lambda).$$

iii) For almost all $\lambda \geq 0$, we have

(9.III.9) $$\sum_{j=-\infty}^{\infty} (\mathfrak{F}_B(G_{j+1}^{L,P})(\lambda) - \mathfrak{F}_B(G_j^{L,P})(\lambda)) = 1.$$

Proof.

From the relations (9.II.1) and (9.II.4) we obtain

$$\frac{1}{C_{g^L}} \int_0^{\infty} \mathfrak{F}_B(g_a^L)(\lambda) \frac{da}{a} = \sum_{j=-\infty}^{\infty} \mathfrak{F}_B(g_j^{L,P})(\lambda),$$
$$= \left(\sum_{j=-\infty}^{J-1} + \sum_{j=J}^{\infty}\right)(\mathfrak{F}_B(g_j^{L,P})(\lambda)) = 1.$$

The relation (9.III.4) gives (9.III.7) and (9.III.8). We deduce the relation (9.III.9) from (9.III.8) and (9.II.4).

Let $\{G_J^{L,P}\}_{J \in \mathbb{Z}}$ be a scale discrete L-scaling function associated with the Bessel operator. We consider for all $J \in \mathbb{Z}$ and $y \geq 0$, the function $G_{J,y}^{L,P}$ defined for almost all $x \geq 0$, by

(9.III.10) $$G_{J,y}^{L,P}(x) = T_x(G_J^{L,P})(y),$$

where T_x, $x \geq 0$, are the generalized translation operators given by the relation (4.II.3).

Some properties of this function are given by the following proposition.

Proposition 9.III.3.

i) The function $G^{L,P}_{J,y}$ belongs to $L^2([0,+\infty[, d\mu_\alpha)$ and we have

$$\|G^{L,P}_{J,y}\|_{2,\alpha} \leq \|G^{L,P}_{J}\|_{2,\alpha}. \tag{9.III.11}$$

ii) We have for almost all $\lambda \geq 0$:

$$\mathfrak{F}_B(G^{L,P}_{J,y})(\lambda) = j_\alpha(\lambda y)\mathfrak{F}_B(G^{L,P}_{J})(\lambda). \tag{9.III.12}$$

$$|\mathfrak{F}_B(G^{L,P}_{J,y})(\lambda)| \leq \frac{1}{|C_{g^L}|}\int_0^\infty |\mathfrak{F}_B(g^L)(\lambda)|\frac{d\lambda}{\lambda}. \tag{9.III.13}$$

Proof.

We deduce these results from the relations (9.III.10), (4.II.8), (5.II.8) and (9.II.3).

Theorem 9.III.1. Let $\{G^{L,P}_J\}_{J\in\mathbb{Z}}$ be a scale discrete L-scaling function associated with the Bessel operator, which corresponds to the L-wavelet packet $\{g^{L,P}_j\}_{j\in\mathbb{Z}}$ associated with the Bessel operator. For f in $L^2([0,+\infty[, d\mu_\alpha)$ we have

$$\text{i) } f(y) = \lim_{J\to+\infty}\int_0^\infty f(x)G^{L,P}_{J,y}(x)d\mu_\alpha(x). \tag{9.III.14}$$

The convergence is in $L^2([0,+\infty[, d\mu_\alpha)$.

$$\text{ii) } f(y) = \int_0^\infty f(x)\, G^{L,P}_{J,y}(x)d\mu_\alpha(x) + \sum_{j=J}^{\infty} \Psi^{L,P}_{g^L}(f)(j,y). \tag{9.III.15}$$

The series converge in $L^2([0,+\infty[, d\mu_\alpha)$.

To prove this theorem we need the following lemma.

Lemma 9.III.1. Let g^L be a L-wavelet associated with the Bessel operator in $L^2([0,+\infty[, d\mu_\alpha)$. For all $\varepsilon > 0$, the function $K^L_{\varepsilon,\infty}$ defined for almost all $\lambda \geq 0$, by

$$K^L_{\varepsilon,\infty}(\lambda) = \frac{1}{C_{g^L}}\int_\varepsilon^\infty \mathfrak{F}_B(g^L_a)(\lambda)\frac{da}{a}, \tag{9.III.16}$$

belongs to $L^\infty([0,+\infty[, d\mu_\alpha)$ and we have for almost all $\lambda > 0$,.

$$\lim_{\varepsilon\to 0^+} K^L_{\varepsilon,\infty}(\lambda) = 1. \tag{9.III.17}$$

Proof.

By a change of variables we obtain for almost all $\lambda > 0$:

$$K^L_{\varepsilon,\infty}(\lambda) = \frac{1}{C_{g^L}} \int_\varepsilon^\infty \mathfrak{F}_B(g^L)(u) \frac{du}{u}.$$

Thus from the relation (9.I.1) we have

$$|K^L_{\varepsilon,\infty}(\lambda)| \leq \frac{1}{|C_{g^L}|} \int_0^\infty |\mathfrak{F}_B(g^L)(u)| \frac{du}{u} < +\infty.$$

Then the function $K^L_{\varepsilon,\infty}$ belongs to $L^\infty([0,+\infty[, d\mu_\alpha)$.

On the other hand from the relation (9.I.2) we have for almost all $\lambda > 0$:

$$\lim_{\varepsilon\to 0^+} K^L_{\varepsilon,\infty}(\lambda) = \lim_{\varepsilon\to 0^+} [\frac{1}{C_{g^L}} \int_{\lambda\varepsilon}^\infty \mathfrak{F}_B(g^L)(u) \frac{du}{u}] = 1.$$

Proof of Theorem 9.III.1.

i) From the relation (9.III.10) and Proposition 4.III.5 we have for $y \geq 0$:

$$\int_0^\infty f(x) G^{L,P}_{J,y}(x) d\mu_\alpha(x) = f *_B G^{L,P}_J(y).$$

But from Lemma 9.III.1 the function $\mathfrak{F}_B(f).\ K^L_{r_J,\infty}$ is in $L^2([0,+\infty[, d\mu_\alpha)$, then from Theorem 5.IV.4 and Proposition 9.III.1, the function $f *_B G^{L,P}_J$ belongs to $L^2([0,+\infty[, d\mu_\alpha)$ and we have

(9.III.18) $$\mathfrak{F}_B(f *_B G^{L,P}_J) = \mathfrak{F}_B(f).K^L_{r_J,\infty}.$$

Using Theorem 5.IV.3 we have

$$\|f *_B G^{L,P}_J - f\|_{2,\alpha} = \|\mathfrak{F}_B(f *_B G^{L,P}_J) - \mathfrak{F}_B(f)\|_{2,\alpha}.$$

Applying the relation (9.III.18) we obtain

$$\|f *_B G^{L,P}_J - f\|_{2,\alpha} = \|\mathfrak{F}_B(f)\ K^L_{r_J,\infty} - \mathfrak{F}_B(f)\|_{2,\alpha}$$
$$= \|\mathfrak{F}_B(f)(1 - K^L_{r_J,\infty})\|_{2,\alpha}.$$

We deduce the relation (9.III.14) from Lemma 9.III.1 and the dominated convergence theorem.

ii) Using proofs of Theorems 9.II.1 and 9.III.1.i), we obtain for $y \geq 0$:

$$\int_0^\infty f(x) G^{L,P}_{J,y}(x) d\mu_\alpha(x) + \sum_{j=J}^{n-1} \Psi^{L,P}_{g^L}(f)(j,y) = f *_B G^{L,P}_J(y) + f^{L,P}_{J,n}(y).$$

But from the relations (9.II.17), (9.III.18) we have for almost all $\lambda \geq 0$:

$$\mathfrak{F}_B(f *_B G^{L,P}_J + f^{L,P}_{J,n})(\lambda) = \mathfrak{F}_B(f)(\lambda)\ (K^L_{r_n,r_J}(\lambda) + K^L_{r_J,\infty}(\lambda)).$$

Thus

(9.III.19) $$\mathfrak{F}_B(f *_B G_J^{L,P} + f_{J,n}^{L,P})(\lambda) = \mathfrak{F}_B(f)(\lambda).K_{r_{n,\infty}}^{L}(\lambda).$$

From Theorem 5.IV.3, we have

$$\|(f *_B G_J^{L,P} + f_{J,n}^{L,P}) - f\|_{2,\alpha} = \|\mathfrak{F}_B(f *_B G_J^{L,P} - f_{J,n}^{L,P}) - \mathfrak{F}_B(f)\|_{2,\alpha}.$$

Using the relation (9.III.19), we obtain

$$\|(f *_B G_J^{L,P} + f_{J,n}^{L,P}) - f\|_{2,\alpha} = \|\mathfrak{F}_B(f)\,(1 - K_{r_{n,\infty}}^{L})\|_{2,\alpha}.$$

We deduce the relation (9;III.15) from Lemma 9.III.1 and the dominated convergence theorem.

We are now able to prove a pointwise reconstruction formula associated with the function $\{G_J^{L,P}\}_{J\in\mathbb{Z}}$.

Theorem 9.III.2. Let $\{G_J^{L,P}\}_{J\in\mathbb{Z}}$ be a scale discrete L-scaling function associated with the Bessel oparator which corresponds to the L-wavelet packet $\{g_j^{L,P}\}_{J\in\mathbb{Z}}$ associated with the Bessel operator. For f in $(L^1 \cap L^2)([0, +\infty[, d\mu_\alpha)$ such that $\mathfrak{F}_B(f)$ belongs to $L^1([0, +\infty[, d\mu_\alpha)$ we have

(9.III.20) $$\text{i) } f(y) = \lim_{J\to+\infty} \int_0^\infty f(x)G_{J,y}^{L,P}(x)d\mu_\alpha(x), \quad y \geq 0.$$

(9.III.21) $$\text{ii) } f(y) = \int_0^\infty f(x)G_{J,y}^{L,P}(x)d\mu_\alpha(x) + \sum_{j=J}^{\infty} \Psi_{g^L}^{L,P}(f)(j,y), \quad y \geq 0,$$

where, for all $y \geq 0$, the series converge absolutely.

Proof.

i) Let $y \in [0, +\infty[$. From the relation (9.III.12) and Theorem 5.IV.3 we obtain

(9.III.22) $$\int_0^\infty f(x)G_{J,y}^{L,P}(x)d\mu_\alpha(x) = \int_0^\infty \mathfrak{F}_B(f)\,(\lambda)\mathfrak{F}_B(G_J^{L,P})(\lambda)j_\alpha(\lambda y)d\mu_\alpha(\lambda).$$

Using the relations (9.III.5) and (9.III.6), the fact that

$$\forall\, y \geq 0,\ \lambda \geq 0,\ |j_\alpha(\lambda y)| \leq 1,$$

and the dominated convergence theorem, we obtain

$$\lim_{J\to+\infty} \int_0^\infty f(x)G_{J,y}^{L,P}(x)d\mu_\alpha(x) = \int_0^\infty \mathfrak{F}_B(f)\,(\lambda)j_\alpha(\lambda y)d\mu_\alpha(\lambda).$$

Then Theorem 5.II.2 gives the relation (9.III.20).

ii) By applying Definition 9.II.2 and the proof of Theorem 9.II.2, we deduce that the series

$$\sum_{j=J}^{\infty} \Psi_{g^L}^{L,P}(f)(j,y)$$

converge absolutely and we have

$$\sum_{j=J}^{\infty} \Psi_{g^L}^{L,P}(f)(j,y) = \int_0^{\infty} \mathfrak{F}_B(f)(\lambda)\Big(\sum_{j=J}^{\infty} \mathfrak{F}_B(g_j^{L,P})(\lambda)\Big) j_\alpha(\lambda y) d\mu_\alpha(\lambda). \tag{9.III.23}$$

On the other hand by using the relation (9.III.4), the equality (9.III.22) can be written in the form

$$\int_0^{\infty} f(x) G_{J,y}^{L,P}(x) d\mu_\alpha(x) = \int_0^{\infty} \mathfrak{F}_B(f)(\lambda)\Big(\sum_{j=-\infty}^{J-1} \mathfrak{F}_B(g_j^{L,P})(\lambda)\Big) j_\alpha(\lambda y) d\mu_\alpha(\lambda). \tag{9.III.24}$$

Then we obtain from the relations (9.III.23), (9.III.24) and (9.II.4) :

$$\int_0^{\infty} f(x) G_{J,y}^{L,P}(x) d\mu_\alpha(x) + \sum_{j=J}^{\infty} \Psi_{g^L}^{L,P}(f)(j,y) = \int_0^{\infty} \mathfrak{F}_B(f)(\lambda) j_\alpha(\lambda y) d\mu_\alpha(\lambda).$$

We obtain the relation (9.III.21) from Theorem 5.II.2.

Bibliography

[1] **A. ACHOUR, H.ANNABI, A. FITOUHI and K. TRIMECHE** : Multiplicateurs associés à l'orpérateur de Bessel. Prépublication. Faculté des Sciences de Tunis - 1980.

[2] **A. ACHOUR and K. TRIMECHE** : La g-function de Littlewood-Paley associée à un opérateur différentiel singulier sur (0, + ∞). Ann. Int. Fourier. Vol. 33, N° 4, 1983, p. 203-226.

[3] **L. C. ANDREWS** : Special functions of Mathematics for engineers. Second edition. Oxford University Press - Oxford - Tokyo - Melbourne - 1998.

[4] **R. ASKEY** : Theory and applications of special functions. Academic Press, New-York 1975.

[5] **R. ASKEY** : Orthogonal polynomials and special functions, in "Regional Conference Series in Applied Mathematics" Vol. 21, SIAM, Philadelphia, 1975.

[6] **N. BEN SALEM** : Convolution semigroups and central limit theorem associated with a dual convolution structure. J. Theoretical Probability, Vol.7, N° 2, 1994, p. 417-436.

[7] **Yu. M. BEREZANSKY and A. A. KALYUZHNYI** : Harmonic Analysis in Hypercomplex Systems. Kluwer Academic Publishers, 1998

[8] **W. R. BLOOM and H. HEYER** : Harmonic analysis of probability measures on hypergroups, in de "Gruyter Studies in Mathematics" (H. Bauer, J.L.Kazdan, and E. Zehnder, Eds), Vol. 20, de Gruyter, Berlin-New-York, 1994.

[9] **W.R.BLOOM and Z. XU** : The Hardy-Littlewood maximal function for Chébli-Trimèche hypergroups. Contemp. Math. 183, 1995, p. 45-69.

[10] **W. R. BLOOM and Z. XU** : Fourier transforms of Schwartz functions on Chébli-Trimèche hypergroups. Mh. Math. 125, 1998, p.89-109.

[11] **H.CHEBLI** : Sur un théorème de Paley-Wiener associé à la décomposition spectrale d'un opérateur de Sturm-Liouville sur]0, + ∞[.
J. funct. Anal. 17, 1974, p.447-461.

[12] **H. CHEBLI** : Opérateurs de translation généralisée et semi-groupes de convolution.
Lecture Notes in Math. N° 404, p. 35-59, Springer-Verlag .

[13] **H. CHEBLI** : Sur les fonctions presque-périodiques associées à un opérateur différentiel singulier sur (0, + ∞).
Prépublication. Faculté des Sciences de Tunis. 1985.

[14] **F.M.CHOLEWINSKI and D.T.HAIMO** : Classical analysis and the generalized heat equation.
SIAM Rev. 10, 1968, p.67-80.

[15] **F.M.CHOLEWINSKI, D.T.HAIMO and A.E.NUSSBAUM** : A necessary and sufficient condition for the representation of a function as a Hankel-Stieltjes transform.
Studia Math. T. 36, 1970, p. 269-274.

[16] **R. COURANT and D. HILBERT** : Methods of Mathematical Physics. Vol.2. Partial differential equations. Inter-Science Publishers, New-York, 1962.

[17] **C.K.CHUI** : An introduction to wavelets.
Academic Press, 1992.

[18] **A. DACHRAOUI and K. TRIMECHE** : Pseudo-differential operators associated with a singular differential operator on]0, + ∞[.
Preprint. Faculty of Sciences of Tunis. 1997.

[19] **G. DARBOUX** : Sur la théorie générale des surfaces, I.II, Paris, 1914-1915.

[20] **I. DAUBECHIES** : Ten lectures on wavelets.
CBMS-NSF Regional Conference Series in Applied Mathematics. Vol.61, SIAM, Philadelphia, PA, 1992.

[21] J. DELSARTE. Une extension nouvelle de la théorie des fonctions presque-périodiques de Bohr.
Acta Math. 69, 1938, p. 259-317.

[22] J. DELSARTE : Hypergroupes et opérateurs de permutation et de transmutation.
Colloque Nancy; 1956, p. 29-45.

[23] J. DELSARTE and J.L.LIONS : Moyennes généralisées.
Comment. Math. Helv. Vol. 33, Fasc. 1, 1959, p. 59-69.

[24] J.DIEUDONNE : Elements d'Analyse.Tomes 2 et 6.
Gauthier-Villars, Editeurs- Paris. 1968-1975.

[25] N. DUNFORD and J.T.SCHWARTZ : Linear operators. Part I : General theory.
A Wiley-Interscience Publication - John Wiley & Sons-New-York. Chichester. Brisbane-Toronto-Singapore. 1957.

[26] M. DUVAL - DESTIN, M.A.MUSCHIETTI and M. TORRESANI :
Continuous wavelet decomposition, multiresolution and contrast analysis.
SIAM. J. Math. Anal. 24, 1993, p. 739-755.

[27] H. DYM and H.P.Mc KEAN : Fourier series and integrals.
Academic Press, 1972.

[28] A. ERDELYI, et al : Higher transcendental functions, 3 vols.
Bateman Manuscript Project, Mc Graw-Hill, New-York, 1963-1965.

[29] J. FARAUT and K. HARZALLAH : Deux cours d'analyse harmonique -
Ecole d'Eté d'Analyse Harmonique de Tunis, 1984.
Brikhäuser-Boston-Basel-Stuttgart, 1987.

[30] J. FAVARD : leçons sur les fonctions presque-périodiques.
Gauthier-Villards. Paris 1933.

[31] A. FITOUHI : Inégalité de Babenko et inégalité logarithmique de Sobolev pour l'opérateur de Bessel.

C.R.Acad. Sci. Paris, t. 305, Série I, 1987, p. 877-880.

[32] **A. FITOUHI and A. ANNABI** : La g-fonction de Littlewood-Paley associée à une classe d'opérateurs différentiels sur (0, + ∞) contenant l'opérateur de Bessel. C.R.A.S. Paris, t. 303, Série I, N° 9, 1986, p. 411-413.

[33] **A. FITOUHI** : Fonctions propres, équations de la chaleur et des ondes associées à un opérateur différentiel singulier sur (0, + ∞). Thèse d'Etat soutenue le 27-5-1988. Faculté des Sciences de Tunis.

[34] **A. FITOUHI** : Heat "polynomials" for a singular differential operator on (0, + ∞). Const. Approx. 51, 1989, p. 241-270.

[35] **A. FITOUHI and K. TRIMECHE** : J.L.Lions transmutation operators and generalized continuous wavelet transform. Preprint. Faculty of Sciences of Tunis. 1995.

[36] **G.B. FOLLAND** : Real analysis. Modern techniques and their applications. John Wiley & Sons-New-York-Chichester-Brisbane-Toronto-Singapore. 1984.

[37] **W.FREEDEN and U. WINDHEUSER** : Spherical wavelet transform and its discretization. Advances in computational Math. 5, 1996; p. 51-94.

[38] **L. GALLARDO and K. TRIMECHE** : Théorèmes local et de renouvellement pour une convolution généralisée sur la demi-droite. C.R.Math. Rep. Acad. Sci. Canada, Vol. 18, N° 2, 3, 1996, p. 61-65.

[39] **L. GALLARDO and K. TRIMECHE** : L'équation de Poisson et les noyaux de Green associés à un opérateur différentiel singulier sur $\mathbb{R}_+$. C.R.Acad. Sci.Paris, Série I, t. 324, 1997, p. 259-264.

[40] **P. GOUPILLAUD, A. GROSSMANN and J. MORLET** : Cycle octave and related transforms in seismic signal analysis. Geoexploration 23, 1984-1985, p.85-102.

[41] A. GROSSMANN and J. MORLET : Decomposition of Hardy functions into square integrable wavelets of constant shape.
SIAM.J.Math. Anal. 15 (4), 1984, p. 723-736.

[42] A. GROSSMANN, R. KRONLAND-MARTINET and J. MORLET : Reading and understanding continuous wavelet transforms.
In wavelets, J.M. Combes, A. Grossmann, Ph. Tchamitchian, eds, Springer, 1989, p. 2-20.

[43] A. GUICHARDET : Calcul integral.
Armand Colin, 1969.

[44] K. HARZALLAH : Fonctions qui opérent sur les fonctions définies négatives - Problèmes concernant la transformation de Hankel.
Thèse d'Etat soutenue le 11-6-1968. Université de Paris - Faculté des Sciences d'Orsay.

[45] S. HELGASON : The Radon transform.
Progress in Mathematics 5, Birkhauser, Boston 1980.

[46] C.S.HERZ : Analyse Harmonique à plusieurs variables.
Publication du Séminaire de Mathématiques d'Orsay. 4eme année, 1964-1965.

[47] I.I.HIRSCHMAN. JR : Variation diminishing Hankel transform.
J. Anal. Math. 8, 1960-1961, p. 307-336.

[48] L. HÖRMANDER : The Analysis of linear partial differential operator I.
Springer-Verlag Berlin Heidelberg, New-York, Tokyo, 1983.

[49] I.A.KIPRIJANOV and V.V.KATRAHOV : On a class of one-dimensional singular pseudo-differential operators.
Math. USSR Sbornik Vol. 33, N° 1, 1977, p.43-61.

[50] E.L.KOH and A.H.ZEMANIAN : The complex Hankel and I-transformations of generalized functions.
SIAM J. Appl. Math. Vol. 16, N° 5, 1968, p. 945-957.

[51] **T.H.KOORNWINDER** : Jacobi polynomials, II. An analytic proof of the product formula.
SIAM J. Math. Anal.Vol. 5. N°1. 1974, p.125-137.

[52] **T.H.KOORNWINDER** : The continuous wavelet transform. Series in approximations and decompositions. Vol. 1. Wavelets : An elementary treatment of theory and applications. Edited by Tom H. Koornwinder. World Scientific. 1993.

[53] **S. LANG** : Analysis I and II.
Addison Wesley Publishing Company 1968-1969.

[54] **M.N. LAZHARI and K. TRIMECHE** : Convolution algebras and factorization of measures on Chébli-Trimèche hypergroups.
C.R.Math. Rep. Acad. Sci. Canada, Vol. 17, N° 4, 1995, p. 165-169.

[55] **N.N.LEBEDEV** : Special functions and their Applications.
(R.A.Silverman, transl. and ed.), Dover, New-York, 1972.

[56] **P.G.LEMARIE** : Introduction à la théorie des ondeletttes, in "les ondelettes en 1989", P.G.Lemarié (ed.), Lecture Notes in Math. 1438. Springer, 1990. exposé N° 1.

[57] **J.L.LIONS** : Opérateurs de Delsarte et problèmes mixtes.
Bull. Soc. Math. France, t. 84, 1956, p. 9-95.

[58] **J. L.LIONS** : On the generalized radiation problem of Weinstein.
J. Math. Mech. 8, 1959, p. 873-888.

[59] **J.L.LIONS** : Equations différentielles opérationnelles et problèmes aux limites.
Springer-Verlag, Berlin, 1961.

[60] **L.H.LOOMIS** : An introduction to abstract harmonic analysis. New-York 1953.

[61] **D. Ludwig** : The Radon transform on euclidean spaces.
Comm. Pure Appl. Math. Vol. 19, 1966, p.49-81.

[62] W. MAGNUS and F. OBERHETTINGER : Formulas and Theorems for the special functions of Mathematical Physics. Springer, New-York, 1969.

[63] Y. MEYER : I-Ondelettes et opérateurs. II-Opérateurs de Calderon-Zygmund. Actualités Mathématiques-Hermann-Editeurs des Sciences et des Arts, 1990.

[64] M.A.MOUROU : Ondelettes généralisées et applications. D.E.A de Mathématiques soutenu le 15-6-1995. Faculté des Sciences de Tunis.

[65] M.A.MOUROU and K. TRIMECHE : Calderon's reproducing formula associated with the Bessel operator. J. Math. Anal. Appl. 219, 1998, p. 97-109.

[66] M.A.MOUROU and K. TRIMECHE : Inversion of the Weyl integral transform and the Radon transform on $\mathbb{R}^n$ using generalized wavelets. Mh. Math. 126, 1998, p. 73-83.

[67] M.A.MOUROU and K. TRIMECHE : Calderon's formula associated with a differential operator on $(0, +\infty)$ and inversion of the generalized Abel transform. Journal of Fourier Analysis and Applications, Vol. 4, (2), 1998, p. 229-245.

[68] M.A.MOUROU and K.TRIMECHE : Inversion of the Radon transform on $\mathbb{R}^n$ using Daubechies wavelets. Preprint. Faculty of Sciences of Tunis. 1998.

[69] A. E. NUSSBAUM : On functions positive definite relative to the orthogonal group and the representation of functions as Hankel-Stieltjes transforms. Trans. Amer. Math. Soc. 175, 1973, p. 389-408.

[70] F.W.J.OLVER : Asymptotic and special functions. Academic Press, New-York, 1974.

[71] R.S.PATHAK and P.K.PANDEY : A class of pseudo-differential oprators associated with Bessel operators. J. Math. Anal. Appl. 196, 1995, p. 736-747.

[72] R. S. PATHAK and S.K.UPADHYAY : Pseudo-differential operators involving Hankel transform.
J. Math. Anal. Appl. 213, 1997, p. 133-147.

[73] R. S.PATHAK and P.K.PANDEY : Sovolev type spaces associated with Bessel operators.
J. Math. Analy. Appl. 215, 1997, p.95-111.

[74] W. RUDIN : Analyse réelle et complexe.
Masson, 1975.

[75] L.SCHWARTZ : Etude des sommes d'exponentielles.
Publication de l'Institut de Mathématique de l'Université de Strasbourg, Hermann, Paris, 1959.

[76] L. SCHWARTZ : Méthodes Mathématiques pour les Sciences physiques.
Hermann, Paris, 1965.

[77] L. SCHWARTZ : Théorie des distributions.
Hermann, Paris, 1966.

[78] A. SCHWARTZ : The structure of the algebra of Hankel transforms and the algebra of Hankel-Stieltjes transforms.
Can. J. Math. Vol. 23, N° 2, 1971, p. 236-246.

[79] R. SEELEY : An introduction to Fourier series and integrals.
Benjamin, 1966.

[80] I.N.SNEDDON : Special functions of Mathematical Physics and Chemistry. 2nd ed. Oliver and Boyd, Edinburgh, 1961.

[81] K. STEMPAK : La théorie de Littlewood-Paley pour la transformation de Fourier-Bessel.
C.R.A.S. Paris. Série I, t. 303, 1986, p. 15-18.

[82] K. TRIMECHE : Probabilités indéfiniment divisible et théorème de limite centrale pour une convolution généralisée sur la demi-droite.

C.R.Acad. Sc. Paris, Série A, 286, 1978, p. 63-66.

[83] K. TRIMECHE : Transformation intégrale de Weyl et théorème de Paley-Wiener associés à un opérateur différentiel singulier sur (0, + ∞).
J. Math. Pures et Appl. 60, 1981, p. 51-98.

[84] K. TRIMECHE : Transmutation operators and mean-periodic functions associated with differential operators.
Mathematical Reports, Vol. 4, Part.1, 1988, p. 1-282.
Harwood Academic Publishers-London-Paris-New-York-Melbourne.

[85] K. TRIMECHE : Transformation intégrale de Weyl généralisée associée à un opérateur différentiel singulier sur (0, + ∞) et problème de radiation de Weinstein généralisé.
Portugaliae Math. Vol. 47, Fasc. 4, 1990, p. 371-389.

[86] K. TRIMECHE : Fonctions moyenne-périodiques associees a un opérateur différentiel singulier sur (0, + ∞) et développement en série de Fourier généralisée.
J. Math. Pures et Appl. 65, 1986, p. 1-46.

[87] K. TRIMECHE : Inversion of the Lions transmutation operators using generalized wavelets.
Applicd and computational Harmonic Analysis, 4, 1997, p. 97-112.

[88] K. TRIMECHE : Generalized wavelets and hypergroups.
Gordon and Breach Sciences Publishers. 1997.

[89] K. TRIMECHE : Asymptotic expansions and linear wavelet packets on certain hypergroups.
Analysis of Divergence. Control and Management of Divergent Processes.
Editors William O. Bray and Caslav V. Stanojevic. Birkhäuser-Boston-Basel-Berlin 1999. p. 251-295..

[90] K. TRIMECHE : Generalized wavelet packets and application to the inversion of the J.L.Lions transmutation operators.
Preprint. Faculty of Sciences of Tunis. 1998.

[91] G.N. WATSON : A treatise on the theory of Bessel functions -2nd ed.
Cambridge Un. Press, London and New-York. 1966.

[92] A.H. ZEMANIAN : Generalized integral transformations.
Interscience, New-York, 1968.

[93] H.M.ZEUNER : One dimensional hypergroups.
Adv. Math. 76, N° 1, 1989, p. 1-18.

[94] H. M. ZEUNER : The central limit theorem for Chébli-Trimèche hypergroups.
J. Theoretical Prob. Vol. 2, N° 1, 1989, p. 51-63.

[95] H.M. ZEUNER : Limit theorems for one-dimensional hypergroups.
Habilitationss-Chrift-Tübingen, 1990.

INDEX

In this index the first numeral of reference indicates the number of chapter in which the subject is treated, the roman numeral indicates the section inside this chapter and when the last numeral exists, it mentions the subsection.

Absolutely continuous 5.VIII
Addition formula 1.I.6-1.III
Adjoint operator 5.I-7.II.2
Admissibility condition 7.I.1-7.II.1-7.V.
Algebra 3.II.1-5.III.
Almost-periodic functions 5.VIII
Anger function 1.III.
Asymptotic formula 1.I.3-1.III
Automorphism 5.I
Babenko's inequality 5.V.
Banach algebra 5.VIII
Banach space 3.I-3.III-5.VIII
Bernouilli topology 3.I.3
Bessel differential equation 1.I.2-1.III
Bessel F.W.1.I.4
Bessel function of first kind 1.I
Bessel operator 2.I.1-2.II.2-2.III-2.V-2.V.3-2.VI-4.I.1-4.II-4.III.1-4.III.2-4.III.3-4.IV-5.II-5.III 5.VII-5.VIII-6.I-6.II-6.III.1-6.III.2-6.IV-7.II.1-7.II.2-7.III-7.IV-8.I-8.I.1-8.I.2-8.II-8.III-8.IV-8.V-9.I-9.II-9.III.
Bohr almost-periodic functions 5.VIII
Calderon's reproducing formula 7.I.2-7.II.2.7.IV-8.I.2-8.III-9.I-9.II
Center 2.I.1
Central limit theorem 6.IV.
Characteristic function 2.II.1.-3.II.1-4.III.1-5.II-5.IV
Classical continuous wavelet transform 7.I.2-7.IV.7.V.
Classical Fourier transform 1.III-5.I-7.V-8.VI.
Classical wavelet 7.I.1-7.I.2-7.IV-7.V
Closed graph theorem 5.VIII
Coherent states 7.II.2
Continuous L-wavelet transform 9.I
Continuous wavelet transform 7.II.2
Convolution product 3.I.2-3.I.3-3.I.4-3.II.2-4.I.2-7.I.2
Convolution semigroup 6.III.1-6.III.2
D-Stability constants 8.VI
D-Stability conditions 8.VI
D-wavelet 8.VI
D-wavelet transform 8.VI
Darboux equation 2.I.1
Daubechies wavelet 8.VI
Detail 8.V
Detail spaces 8.V
Differential equation 1.II.2-1.III-3.III
Differential operator 2.III
Dilated measure 7.V
Dilation operator 7.I.1
Dirac measure 2.VI-3.I.2-4.III.2-5.VI-6.II
Dispersion 6.I.2-6.IV
Dual 2.II.1-2.V
Dual D-wavelet 8.VI
Dual exponent 2.I.3-5.V

Dual modified wavelet packet 8.III
Dual M-wavelet packet 8.III
Dual M-wavelet packet transform 8.III
Dual pair 3.I.3
Dual Radon transform 2.VI-7.V.
Dual S-wavelet packet 8.IV
Dual S-wavelet packet transform 8.IV
Dual Sonine integral transform 2.V.3-2.V.4.
E-multiresolution analysis 8.V
Euclidean structure 1.III
Even mean-periodic functions 5.VIII
Exponential type 3.II.2-3.II.5
Finite part 2.VI
Fitouhi. A 5.V.
Fourier coefficients 5.VIII
Fourier exponents 5.VIII
Fourier series 5.VIII
Fourier-Bessel transform 5.II-5.III-5.IV-5.V-5.VI-5.VII-5.VIII-6.I.3-7.III-8.VI
Fourier-Bessel-Plancherel transform 5.IV-5.VIII
Fourier-cosine transform 3.II.1-3.II.2-3.II.3-3.II.4-3.II.5-5.II-7.III.
Fréchet space 2.I.2-2.II.1-3.II.2-3.II.5
Gaussian distribution 6.IV
Gaussian function 5.V.
Gegenbauer polynomial 1.I.6-4.IV
Generalized convolution product 4.III.1-4.III.2-4.III.3-5.III-5.V-7.II.2-8.I.1-8.III-8.IV-8.V-9.I-9.II.
Generalized Fourier coefficients 5.VIII
Generalized Fourier exponents 5.VIII
Generalized Fourier series 5.VIII
Generalized mean-periodic functions
Generalized translation operator 4.II-4.IV-5.III-5.VIII-6.I.1-6.II-7.II.1-7.V-8.I.1-8.II-8.III-8.IV-9.I-9.II
Generalized quadratic form 6.I.1
Generating function 1.I.1-1.III
Generating function relation 1.III
Hahn Banach theorem 5.VIII
Haar measure 1.III
Heaviside function 2.VI
Helglotz-Bochner-Weil-Raikov theorem 5.VIII
Hilbert space 5.VI-7.II.2-8.II
Homomorphism 3.II.1-5.VIII
Hyperplan 2.VI
Ideal 5.VIII
Inductive limit 3.I.3
Inductive limit topology 2.II.1-3.II.2-3.II.5
Infinitely divisible probabilities 6.III.2
Integral formula 1.III.
Integral representations 1.II.3-1.III
Inversion formulas 3.II.1-5.II-7.III-7.IV
Isometric 7.I.1-7.II.2
Isometric isomorphism 3.II.3-5.IV
Isometry 5.VI
Isomorphism 5.VIII
J-scale smooth functions 8.V
J^{th} partial reconstruction 8.II
Law-pass filters 8.II
L-wavelet 9.I-9.II-9.III
L-wavelet packet 9.II-9.III-8.I.1-8.III-8.IV-8.V-9.I-9.II
L-wavelet packet transform 9.II
L-wavelet packet member 9.II
Laplacian 1.II.2-1.III-2.I.1-4.I.1.

5.VIII.
Levy-Khintchine formula 6.I.3
Linear wavelet 9.I
Linear wavelet packet 9.II
Lions J.L. 2.I.2
Locally convex space 3.I.3
M-wavelet packet 8.III
M-wavelet packet transform 8.III
Modified wavelet packet 8.III
Multiresolution analysis 8.V
Negative definite function 6.II-6.III.2
Norm 1.II.3-5.IV-5.VIII-8.VI
Normal distribution 6.IV
Normalized Bessel function 1.II-5.I-5.II
Orthogonal group 1.III
Orthogonal projection 7.II.2
P-wavelet packet 8.I.1-8.I.2-8.II
P-wavelet packet transform 8.I-8.I.1-8.II
P-wavelet packet members 8.II
Palet-Wiener theorem 3.II.2
Parseval formula 7.I.2-7.II.2-8.I.1-8.II.
Partial differential equation 1.III-2.I.1
Partial differential operator 1.III
Permutation relation 2.V.3
Plancherel formula 5.IV-7.I.2-7.II.2-8.II-8.III-8.IV.
Plancherel theorem 3.II.3-5.IV
Pointwise reconstruction formula 7.I.2-7.II.2-8.III-9.I-9.II-9.III
Poisson integral representation 1.I.4-1.II.3-1.III.
Poisson S.D.1.I.4
Poisson transform 2.I.1.
Polar coordinates 4.I.1
Positive definite functions 5.VIII-6.II-6.III.1
Lebesgue point 4.IV-7.V
Levy theorem 6.I.3
Probability measure 3.I.3-5.VIII-6.IV
Product formulas 1.I.7-1.II.4-1.III-3.I.1-4.I.
Pseudo-differential operator 5.VIII
Quadratic form 6.I.1
Radial classical wavelet 7.V
Radial functions 1.III-2.VI-4.I.1-4.I.2-4.III.1-5.I-8.VI
Radial part 1.II.2
Radiation problem 2.VI
Radius 2.I.1
Radon measure 3.I.3
Radon transform 2.VI-7.V-8.VI
Reconstruction formula 8.I.1-8.II-8.III-8.IV-9.I
Riemann-Lebesgue Lemma 5.II
Riemann-Liouville integral transform 2.I.1-2.I.3-2.II.1-2.IV-2.VI-5.III-5.VIII
Riesz convexity theorem 5.V
Riesz M. 2.I.2
S-wavelet packet 8.IV-8.V
S-wavelet packet transform 8.IV
Scalar product 1.II.3-1.III-2.VI-5.I-5.VIII-8.II-8.VI-7.II.2-7.V
Scale discrete L-scaling function 9.III
Scale discrete scaling function 8.II-8.III-8.IV
Scale sequence 8.I-8.VI-9.II-9.III
Scale spaces 8.V
Schwartz L. 5.VIII
Seminorms 2.I.2-2.II.1-2.III-3.I.3-3.II.2-3.II.5
Series representation 1.III

Pre-image 5.VIII
Prehilbertian space 5.VIII
Sonine N.J. 1.II.4
Sonine transform 2.V.4
Sonine's first integral 1.I.5-1.II.3
Spectrum 5.VIII
Sphere 2.I.1
Spherical average 2.I.1
Stability conditions 8.IV
Stability constants 8.IV
Struve function 1.III
Surface measure 2.VI
Symbol 5.VIII
Tempered distributions 3.I.4
Three-term recurrence formula 1.I.2
Topological automorphism 7.I.1
Topological dual 2.IV-2.VI-3.I.4
Topological isomorphism 2.I.2-2.II.2-2.III-2.V.1-2.V.3-2.V.4-2.VI-3.II.2-3.II.5-5.III-7.III
Translation operator 3.I.1-3.I.3-5.III-5.VIII-6.I.1-7.I.1
Transmutation operator 2.I.2-2.II.2-2.III-2.V.1-2V.3
Singular support 3.II.5-5.VII
Smoothing 8.II
Sonine integral transform 2.V.1-2.V.2-2.V.4.
Triangle 1.II.4
Uniform norm 3.I.3
Unit sphere 1.II.3-1.III-2.VI-4.I.1
Usual convolution product 2.VI-7.V
Usual Schwartz space 8.VI
Vague convergence 6.I.3-6.II
Vague topology 3.I.3
Vaguely continuous 6.III.1
Variance 6.I.2
Wave equation 2.I.1
Wavelet 7.II.1-7.II.2-7.IV-8.I-8.I.1
Wavelet packet 8.I.1-8.V
Weak convergence 6.I.3-6.IV
Weak topology 3.I.3
Weakly oscillating 7.V
Weber function 1.III.
Weyl integral transform 2.II.1-2.II.2-2.III-5.II-5.III
Weyl transform 2.IV-2.VI-5.VII
Young's inequality 5.V.